INSECT CHEMORECEPTION
FUNDAMENTAL AND APPLIED

Insect Chemoreception

Fundamental and Applied

by

Michael F. Ryan

Department of Zoology, University College Dublin

KLUWER ACADEMIC PUBLISHERS

DORDRECHT / BOSTON / LONDON

A C.I.P. Catalogue record for this book is available from the Library of Congress.

ISBN 1-4020-0270-X

Published by Kluwer Academic Publishers,
P.O. Box 17, 3300 AA Dordrecht, The Netherlands.

Sold and distributed in North, Central and South America
by Kluwer Academic Publishers,
101 Philip Drive, Norwell, MA 02061, U.S.A.

In all other countries, sold and distributed
by Kluwer Academic Publishers,
P.O. Box 322, 3300 AH Dordrecht, The Netherlands.

Printed on acid-free paper

Printed in the Netherlands.

DEDICATED TO MY FAMILY

PREFACE

In this time of edited volumes when the list of individual contributors may reach double figures, it is appropriate to question the usefulness of a volume, with such a broad scope, by a single author. The answer is simple. For years he has believed that the rather sharp distinction between fundamental and applied aspects of this discipline, has ill-served the significance of each; and has diminished the incidence of fruitful synergies. Yet the need for these was never greater, and this case may be developed by a single author with experience of each aspect.

The inclusion of a Chapter on Genetic Engineering may raise some doubts, but it is enabled by the chosen title "Chemoreception", as distinct from Chemoperception: the latter implies detection of a chemical, followed by a behavioural response. But the former broader category subsumes Chemoperception and allows for the reception of a chemical toxin so potent as to prelude a behavioural or physiological response, other than death. Accordingly, chemical toxins are a legitimate inclusion. In which event, their delivery through a GM plant is as appropriate for study as their application in a spray.

Following a Chapter devoted to an Introduction and Overview, Part I comprises Chapters 2, 3, 4, 5, and 6 respectively, providing a fundamental framework. Part II comprising Chapters 7, 8, 9, and 10 examines applied aspects. For internal consistency, the sequence followed in Part I in regard to plant chemicals and pheromones (Chapters 2 and 3 respectively), is also followed in Part II (Chapters 7 and 8 for plants and Chapter 9 for pheromones). However, as progress in genetic engineering for insect pest control (Chapter 10) is less with insects than with plants, Chapters 7, 8, and 10 may be read as a continuum on plants.

Chapter 10 brings to the fore the Third World dilemma of feeding, in the future, vastly more people from less land and poorer-quality water. The problem is most acute in sub-Saharan Africa where, from 1970 onward, food production grew at about half the rate of population growth, entailing decreased per capita food production. Undernourished people in this region are estimated to number some 300 million or half the world's total. In seeking to find a balance between GM-based and sustainable ecological solutions, this Chapter explicitly calls for a renewed commitment from the West, especially from the young researcher. Shall a coalition develop?

My most grateful thanks are due to the many who encouraged or helped, especially President Art Cosgrove, University College Dublin (UCD), and my colleagues at the Institute of Environment and Life, Department of Zoology, Coimbra University, Portugal, where my presence was enabled by the PRAXIS XXI programme of the EU, and by the Foundation for Science and Technology, Lisbon.

A very special debt of gratitude is owed to Professor Maria Susana de Almeida Santos, Leader of the Nematology Group, whose kind hospitality sustained this enterprise at difficult and decisive moments. She also, with her colleague Professor Isabel Abrantes, critically read the manuscript leading to significant improvements and the avoidance of countless solecisms.

The hand-drawn figures are by William Clarke, UCD except for Figure 33 by Professor Mary Behan, and Figure 39 by Dr. Peter Daly, both originally of UCD. Figures 18, 32, 35, 36, 37, 38, 72 and 74 are by Fernando Correia, Coimbra University; chemical structures and all remaining diagrams are by Dr. Wen-Yuan Chung, UCD. At Coimbra University, the species index was compiled by Sofia dos S. da R. Costa, and the entire manuscript was prepared and formatted with unforgettable insight, patience, and skill by M. Clara Vieira Santos.

MFR

September 2001

TABLE OF CONTENTS

PART II – APPLIED ASPECTS

CHAPTER 1

INTRODUCTION AND OVERVIEW

FUNDAMENTAL ASPECTS

Historical

From early times man has studied insects for their intrinsic interest and for their economic and medical significance. Early naturalists were exercized by the behavioural repertoire of insects, solitary and social. The Sacred Beetle or Scarab was known to and emblazoned by the Egyptian civilisations of the Rameses. Pharaonic accounts claimed that the Scarab rolled the ball of soil from east to west, the direction in which the world turns; immersion in the sacred waters of the Nile elicited emergence of the adult beetle. Such accounts were transmitted without verification through the millennia until the eighteenth century during which ecclesiasticism and conventional wisdom were subjected in Europe to severe scrutiny.

J.H. Fabre used direct observations to correct long standing myths regarding the Scarab. He confirmed that a female dung beetle buried her soil-pelleted egg underground, but when he disinterred the egg she showed no further interest in it. This elicited the observation that "Instinct sees the future and knows nothing of the past" [1.10]. Subsequent exploration of instinct in insects and of the senses associated with its expression provided the substratum for much fundamental data on chemoreception.

Also of interest were the endeavours of social insects such as species of wasps, ants, bees and termites each working through intraspecific cooperation. There is a delightful literature with vivid if somewhat anthropomorphic descriptions of such cooperativity. Nor was such interest confined to naturalists: the early Portuguese explorers described ants as the "Kings of Brazil" and "owners of the Amazon Valley". The honeybee in 1609 enjoyed the epithet "Sacred Bee", Napoleon I subsequently taking it as his personal emblem.

Insect phylogeny

Insects belong to the Phylum Arthropoda or jointed-limbed invertebrates the origins of which are in doubt. There is agreement that the immediate ancestors are annelids (worms) but at issue is whether this origin is mono- or diphyletic. The fossil record indicates the complexity of arthropod radiation. The three extant subphyla are: chelicerates (horseshoe crabs such as *Limulus*, spiders, scorpions, mites and ticks); crustaceans (including crabs and lobsters); and uniramеs (millipedes, insects and related groups). In the monophyletic view, the latter two subphyla are frequently united in the Mandibulata (jointed jaws). The polyphyletic view traces crustaceans from polychaete-type annelids and the uniramеs from oligochaete (earthworm)-like ones. This issue has been addressed by comparing ribosomal RNA from representative species. There was no evidence for a diphyletic origin and the validity of the Mandibulata grouping was sustained, although the millipedes occupied an unexpectedly primordial position in the phylogenetic tree [1.11].

Insect systems

Arthropods include the first truly terrestrial invertebrates of which insects comprise the largest class; the principal characters, essential for access by non-biologists to various Chapters, are outlined sparingly as follows. Characteristics contributing to insect success include: the ability of the epidermis to secrete a tough cuticle that functions as an exoskeleton; the evolution of jointed limbs serially repeated along the body; and the development of a haemocoel or blood-filled body cavity such that the main organs are bathed in blood (the true coelom being restricted to the cavity of the excretory organs and reproductive system). Such organ systems display quite advanced levels of organization [1.30].

The cuticle consists essentially of the polyglucosamine chitin, that may be impregnated by salts, proteins, and phenolic deposits, overlain by impermeable, non-chitinous layers. Tanning of the proteins by oxidized phenols represents sclerotization. This is most pronounced in insects in which the epicuticle comprises a double layer of protein and lipid, the latter conferring unwettability and the whole decreasing water loss, thus facilitating success in a terrestrial milieu.

Water-retention is also facilitated by the excretory system consisting of Malpighian tubules arising as outgrowths from the posterior gut. The distal part of each tubule pours out nitrogenous salts that, in conjunction with carbon dioxide, precipitate urates as crystalline uric acid that is then excreted; water and salt are resorbed from the proximal part of the tubule and from the rectum.

Insects are the first flying invertebrates and their wings are confined to the meso- and metathorax. Each wing consists of an upper and a lower membranous layer as extensions of the cuticle supported by a branching framework of veins, each a tubular extension of the haemocoel; an extension of the trachea and a nerve complete the arrangement. Wing membranes are secreted by the underlying epidermis and the anterior or mesothoracic wings are more heavily sclerotized than the larger metathoracic wings that are membranous. The anterior ones may function as covers or elytra.

The insect nervous system represents a major advance over that in the Annelida, probably stemming from the need to coordinate complex appendages. Such control is to some extent self-regulating, i.e. not brain-dependent, and is vested in the sympathetic nervous system so, a decapitated insect can still walk. The brain's significance as an inhibitor is apparent from the fact that a decapitated insect will attempt to perform conflicting functions simultaneously.

The insect nervous system consists of a dorsal brain and a ventral double chain of ganglia, connected by longitudinal and transverse commissures. The anterior three pairs of ganglia fuse to form the suboesophageal ganglion (SOG), nerves from which supply the mouth parts, and circumoesophageal connectives unite the brain to this ganglion. From it double commissures connect with the prothoracic ganglia, followed by the meso-, metathoracic, and six abdominal ganglia; the last of these represents a number of fused ganglia. From each pair of ganglia segmentally-arranged nerves run to various parts of segments to form the peripheral nervous system.

The brain consists of three pairs of closely-fused ganglia that supply eyes, antennae and labrum; the other mouthparts are innervated from the suboesophageal ganglia. The sympathetic nervous system comprises: an oesophageal system supplying the heart and foregut; a ventral system represented by a longitudinal nerve, between longitudinal connectives, dividing in each segment to supply the spiracles; and a caudal system arising from the last abdominal ganglion and supplying the hindgut and reproductive system (Fig. 1).

The hormonal system of insects controls metamorphosis and is quite sophisticated. Apterygota or wingless forms exhibit slight metamorphosis in which development proceeds from egg to nymphal stages, that differ only from the succeeding adult in terms of size and sexual maturity. Included in the Pterygota or winged forms are: the Exopterygota that also exhibit relatively slight metamorphosis and externally developing wing pads; and the Endopterygota that display pronounced metamorphism with immature stages quite unlike the adult. Essentially, the corpora allata of the brain produce the prothoracicotropic hormone (PTTH) that stimulates the prothoracic glands to produce ecdysone, the hormone that elicits moulting and pupal formation. Other brain cells produce a third or juvenile hormone (JH) which modulates the effect of ecdysone. Present in large concentrations, JH suppresses ecdysone thus allowing increased larval growth; decreased concentrations allow ecdysone to initiate pupal development; and its complete absence allows adult development (Fig. 2). In effect, JH regulates the genes for adult development [1.30, and references therein].

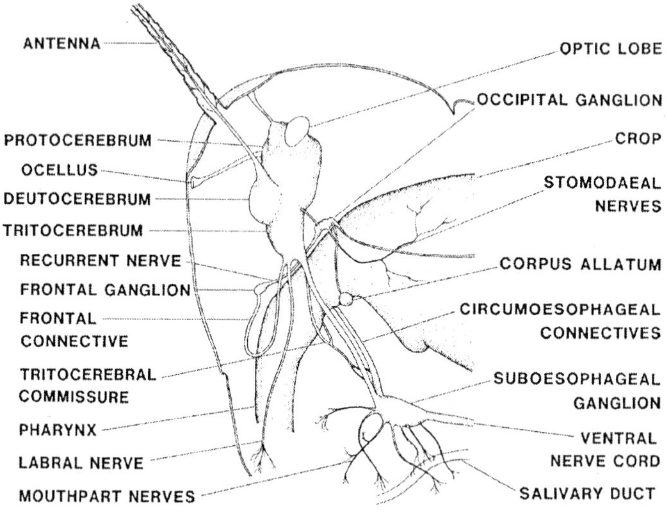

Figure 1. The insect brain and associated structures.

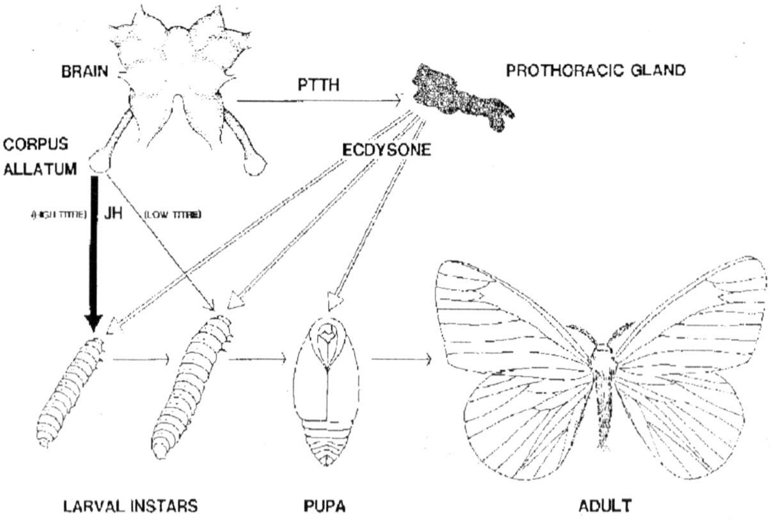

Figure 2. Hormonal regulation of insect metamorphosis.

Insect senses

The early naturalists recognized the significance of each of the five senses: vision, hearing, touch and the chemical senses, olfaction and taste, providing evidence that the latter two were of cardinal significance for insects. In 1670, Wray isolated formic acid, the simplest fatty acid, from a distillate of formicine ants. It was also established in that era that population densities in bee hives and ant communities reached orders of 50,000 and 500,000 respectively. In the late eighteenth and early nineteenth centuries observations by investigators such as Latreille, Forel, Huber and Sir John Lubbock began a focus on the primacy of chemical signals. Thus, Huber discovered slavery in ants and, while walking in the environs of Geneva in 1804, observed a full scale attack involving obvious communication within the ranks of the attackers and also between the defenders in a nest. Members of such colonies readily recognized each other and distinguished sibs from strangers. This led to the suggestion by Gélieu that each nest had a sign or password that gave admission to sibs and prevents strangers from entering. Other entomologists favoured the existence of a distinctive smell [1.20, and references therein]. It fell to Lubbock, parliamentarian, originator of the idea of the Bank Holiday, president of both the Linnean Society and The Institute of Bankers, to perform some decisive experiments.

In 1874, Lubbock ascertained that noises, loud, shrill, piercing or startling to the human ear, elicited no perceptible effect on ant behaviour. Furthermore, he established that ants do not communicate by sound the location of food sources by depositing honey on one of six small pillars of wood standing on a board 12 inches below the nest. Three ants put to the honey and allowed to feed were then confined there. The number of ants (seven) visiting the honey-rich pillar from the board did not exceed those visiting the other, untreated pillars. However, when the fed ants were allowed return to the nest, 54 visited in 45 min. In 1875, he introduced worker ants to a batch of larvae of a species appropriate for enslavement, providing, as a route to the workers hive, a series of narrow paper strips with movable strips acting as interconnecting bridges (Fig. 3). Switching the movable strips, after the ant had returned to the hive, showed unequivocally that investigating sibs tracked the trail of the first ant, even when this led them astray. Thus, they were not following directions to the larval source. After the first description of antennal sensory structures by Hicks, Lubbock [1.20] described their subcuticular structure in an outline drawing (Fig. 4) that foreshadowed many subsequent morphological studies.

It took a further 85 years for the investigation of insect-produced stimuli to achieve isolation, purification, and identification of active chemicals. Designated as pheromones, they were defined as substances which are secreted to the outside by an individual and received by a second individual of the same species in which they release a specific reaction, for example, a definite behaviour or developmental process. The sex attractant, (*E*)-10, (*Z*)-12-hexadecadien-1-ol or bombykol was purified from half a million abdomens of the female silk moth *Bombyx mori* [1.17]. Pheromones may have been primordial signals employed in animal communication, and as cells within the metazoan body may communicate through hormones, then a lineal evolutionary relationship between hormones and pheromones has been proposed [1.14]. This was reflected in the early use of the term ectohormone, although this was subsequently replaced by pheromone [1.17].

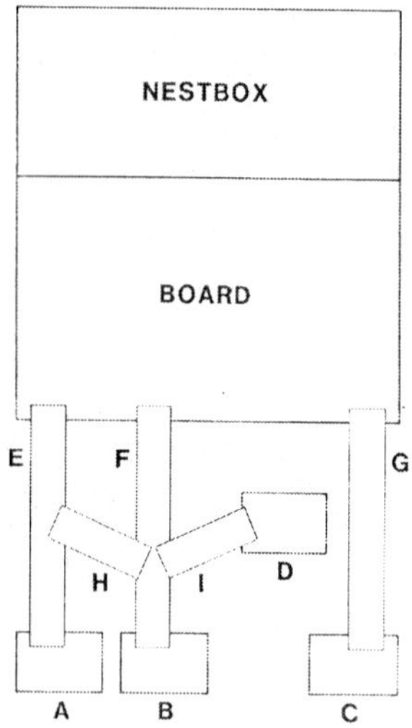

Figure 3. Lubbock's apparatus (redrawn) to establish that the presence of slave ants was signalled by chemical trails. A, B, C and D are pieces of glass connected with the board by narrow strips of paper. E, F and G: H and I are movable paper strips. Larvae suitable for enslavement were placed at C and D and a worker ant was introduced at D. After its return to the nest, other workers visited only D or other sites that terminated trails laid on the movable bridges.

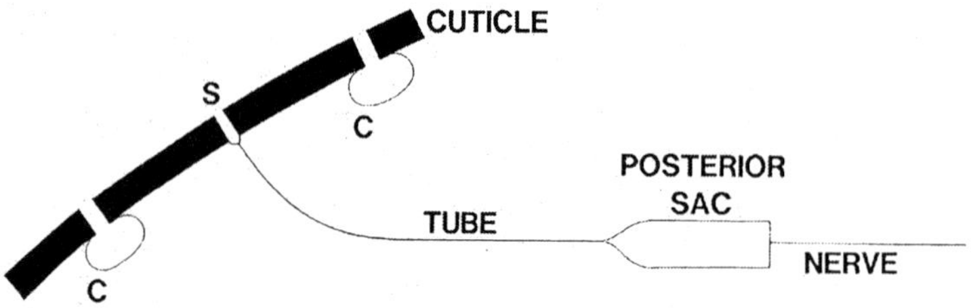

Figure 4. Lubbock's diagrammatic representation (redrawn) of a section through the terminal portion of the antenna of ant Myrmica ruginodis. *C, cork-shaped organ first reported by Hicks and subsequently by Forel; S, external chamber of stethoscope-shaped organ, the tube of which is, in fact, the axon.*

Early conceptual investigations of pheromones focussed on their information content. The total rate of spatial information transmitted in the fire ant trail is 0.1 to 5.0 bits/s, comparable to the transfer rate from the waggle dance of the honeybee. In principle, patterned transmission of pheromones through amplitude and frequency modulation could convey 100 bits, equivalent to approximately 20 words of English/second. Also from first principles came the rationale that most pheromone classes would have five to 20 carbons and molecular weights of 80 to 300. Relatively few kinds of molecules are readily synthesized below the lower limit; above it there is a rapid increase in molecular diversity and olfactory efficiency. Approaching the proposed upper limit, molecular diversity becomes so large that further increases confer no advantage but incur the disadvantage of increasing metabolic expense [1.31]. Subsequently, investigators tended to focus in the following order on: identification of the chemical components of pheromones; mechanisms governing their biosynthesis release, and evolution. Pheromones also served as investigative tools to reveal the neurophysiological and molecular mechanisms underlying the perception of odour.

Plant/insect relationships

Social insects also served to provide early evidence of the close and evolving association between plants and insects. Darwin wrote:

> Thus can I understand how a flower and a bee might slowly become, simultaneously or one after the other, modified and adapted to each other in the most perfect manner by the continued preservation of the individuals which presented slight deviations of structure mutually favourable to each other [1.6].

This observation anticipated the recognition of plant/insect coevolution as a distinct and significant area of investigation.

The fragrance of flowers was readily understood as attractive to bees, but the concept of chemical attraction was rather slowly extended to the relationship between solitary insects and their host plants, especially crops. It was almost 100 years after Darwin's comment that this chemically-mediated relationship was singled out for attention at the symposium on "Physiological Relationship between Insects and Their Host Plants" held at the International Congress of Entomology in Amsterdam, 1951. Another influential meeting was "Insect and Foodplant" in Wageningen, 1957. Fraenkel was associated with the widely accepted claim that host preference in phytophagous insects represented "the very heart of agricultural entomology". In lineal descent from this concept was his seminal paper "The Raison d'Être of Secondary Plant Substances" with the subheading "These odd chemicals arose as a means of protecting plants from insects and now guide insects to food" [1.12].

For this analysis Fraenkel ranged over a selection of plant families including Cruciferae, Umbelliferae, Leguminosae, Solanaceae, Moraceae and Gramineae. He drew attention to the distribution of sinigrin among the Cruciferae, to the quantitative relationship between glucoside content of food and the feeding response of the diamondback moth *Plutella maculipennis*, and to the necessity of treating non-cruciferous plants with sinigrin before *Plutella* would eat them. Accordingly, he claimed to:

> clearly demonstrate the function of secondary substances in plants as a means of repelling or attracting insects.

He also foresaw the applicability of this concept to plant relationships with other pest groups including mites, bacteria, and fungi. He acknowledged in full the existence of an earlier assessment by the German botanist Stahl that Fraenkel translated as:

> We have long been accustomed to comprehend many manifestations of the morphology [of plants], of vegetative as well as reproductive organs, as being due to the relations between plants and animals, and nobody, in our special case here, will doubt that the external mechanical means of protection of plants were acquired in their struggle [for existence] with the animal world. The great diversity in mechanical protection does not appear to us incomprehensible, but is fully as understandable as the diversity in the formation of flowers. In the same sense, the great difference in the nature of chemical products [excreta], and consequently of metabolic processes, are brought nearer to our understanding, if we regard these compounds as means of protection, acquired in the struggle with the animal world. Thus, the animal world which surrounds the plants deeply influenced not only their morphology, but also their chemistry [1.12].

This rationale of Fraenkel's, met with less than universal acceptance, and dissension in various reviews, stimulated a clarification of his position in regard to the importance of nutrients. The issue was not whether they play a role in host selection but whether they can explain it. His examination of the literature published since his 1959 paper indicated that nutrients failed this critical test. Drawing on studies investigating such insect responses as, induced feeding, oviposition, and sensory responses, Fraenkel showed the critical importance of secondary plant substances in example after example. The roles of hypericin, cucurbitacins, coumarins, volatile terpenes, and alkaloids are all identified, as is the fact that many of these compounds would be toxic or repellent [1.13]. The concept of the primacy of chemical signals was thus firmly established but more recent data have redrawn attention to the significance of plant nutrients for insects. The following two years brought a masterly and seminal categorization of the interactions between species [1.29] and the establishment of chemical ecology as the discipline devoted to their study [1.26]. Fundamental aspects of insect chemoreception are considered in regard to: plants in Chapter 2, pheromones in Chapter 3, sensory structures in Chapter 4, and neurophysiological and molecular mechanisms in Chapters 5 and 6 respectively.

APPLIED ASPECTS

The practical significance of insects was readily apparent to the scribe who wrote in the time of Rameses II:

> Worms have destroyed half of the wheat, and the hippopotami have eaten the rest; there are swarms of rats in the fields and the grasshoppers alight there.

It is salutary to reflect on the extent to which such difficulties have been overcome almost 3,500 years later.

It seems likely that, from prehistoric times, man has been troubled by insects, especially mosquitoes, flies, fleas and lice. Furthermore, man's profoundly social nature must have led to population concentrations that enhanced the scope for such infestations and for the transmission of diseases. The mosquito and tsetse fly transmit the protozoa causing malaria and sleeping sickness respectively, diseases that have afflicted up to 1,000 million people at a given time. Some flies elicit river blindness, others are vectors for bacteria that cause enteritis a major mortality factor for children in underdeveloped regions; additional insect-borne diseases include bubonic plague, dengue, typhus, yellow fever, and filariasis.

Cultivation of food plants and storage of the subsequent harvest would have provided favourable conditions for the emergence of phytophagous and stored product pests respectively. Man's crops are attacked in the field by locusts, an enormous array of Lepidoptera, and by Diptera and Coleoptera among others. Major sources of food and fibre thus damaged include cereals, rice, vegetables and fruit as well as cotton and timber. Insect pests were estimated to elicit crop protection costs of more than U.S.$6 billion p.a. world-wide and more than $400 million was spent against lepidopteran pests in the U.S.A. alone (1990 values). It must be acknowledged that total alleviation of these losses would not necessarily be beneficial as the resultant glut would bring many producers to ruin. However, such an eventuality is not in prospect.

First generation insecticides

About 1850, two important natural insecticides were introduced: rotenone from the root of the *Derris* plant, and pyrethrum from the flower heads of species of *Chrysanthemum*, both in use today (see below). Paris Green was used in the control of Colorado potato beetle *Leptinotarsa decemlineata*, in the State of Mississippi in 1867. Later shown to be hazardous to plants and animals, due to the inconsistency of different preparations, it was ultimate by used as a bait. In 1912, calcium arsenate was used to control the cotton boll weevil *Anthonomus grandis*, on cotton in the U.S.A. However, the extensive application of arsenical insecticides to fruit and vegetables proved to be dangerous. By 1956, 35% of accidental deaths associated with pesticides were caused by arsenicals necessitating their replacement.

Second generation insecticides

During World War I, the manufacture of explosives created a spin-off industry in synthetic insecticides. For example, *p*-dichlorobenzene (PDB) was recommended for control of the peach tree borer *Sanninoidea exitiosa*, by the U.S. Bureau of Entomology. The dinitrophenols, including derivatives of aromatic hydrocarbons, were also discovered in advance of any systematic work on insecticidal properties. These compounds were reported as ovicidal against the purple thorn moth *Selenia tetralunaria*.

Organochlorines

First prepared by Othmar Zeidler in 1874, the powerful insecticidal properties of 1,1,1-trichloro-2,2-bis(*p*-chlorophenyl) ethane (DDT) were not discovered until

1939 by Paul Müller of the Swiss Company, Geigy. These stimulated development of analagous organochlorine compounds comprising three main families related to DDT, cyclodienes including chlordane (Fig. 5), and lindane (hexachlorocyclohexane) respectively. The body louse *Pediculus humanus* is the vector of epidemic typhus and epidemic relapsing fever, diseases typical of population disruption and refugee camps. A typhus epidemic, threatening the whole population of Naples in 1943, was controlled by effective delousing using DDT. Also during World War II, DDT was used in the field to control lice, flies, and mosquitoes, and the availability of this compound to only the Allied powers minimized their casualties from insect-transmitted diseases. The main advantages of DDT seemed to be: stability; persistence of insecticidal action; cheapness of manufacture; low mammalian toxicity; and a wide spectrum of insecticidal activity. A poison of the nervous system, it apparently blocks the sodium channel [1.21].

Organophosphates

During World War II, German scientists under the direction of Schrader investigated the potent nerve gases, tabun and sarin for potential use in warfare, leading in 1941 to the preparation of octamethyl pyrophosphoramide known as schradan. Insecticidally-active organophosphate compounds (Fig. 5) comprise three classes, aliphatic, phenyl and heterocyclic derivatives.

Carbamates

In 1951, Ciba Geigy introduced carbamate insecticides, for example, Temik (Fig. 5). The most generally effective member of the group is carbaryl or Sevin, a contact insecticide with slight systemic properties. Effective against insect pests of fruit, vegetable and cotton, it biodegrades and is sometimes used as a DDT-substitute. Like the organophosphates, carbamate insecticides function as inhibitors of the enzyme acetylcholinesterase [1.21, and references therein].

Pyrethroid insecticides

Pyrethrums are extracted from the dried flowers of *Chrysanthemum cinerariaefolium* (=*Pyrethrum cinerariaefolium*), a member of the Compositae. Originating in Persia this plant was introduced into Europe in 1828, to the U.S.A. in 1876, and then to Japan, Africa and South America. By 1939, imports of pyrethrum to the U.S.A. reached 13.5 million pounds and by 1941 Japan was the major producer. Production declined after the Second World War with the advent of new synthetic insecticides. The demand world-wide for the pyrethrum flower remains in excess of 25,000 tonnes annually and is satisfied by the estimated 150 million flowers still hand-harvested in Kenya, Tanzania and Ecuador. There are six active constituents : pyrethrin I, jasmolin I, cinerin I, pyrethrin II, jasmolin II, and cinerin II (Fig. 6). The extracts were mainly used in domestic insect sprays against, for example, flies and mosquitoes but instability in air and light rendered natural pyrethrins of little value in crop protection. The first synthetic pyrethroid allethrin (Fig. 6) proved as effective as natural pyrethrin against the housefly *Musca domestica*, but instability precluded a practical use. Subsequently, resmethrin and bioresmethrin were synthesized, by Elliott at Rothamsted Experimental Station in 1967, with LD_{50} (dose lethal to 50%) values of 6 and 4 ppm respectively, when applied topically to the adult male desert locust *Schistocerca gregaria*. The same group synthesized permethrin as the first potent, photostable pyrethroid, which led to the development of a range of additional, potent and photostable molecules [1.9]. These have established pyrethroids as one of the principal categories of commercial insecticides accounting for about 25% of the global insecticide market.

Organochlorines

D.D.T

1,1,1-trichloro-2,2-bis(p-chloro-phenyl) ethane

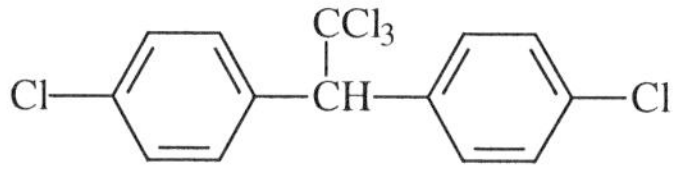

Chlordane

1,2,4,5,6,7,8,8-octachloro-3a, 4,7,7a-tetrahydro-4,7-methanoindane

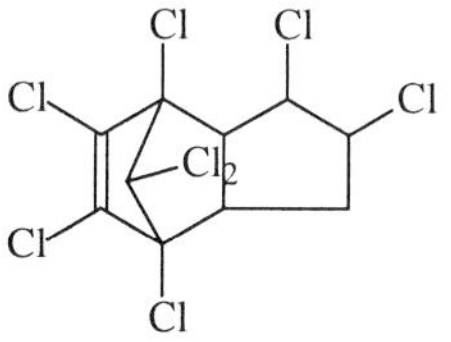

Organophosphates

Malathion

Diethyl mercaptosuccinate, s-ester o,o-dimethyl phosphorodithioate

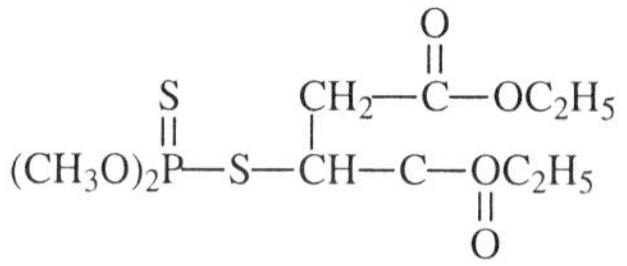

Carbamates

Aldicarb

(Temik)

2,methyl-2-(methylthio) propionaldehyde o-(methylcarbamoyl) oxime

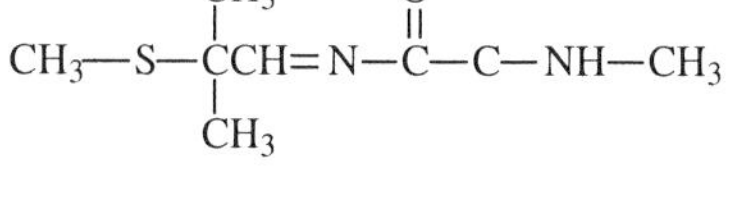

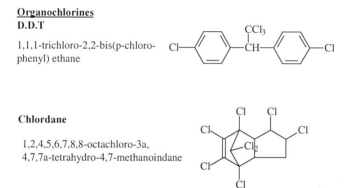

Figure 5. Structures of some synthetic insecticides.

Pyrethroids are neurotoxins affecting axons, sensory organs, neurosecretory neurons, and presynaptic nerve terminals. A general effect is increased neuronal firing followed by conduction block, and a specific target site is the voltage-dependent sodium channel that opens and shuts during the generation of the action potential. Pyrethroids seem to delay the closure of a proportion of such channels, allowing sodium ions to enter the neuron at the end of the action potential.

This creates a tail current that may decay rapidly if elicited by a heterogeneous group designated Type I pyrethroids, or slowly if elicited by cyano-containing pyrethroids Type II. The former produces repetitive discharges and the latter prevents achievement of the threshold potential; the effect is the same, a conduction block [1.25].

Pyrethrins
Pyrethrin I

(1R)-trans-2,2,dimethyl-3-(2-methylprop-1-enyl) cycloptopane-carboxylic acid

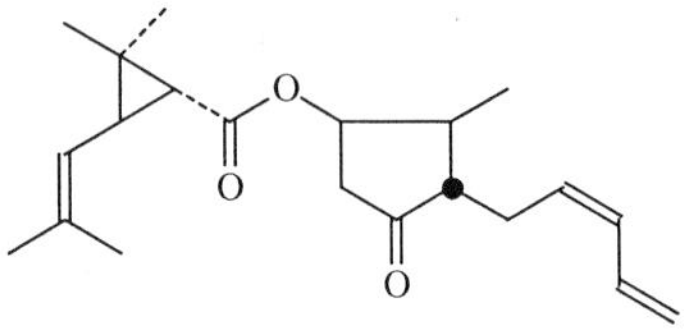

Cinerin I

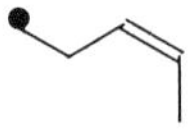

Jasmolin I

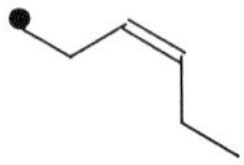

Pyrethrin II

(1R)-trans-2,2,dimethyl-3-(2-methylprop-1-enyl) cycloptopane-carboxylic acid

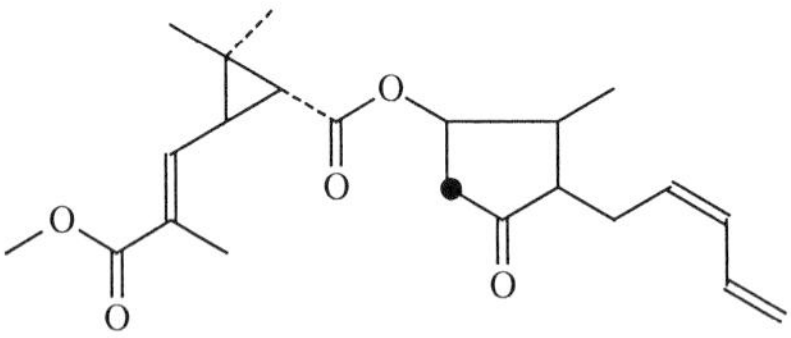

Cinerin II

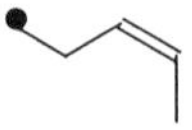

Jasmolin II

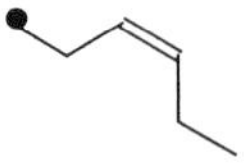

Pyrethroids
Allethrin

2-methyl-4-oxo-3(2-prenyl) 2-cyclo-penten-1-yl

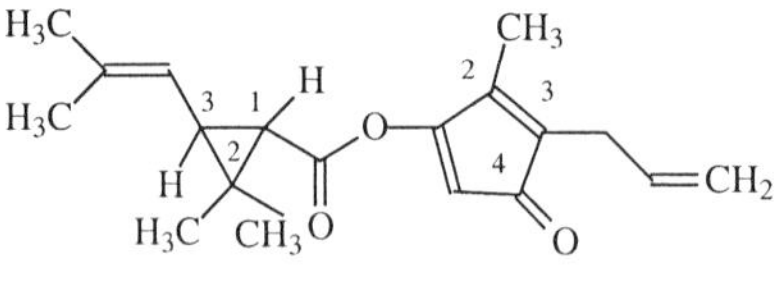

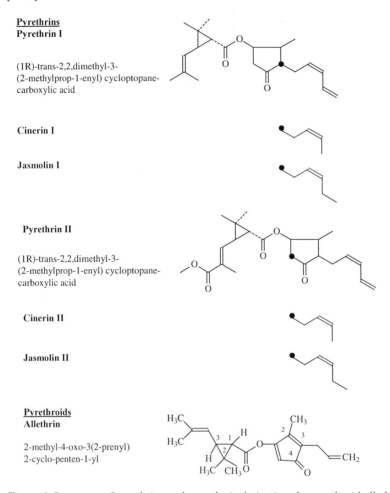

Figure 6. Structures of pyrethrins and a synthetic derivative, the pyrethroid allethrin.

Avermectins

Avermectins are a group of potent, broad-spectrum antiparasitic and insecticidal agents that inhibit neurotransmission by potentiating the release and binding of γ-aminobutyric acid (GABA) in nerve synapses, thus blocking GABA-mediated transmission of nerve impulses. Isolated from mycelia of a novel actinomycete *Streptomyces avermitilis*, avermectins are derivatives of pentacyclic 16-membered lactones. They separate into major components A_{1a}, A_{2a}, B_{1a}, B_{2a} and minor components A_{1b}, A_{2b}, B_{1b} and B_{2b}. Commercially-used ivermectin is a mixture of at least 80% 22, 23 dihydro-avermectin B_{1a} and not more than 20% 22, 23 dihydro-avermectin B_{1b} (Fig. 7). Ivermectin concentrations of 0.1-1.6 ppm elicit 100% mortality of the tsetse fly as adult male and female respectively [1.19].

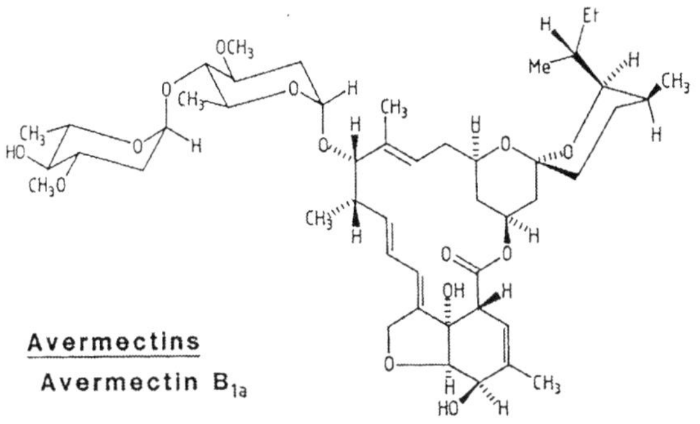

Figure 7. Structure of avermectin B_{1a} – the principal constituent of ivermectin.

Chitin inhibitors

Another category of insecticides takes effect by inhibiting chitin synthesis. A polymer of *N*-acetylglucosamine, chitin is synthesized from uridine diphosphate (UDP) *N*-acetylglucosamine wherein units of *N*-acetylglucosamine are transferred to the growing chain of chitin. The reaction is controlled by a chitin transferase, inhibition of which leads to an accumulation of UDP-*N*-acetylglucosamine. Treated insects, especially larvae, are unable to shed the old cuticle, and the new cuticle is distorted and deformed by lacking the appropriate lamellar structure.

Third generation insecticides

As indicated earlier, JH released from the corpora allata, governs larval growth and development and its absence facilitates development to the adult stage. Applied effectively JH prolongs larval life and may elicit growth abnormalities that are usually lethal; applied to insect eggs it blocks hatching. However, the effect may be offset by prolongation of larval feeding activities that may be deleterious to the host crop. Accordingly, their use is restricted to insects that are pests only as adults.

Balsam fir wood contains a steroid, juvabione, structurally similar to JH, that inhibits development of the European bug *Pyrrhocoris apterus*. A search for other, naturally occurring, JH mimics in plants has not proved very fruitful. Attention has focussed on synthetic mimics of JH leading to the registration in the U.S.A. of methoprene to control mosquitoes. As this and related compounds specifically affect the insect endocrine system, they have low mammalian toxicity and seem to have little effect on non-target aquatic species such as mayflies and water beetles. This represents, in principle, a significant advance over other synthetic insecticides, but disadvantages include: critical significance of the timing of the application; lack of persistence in crop situations; and cross-resistance elicited by previous treatment with conventional insecticides. In the event, such compounds have had little impact in the market place.

Fourth generation insecticides

Essentially, these are the behaviour-modifying chemicals such as antifeedants and pheromones. The former could protect fresh plant growth if they exhibit high potency and broad-spectrum effects combined with systemic action. To date no antifeedant fulfils all these requirements but azadirachtin, from the neem tree used by subsistence farmers, is one of the most promising (see Chapter 7).

Pheromones are potent and stereospecific effectors of insect behaviour and so, at first glance, seem to offer greater potential for manipulation of insect behaviour. Typically, but not exclusively, they influence mating behaviour of the male but their efficacy in direct pest control, through permeating the atmosphere and confusing the male as to the location of the female, is uneven. An alternative use is in monitoring pest population build-up to improve the timing of insecticide applications (see Chapter 9).

The use of synthetic insecticides was a major contributing factor to the success of the agricultural revolution (Fig. 8), viewed by some scientists as among the most important single steps taken by the human race. Overall, scientific and technical improvements have provided a steadily increasing food harvest such that in the quarter century from 1960, world grain harvests doubled to 1.66 billion metric tonnes, although the acreage increased by only 11%. In Europe, wheat yields tripled as did maize yields in the U.S.A.; rice yields doubled in Asia and many countries chronically short of food become self-sufficient, especially in grain. India, stricken by famine in 1965-67, became a food exporter. All of this has not abolished hunger in the world but the problem is more one of local than of total production. Interestingly, the U.S.A., the intellectual source of many of these advances, produced insufficient grain for its needs in 1988 [1.22].

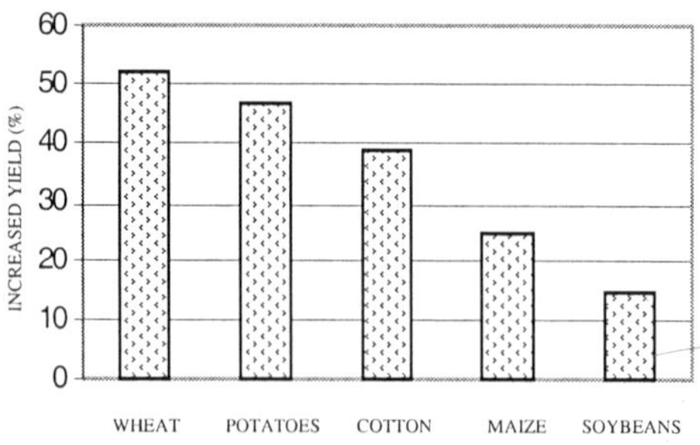

Figure 8. Increased yield of various crops attributable to treatment with insecticides. Compiled from Young [1.32].

ECOLOGICAL CONSEQUENCES

Concomitant with these enormous benefits, insecticides are associated with a range of ecological and medical problems especially: pest resistance; mortality of innocuous and beneficial species; and contamination of food. The social and environmental cost of pesticide usage was estimated at some US$800 million in 1980 [1.22].

Insect resistance

Exposure to insecticides elicits genetic selection for insects that are tolerant, either genetically or behaviourally. Over time this may render the insecticide ineffective necessitating the use of another with usually a different mode-of-action. This becomes an expensive treadmill especially through the need to meet the exacting safety standards of the Food and Drug Administration (FDA) and of the Environmental Protection Agency (EPA) in the U.S.A. One result of these high costs has been the extinction of many small companies that tested prototype compounds. Thus, the range available to the multinational companies is much restricted. This in turn impacts on the rate at which they can produce new, commercially-viable compounds and formulations. Specifically, marketing a new insecticide is the culmination of screening 15,000 compounds, up to 10 years research, with development costs of US$ 20-40 million (1989 values) [1.18]. One response has been to coordinate activities designed to prolong the useful life of an insecticide. Thus, the Pyrethroid Efficiency Group monitors the onset of resistance and advises on strategies to manage it through collective voluntary programmes.

Organochlorine insecticides elicit two different types of resistance mechanisms: one to DDT and analogues like 1,1-dichloro-2,2-bis-(4-chlorophenyl) ethane (DDD) and methoxychlor; the other, known as cross resistance, to cyclodiene derivatives and lindane. Among insects of agricultural importance more than 119 species have developed resistance to insecticides with the highest number among Hemiptera, 33, followed by Lepidoptera (as caterpillars), 29, and Coleoptera, 23. Significant crops threatened by pest resistance include, in the U.S.A., cotton by *A. grandis*, and by the cotton bollworm (= corn earworm, = tomato fruitworm) *Helicoverpa* (=*Heliothis*) *zea*. Cotton in Egypt is also threatened by the cotton leafworm *Spodoptera littoralis* [1.18].

Resistance is not confined to organochlorine, organophosphate and carbamate insecticides as there is laboratory evidence that insects can develop resistance to juvenile hormone mimics. Insect resistance frequently derives from enzymes that detoxify the insecticide. Enhanced production of the enzyme may increase detoxification through phosphorylation and inhibition of the insecticide. Alternatively, target enzymes such as acetylcholinesterase may mutate to exhibit decreased sensitivity to the insecticide. More than one breakdown mechanism may co-exist within the same insect, in which event resistance is multiplicative.

Resistance to organophosphorous insecticides is also derived from genes controlling enzyme production. Technological advances exploited to detect resistance include the use of microassays, in microtitre plates, for acetylcholinesterase and malathion carboxylester hydrolase as examples. Each microassay allows 30 readings from a single mosquito and may be conducted in the field or in the laboratory; the advent of hybridoma technology has facilitated the creation of monoclonal antibodies sensitive to resistant enzymes [1.5].

Mortality of beneficial and innocuous insects and of wildlife

Insecticides may leave toxic deposits, lasting from several weeks to years, that may kill beneficial insects with consequences such as target pest resurgence and secondary pest outbreaks.

Target pest resurgence
This occurs when the insecticide not only kills the pest population but also its natural enemies. For example, monocrotophos, an insecticide Federally-registered for bollworm control, destroyed bollworm predators eliciting a resurgence of the pest. Predators such as ladybirds and hoverflies are often killed by DDT sprays. Also organochlorines are very toxic to predatory carabid beetles such that a light dusting of dieldrin around the cabbage plants, to protect them from cabbage root fly *Delia radicum,* may induce more damage by killing predatory carabid beetles, such as *Bembidion lampros*, that eat pest eggs [1.28].

Secondary pest outbreak
This occurs when the insecticide kills not only the pest but also creates a new pest. For example, DDT sprays applied to control apple pests resulted in the fruit tree red spider mite becoming a new pest. Prior to the use of this insecticide the mite had led an inconspicuous existence on fruit trees. Both foregoing phenomena contribute directly to insecticide resistance by eliciting more frequent treatments that speed genetic selection for resistance, and lead increasingly to a "pesticide addiction" from which it is difficult to withdraw.

Mortality of wild life
One of the most telling arguments against the continued use of DDT and other organochlorines derives from their effects on wild life. Seed dressed with these compounds kills birds directly. In addition, pesticide run-off into streams concentrates in tissues of plankton at 3.0×10^{-4} ppm, in aqueous invertebrates at 1.0×10^{-3} ppm, and in fish at 5.0×10^{-1} ppm. This is fatal for many fish species and a further concentration step yields doses that adversely affect calcium metabolism in predatory birds. Thus, their egg shells become so thin that they fracture in the nest, leading to significant population decline by this indirect poisoning; land birds preying on small mammals are similarly affected. The outcry in the 1960's following population decline of species such as the peregrine falcon contributed to the

withdrawal of organochlorine pesticides from agricultural use in most of the Western World.

The exigencies of life in the Third World means that the advantages of significant potency and cheapness associated with such formulations are not foregone so readily. Although officially banned from farm use in Brazil since 1985, lindane is readily available under-the-counter and is the insecticide of choice for Brazilian cocoa farmers. Uncontrolled applications of organochlorines on cocoa ultimately impact on the developed world in the form of contaminated chocolate. This is not subject to statutory control in the U.K., as it is not considered to be a basic item of diet [1.15]. Unrestricted by dose constraints, such treatments significantly decrease the total insect population, thus threatening insectivorous mammals. The disappearance of the endemic shrew *Crocidura thomensis*, from much of the equatorial island of São Tomé, (Gulf of Guinea) is associated with insecticide sprayings of the early 1970's [J. Dutton, University of East Anglia, personal communication, 1993].

Contamination of food

In the relatively recent past humans regularly ingested small amounts of chlorinated pesticides persistent on foodstuffs. Of 28 restaurant meals analyzed in 1954, 24 contained at least 55 μg of DDT and 28 μg of DDE, the dichloroethylene derivative. Meat was contaminated by DDT but vegetables and fruit were apparently free, although amounts below 1 μg could not be detected. Two hundred and sixteen composite samples, collected from cooked food during 1964-65, were contaminated by pesticides but at a low level e.g. 0.001 ppm DDT, DDE, DDD, dieldrin, lindane, benzene hexachloride, aldrin and endrin. In 1961, contamination of milk residues was almost universal as judged from 4000 samples taken from widespread locations in the U.S.A.: 91% contained DDT, 43% DDE, 29% methoxychlor, and 12% had DDD. As a result it was suggested that DDT should not be used in or around dairy barns. The problem is exacerbated by the long persistence of DDT, typically up to 10 months in the bovine body [1.21].

Forty men with intensive and prolonged occupational exposure to DDT ingested 200 times more DDT than the average. However, no ill effects were noted apart from minor skin and eye irritations caused by DDT dust that cleared up when exposure was terminated. In a Romanian study, workers inhaling DDT at 21 mg/kg body weight as well as 6 mg/kg lindane daily for 30 days showed no ill effects. Volunteers fed 3.5 or 35 mg/man/day for one year, or equivalent to 20 and 200 times regular intake, followed by similar consumption for 27 months were observed for a further year. DDT storage in adipose tissue equilibrated after one year and no metabolic effects were detected. Specifically, production workers gave no evidence of ill-health with fat levels of 600 ppm: mean levels in the population at large in Europe and Asia are 4 and 17 ppm, respectively [1.7]. Present-day consumer resistance to pesticide residues seems to stem from a seeping fear of the infiltrating tentacles of cancer.

Cancer

Up to 80% of human neoplasms may depend upon environmental factors, and given that many chemicals elicit cancer in experimental animals, pesticide management practices seem relevant (Table 1). A comprehensive epidemiological study was made on cancer death rates for populations in agricultural areas in the southeastern U.S.A. for the period 1950 through 1975. A multiple regression model included as variables specific crops requiring heavy use of pesticides. This was adjusted for socio-demographic factors, contributing to known geographical variations in cancer mortality, within each 5-year interval from 1950-1975. In white males the percentage of harvested land in cotton or in vegetables was significantly associated with increased mortality from total cancer. Similar analyses for total cancer in white females and for lung cancer in both males and females showed some positive correlation with cotton and with vegetable production [1.8].

Table 1. Estimated quantification (%) of the causes of cancer (all types)[†]

Cause	Diet	Tobacco	Sexual Behaviour	Alcohol	Occupation	Geo - physical	Pollution	Industrial	Medicines
Cancer (%)	35	30	7	4	3	3	2	1	1

[†] Compiled from 1.3

The following pesticides are produced in quantities greater than 500 tonnes/year: lindane, DDT, dieldrin, carbon tetrachloride, polychlorinated biphenyls (PCBs), propylene oxides, 1,1-dimethyl hydrazine, trichloroethylene, benzylchloride, amitrole. Some of the foregoing compounds, identified in surveys of human adipose residue, have proven carcinogenic in experimental animals, and DDT was associated with a liver cancer in man. In contrast, are data for total DDT and dieldrin residues in adipose tissue compared from 122 cancer cases and 122 controls matched for age, race, sex and social class. The cases comprised 24 primary lung cancers, 14 gastrointestinal cancers, 29 breast cancers, and 15 generalized metastatic carcinomas. The mean total DDT was 8 mg/kg in cancer cases and 7.8 mg/kg in control, a difference that was not significant at the 5% level [1.8]. It might be expected that such early studies dating from 1975 would be clarified by more recent results.

In 1993, an epidemiological study of 58 women in New York indicated that breast cancer among women with the highest pesticide residues in blood was four-fold higher than in women with the lowest levels. The pesticide was DDE a breakdown product of DDT. A similar 1994 investigation in California used a more comprehensive sample of 150 women (50 white, 50 black, 50 Asian), who developed breast cancer and 150 matched controls. Furthermore, the second study used blood taken prior to the 1972 ban on organochlorine pesticides, when women were exposed to far higher levels than present, and with DDE levels four-to five-fold higher than in the New York study. There was no association between the pesticide and cancer [1.27]. Such data leave the link between organochlorines and cancer as tenuous indeed.

In fact, cancer of the stomach, a front-line organ in human interaction with ingested residues, has been falling steadily since the 1930's (Fig. 9). This might be associated with the elimination by pesticides of mycotoxins from food, or with decreased use of nitrate preservatives that, in turn, is linked to the wide availability of refrigeration and of fresh fruit and vegetables. Furthermore, there has been no general increase in mortality from liver tumours in the U.S.A. since long-lasting pesticides were introduced. In the population under age 65, there was a non-significant increase in mortality from liver cancer among women and a non-significant decrease in men.

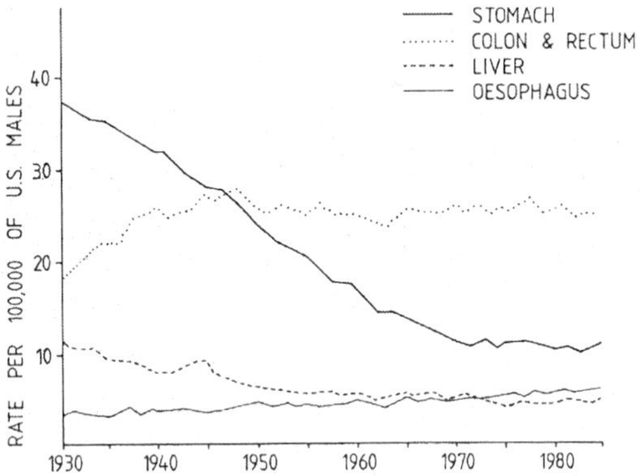

Figure 9. Death rates from cancer in U.S.A. males. Redrawn from data originally from U.S. Department of Health and Human Resources.

The present position

It is obvious from the foregoing that the first flush of scientific achievement at the potency and efficacy of insecticides has long been replaced by a sense of unease in the scientific and wider community. For many this process dates from the publication in 1962 of *Silent Spring* by Rachel Carson. Although polemical, this account of the environmental consequences of the use and misuse of pesticides represented a turning point. For the first time the insecticide industry, hitherto a spearhead of the agricultural revolution, was placed on the defensive and many young biologists sought investigate possible alternatives such as the use of predators and parasitoids in biological control. This was subsequently modified to the use of such natural enemies in conjunction with pesticides in integrated control, subsequently subsumed in the concept of integrated pest management (IPM). Other alternatives include the use of: insect-resistant crop cultivars; pheromones; naturally occurring insecticides and insect repellents.

An entirely new dimension has been added to such efforts by the views in the U.S.A. of the college-educated layman. Alarm at *Silent Spring*, associated hostility to the insecticide industry and especially to indiscriminate aerial spraying, was fed by distrust of big government dating from the Vietnam War, to form a growing constituency of doubt and rejection. The apparent association, however loose, of pesticide residues with cancer elicited a new sociological phenomenon - the chemophobic local community that led to a market for organically-grown foods.

In Europe, especially The Netherlands and West Germany, the educated young also expressed rejection of a range of perceived risks including: nuclear weapons; waters polluted by heavy industry and by nitrate run-off from farms; and insecticide-contaminated food. One result was a politically significant Green Movement.

Risk assessment

The raison d'être of publicized resistance to pesticide residues merits closer scrutiny as it almost certainly involves the perception of risk. Risk estimation entails an assessment of the size, probabilities and consequences of negative events, and varies greatly between informed and ill-informed individuals and groups. Thus, rankings by representatives from two U.S.A. groups (League of Women Voters, and college students) of the top five physical hazards correctly included smoking (150,000 actual deaths) and handguns (17,000); only one group recognized motor vehicles (50,000) and neither included electric power (14,000); both incorrectly included nuclear power (100) and pesticides (0) [1.4]. Similarly, a ranking by the EPA of environmental problems, revealed after peer review, that the public's greatest worries such as hazardous waste disposal and oil spills, were trivial by comparison with actual dangers from: radar; stratospheric ozone depletion; air pollution; climatic change; habitat alteration; and species extinction [1.24]. Such data leave little doubt that the opinions of the educated non-scientist may be ill-founded on scientific matters. Yet such opinions may set the political context within which scientists have to operate.

A contributing factor to such perceptions of risk may derive from extrapolations. Quantifying a carcinogenic dose of say C for a chemical may lead to the assumption that a daily dose of 1×10^{-3} C ingested over 1×10^{3} days will be carcinogenic. Accordingly, such a carcinogen would be, over a sufficiently long time interval, unsafe at any dose. This found expression in a 1993 U.S.A. court ruling that the 1958 Delaney Clause or Amendment, requiring any level of a certified chemical carcinogen to be removed from food for human consumption, be strictly applied. Objectors claimed this would ban pesticides and increase food costs, ruin growers and processors, and enhance soil erosion as no-till agricultural practices would decline, although the latter claim was disputed [1.23]. Enacted in 1958, this clause neglects the concept of thresholds, i.e. levels below which a chemical has no effect or, like oxygen, has a beneficial one. There is an obvious need to establish if there are thresholds for the effects of carcinogens encountered by humans, and if these are affected by carcinogen interactions. In the event, the Delaney Clause was removed by enactment in August 1996 of a new law, the Food Quality Protection Act, driven by a court ruling that enforced the Delaney Clause and that would have banned outright the use of pesticides. This law sets a threshold of 1 ppm for known carcinogens or toxins, and provides for a review of regulated chemical and other pesticides within 10 years, at which time all such compounds must fulfil the new standard [1.16].

A second element in the disposition to pesticide residues is the notion that unprotected food is intrinsically healthy, wholesome, and risk-free. After all, it is natural. A powerful series of papers by Ames and coauthors points out that this entirely overlooks the elaborate range of plant chemicals, associated with plant defence, that are carcinogenic (Table 2). It is estimated that human diet has 10,000-fold more natural than synthetic pesticide (1.5 g/day compared with 150 μg/day). Three cups of coffee equivalent to 13 g roasted coffee contain some 750 mg chlorogenic acid, 300 mg caffeine, and 24 mg caffeic acid; a mutagen, an inhibitor of DNA repair, and a rodent carcinogen respectively. An additional several hundred milligrams of toxins are ingested daily from plant phenolics, with flavonoids and glucosinolates contributing a similar amount. Potato and tomato would add a further one hundred milligrams as would saponins from legumes [1.2].

The common constituent of brassicas, allyl isothiocyanate, is a mutagen and as a carcinogen elicited bladder papillomae in the male rat, an extremely rare condition in controls; sinigrin (allyl glucosinolate, up to 1,500 ppm in brussel sprouts) is cocarcinogenic in the rat pancreas. 5- and 8-Methoxypsoralens, light-activated mutagens that when topically applied elicit skin cancer in rats, represent 32 ppm of cooked parsnip [1.2]. Applying the "carcinogens-are-unsafe-at-any-dose" rationale would lead to banning these and indeed most other vegetables from the human diet. Yet a diet rich in vegetables seems to confer protection against cancer and it is claimed that such plant constituents as vitamin E, carotene, selenium, glutathione, and ascorbic acid are anticarcinogenic [1.1]. In addition, the human race will have inherited, from hominid and earlier ancestors, genes facilitating the detoxification of plant carcinogens. However, these arguments are not sufficient to relax vigilance in regard to pesticides, or other industrial products such as halogenated hydrocarbons, that our ancestors have not encountered in food. Against this background, it seems that chemical insecticides will continue in general use for a very long time, at least on the basis of continuing control of their residues on food.

Table 2. Rodent carcinogens naturally occurring in food plants[†]

Compound	Food Plant	Conc. (ppm) In Food
Sinigrin when converted to allyl isothiocyanate	Brassicas	12-1560
Limonene	Carrots, Citrus	31
5- and 8-Methoxypsoralens	Parsnip	32
Caffeic acid	Coffee	1800

[†] Data from 1.2 which gives extended list with mutagenicity e.g. allyl isothiocyanate is mutagenic at 0.0005 ppm or $2x10^5$-fold less than the concentration of the percursor sinigrin in cabbage.

This is not to argue for the perpetual use of insecticides. Their concentration in food chains with harmful effects on predators is acknowledged above; also serious questions arise regarding their long-term effectiveness in terms of pest resistance. Since the 1940's, there has been a 33-fold increase in the tonnage of chemical pesticides applied to croplands in the U.S.A. (Fig. 10) with a 10-fold increase in their toxicity. But losses assignable to insects, weeds and fungi have increased from 31 to 37% due to pest resistance and to the ideal conditions created for pests by monocultures [1.22]. So a long-term perspective is likely to include greater rotation of crops and increasing use of non-pesticidal control, especially in judicious integration with pesticides, i.e. IPM [1.22].

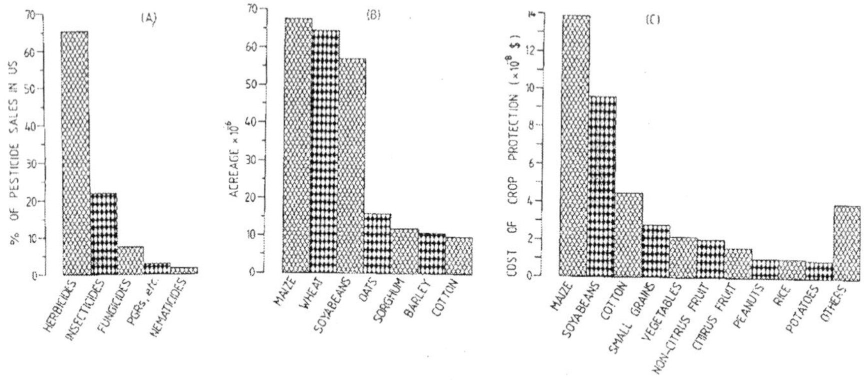

Figure 10. Basic statistics (1994 values) of pesticide usage in the U.S.A. in terms of: (A) pesticide type; (B) crop acreages protected; and (C) cost. Compiled from various sources.

There has been a 50% decrease in pesticide usage in Denmark, Sweden, The Netherlands, and the Canadian province of Ontario; Sweden aims for a further 50% decrease without decreasing crop yield or appearance. Indonesia has decreased pesticide use by 65% but rice yields have increased by 12% [1.22].

Applied aspects of insect chemoreception are considered in regard to: naturally occurring insecticides and repellents (Chapter 7); insect-resistant crop cultivars (Chapter 8); and the use of pheromones (Chapter 9). A further and very exciting possibility is represented by insect control through genetic engineering of natural toxins into crop species (Chapter 10), thus completing the scope of this book.

REFERENCES

1.1 Ames, B.N. (1983) Dietary carcinogens and anticarcinogens. *Science* **221**, 1256-1264.

1.2 Ames, B.N., Profet, M. and Gold, L.S. (1990) Dietary pesticides (99.99% all natural). *Proc. Natl. Acad. Sci.* **87**, 7777-7781.

1.3 Anon (1986) *Cancer Control Objectives for the Nation*; 1985-2000; NCI Monograph; US Department of Health and Human Cervices. U.S. Government Printing Office. Washington, DC.

1.4 Berry, C.L. (1990) The hazards of healthy living - the agricultural component. *Proc. Br. Crop. Pr. Con.* **1**, 3-13.

1.5 Brown, T.M. and Brogdon, W.G. (1987) Improved detection of insecticide resistance through conventional and molecular techniques. *Ann. Rev. Entomol.* **32**, 145-162.

1.6 Darwin, C. (1859) *On the Origin of Species*. John Murray, London.

1.7 Davies, J.E., Barquet, A., Morgade, C. and Raffonelli, A. (1975) Epidemiologic studies of DDT and dieldrin residues and their relationships to human carcinogenesis. In *Proceedings of International Symposium, Recent Advances in the Assessment of the Health Effect of Environmental Pollution* (EUR 5360), Vol. 2, Luxembourg, Commission of the European Communities, pp. 695-702.

1.8 Durham, F.W. (1983) *Pesticides and Human Health*. Academic Press, New York.

1.9 Elliott, M. and Janes, N.F. (1978) Synthetic pyrethroids - a new class of insecticide. *Chem. Soc. Rev.* **7**, 473-505.

1.10 Fabre, J.H. (1918) *The Sacred Beetle and Others* (Eng. trans.). Hodder and Stoughton, London.

1.11 Field, K.G., Olsen, G.J., Lane, D.J., Giovannoni, S.J., Ghiselin, M.T., Raff, E.C., Pace, N.R. and Raff, R.A. (1988) Molecular phylogeny of the animal kingdom. *Science* **239**, 748-753.

1.12 Fraenkel, G.S. (1959) The raison d'être of secondary plant substances. *Science* **129**, 1466-1470.

1.13 Fraenkel, G. (1969) Evaluation of our thoughts on secondary plant substances. *Entomol. exp. appl.* **12**, 473-486.

1.14 Haldane, J.B.S. (1955) Animal communication and the origin of human language. *Sci. Prog.* **43**, 385-401.

1.15 House, R. (1990) The cocoa slaves of Brazil. *Int. Ag. Dev.* **10**, 10-14.

1.16 Hoyle, R. (1996) Delaney clause knocked out by surprise compromise act. *Nature Biotech.* **14**, 1056.

1.17 Karlson, P. and Butenandt, A. (1959) Pheromones (Ectohormones) in insects. *Ann. Rev. Entomol.* **4**, 39-58.

1.18 Knight, A.L. and Norton, G.W. (1989) Economics of agricultural pesticide resistance in arthopods. *Ann. Rev. Entomol.* **34**, 293-313.

1.19 Lasota, J.A. and Dybas, R.A. (1991) Avermectins, a novel class of compounds: implications for use in arthropod pest control. *Ann. Rev. Entomol.* **36**, 91-117.

1.20 Lubbock, J. (1915). *Ants, Bees, and Wasps*. Kegan Paul, Trench, Trübner & Co. LTD. London.

1.21 O'Brien, R.D. (1967) *Insecticides - Action and Metabolism*. Academic Press, New York.

1.22 Pimentel, D. (1991) *Handbook of Pest Management in Agriculture*. CRC Press, Boca Raton.
1.23 Pimentel, D. and Pimentel, M. (1993) Pesticides and the Delaney Amendment. *Science* **260**, 1409-1410.
1.24 Roberts, L. (1990) Counting on science at EPA. *Science* **249**, 616-618.
1.25 Soderlund, D.M. and Bloomquist, J.R. (1989) Neurotoxic actions of pyrethroid insecticides. *Ann. Rev. Entomol.* **34**, 77-96.
1.26 Sondheimer, E. and Simeone, J.B. (1970) *Chemical Ecology*. Academic Press, New York.
1.27 Taubes, G. (1994) Pesticides and breast cancer: no link? *Science* **264**, 499-500.
1.28 Van Den Bosch, R. and Messenger, P.S. (1973) *Biological Control*. International Textbook Company, Aylesbury, UK.
1.29 Whittaker, R.H. and Feeny, P.P. (1971) Allelochemics: Chemical interactions between species. *Science* **171**, 757-770.
1.30 Wigglesworth, V.B. (1972) *The Principles of Insect Physiology*. Chapman and Hall, London.
1.31 Wilson, E.O. and Bossert, W.H. (1963) Chemical communication among animals. *Res. Pro. Hormone Res.* **19**, 673-716.
1.32 Young, A.L. (1987) Minimizing the risk associated with pesticide use: An overview. In Ragsdale, N.N. and Kuhr, R.J. (eds.) *Pesticides: Minimizing the Risk*. ACS Symposium Series 336, American Chemical Society, Washington, DC, pp. 1-11.

PART I

FUNDAMENTAL ASPECTS

CHAPTER 2

PLANT CHEMICALS

Two major chemical resources for insects are food and pheromones. Without the latter insects will not reproduce, but without the former they will not live long enough to try. So it is reasonable to open the examination of chemical stimuli with an account of food constituents. To condense this into a single chapter is a daunting task given that a recent account of plant-insect interactions extends to five volumes [2.13], and a recent consideration of behaviour in host-plant selection merited a separate text [2.14], as did one category of behaviourally-significant chemicals [2.44], the evolutionary approach to chemical ecology [2.46], and the influence of microbes [2.6]; an up-date on the interaction of herbivores with secondary plant metabolites required two volumes [2.47]. Accordingly, a single Chapter can do little more than provide an intellectual framework within which significant components may be picked out and linked.

Feeding in phytophagous insects involves a sequence of stereotyped behavioural components: (1) host plant recognition and orientation; (2) initiation of feeding (biting or piercing); (3) maintenance of feeding; and (4) cessation of feeding, followed usually by dispersal. Each innate behavioural component may occur only in response to a particular cue usually a chemical.

Phytophagous insects represent three categories in terms of the specificity of their host plant range: monophagous or species restricted to plants of a single species or at most a few closely related ones; oligophagous or species that feed on plants within one family or members of closely related ones and including "disjunct oligophagy" where the connection between the hosts is not understood; and polyphagous or species that utilize host plants from more than one botanical order. Host plant specificity may be further subdivided in terms of the plant parts that are utilized (leaf, stalk, and root) and by differences in the feeding specificities of the larval and adult stages of the insect. Thus, in the northern corn rootworm *Diabrotica longicornis*, larvae are monophagous, but adults are polyphagous.

Monophagous insects are significant because the combination of dense populations and restricted host range may inflict considerable damage. In contrast, polyphagous insects spread their feeding activity thus causing less apparent damage to any particular host. However, the combined losses from a polyphagous species, such as the locust, may exceed those from a monophagous species attacking one host within the range.

This might suggest that most green plants offer a readily available supply of highly suitable food for insects. But it is noteworthy that only nine of the twenty-nine insect orders use seed plants in a major way, despite the ubiquity of green plants, leading to the suggestion that herbivory entails overcoming hurdles such as nutrition, attachment, desiccation and host finding [2.55]. In order to survive, every plant species must defend itself against most phytophagous insects and pathogens. Such defence rests on two principal mechanisms: morphological/physical, and chemical/biochemical of which only the latter is of interest here.

CHEMICAL DEFENCE BY NUTRIENTS

Plant nutrient levels affect growth, development, survival and fecundity of insects. For example, mean percentage, N, P and K is significantly larger in susceptible germplasms of maize *Zea mays*, as compared with those resistant to the rice stem borer *Chilo zosellus*. In the cabbage aphid *Brevicoryne brassicae* changes in the concentration of only one amino acid (threonine) explain much of the variation in the insect's fecundity. Other results discounted the importance of nutrients in determining feeding choices by insects. For example, host and non-host plants for the cotton boll weevil *Anthonomus grandis*, all contain the amino acids required by the pest and quantitative differences were insufficient to distinguish between host and non-hosts. In addition, it was believed that as most species of insects do not differ greatly in their qualitative requirements for nutrients, once the host plant satisfies nutritional requirements, nutrients are unlikely to play more than a minor role in host plant specificity [2.50, and references therein].

However, studies of interactions between tundra plants (deciduous shrub, evergreen shrub, deciduous forb, tussock graminoid, moss and lichen) and eight generalist herbivores (four vertebrates and the following four lepidopteran larvae: *Polia* spp., *Apentesis* spp., *Parasemia parthenost* and *Gynaephora rossii*) represented a turning point [2.18, 2.20]. Specifically, feeding preferences were strongly and positively correlated with levels of N, P and K, and were strongly and negatively correlated with levels of cellulose and lipid. This was interpreted as indicating that tundra generalist herbivores select for a diet high in protein and P; and low in fibre, resins, and terpenes. However, factors not studied include: leaf morphology; hairiness; trichomes; saponins; and previous feeding experience of the herbivores. The data led to the suggestion that browsing by herbivores may be influenced by C/nutrient ratios in the plant [2.20]. Resource availability from the environment was proposed as the factor principally affecting plant defence in terms of investment and type. Thus, limited resources favour plants with inherently slow growth rates that, in turn, favour large investments in antiherbivore defence. Leaf lifetime, also affected by resource availability, influences evolution of mobile or immobile defences. It may be questionable if this general approach involving polyphagous species is also significant for monophagous or oligophagous ones.

Some recent data defining resistance of *Z. mays* to the fall armyworm larva *Spodoptera frugiperda* are relevant. Free amino acids were more useful than any other factor, including toxins, in defining leaf-feeding resistance in this relationship. Ratios of essential amino acids (arginine, histidine, isoleucine, lysine, methionine, phenylalanine, valine, threonine, tryptophan and leucine) were similar for susceptible (S) and resistant (R) lines. But differences existed in ratios of non-essential amino acids: asparagine and proline were higher in S lines; and tyrosine and asparagine, the only ones non-stimulatory for feeding, were higher in R lines. Amino acid extracts and reconstituted mixtures of S lines stimulated more feeding than R lines. However, larvae grew as well on diets based on S and R mixtures, perhaps because levels and ratios of essential amino acids were similar. It was suggested that the feeding preference for the amino acid composition of S lines, coupled with a larger hemicellulose content of R lines, provided the best available explanation for resistance to feeding by *S. frugiperda* larva [2.29].

In contrast to the primary substances, proteins, carbohydrates, nucleic acids and fats, defence-related compounds are categorized under secondary plant substances. More than 80% of all known natural products are of plant origins and some 30,000 plant secondary compounds figure in relationships between plants and disease organisms, and between grazing animals and food. Major chemical classes involved in plant-animal interactions are: (1) phenolics, e.g. simple phenols, flavonoids including tannins and quinones; (2) nitrogenous compounds, e.g. alkaloids, amines, "non-protein" amino acids, cyanogenic glycosides, glycosinolates; (3) terpenoids, saponins, limonoids, cucurbitacins, cardenolides, carotenoids; (4) other compounds, e.g. polyacetylenes [2.64, and references therein]. In 1971, the term "secondary plant chemicals" was replaced by allelochemics, used interchangeably with allelochemicals, and defined as: non-nutritional chemicals produced by an individual of one species that affect the growth, health, behaviour, or population biology of another species [2.66]. An example was the suppression of the growth or occurrence of some higher plants by chemicals released from another higher plant, or allelopathy.

CHEMICAL DEFENCE BY ALLELOCHEMICALS: STRUCTURAL CATEGORIES

Phenolics

Plant polyphenols or vegetable tannins comprise three principal subdivisions: proanthocyanidins; glycosylated flavonols, and hydroxycinnamoyl esters. Their biosynthesis in plants produces compounds with a molecular weight ranging from 500 to ca 3,000 and may be related to a plant's ability to synthesize lignin. Proteins in small concentrations may interact with polyphenols by hydrogen binding to form a hydrophobic surface layer leading to aggregation and precipitation. Proteins in large concentrations crosslink using the polyphenols as bridges followed also by precipitation. Polyphenol formation may facilitate protein precipitation without the need for large concentrations of phenol, potentially damaging to the plant.

Tannins account for at least 5% of oak leaf dry weight and their inhibitory effect on winter moth larva *Operophtera brumata* is dose-dependent. The proposed mechanism entails binding by tannins to leaf protein, thus denying leaf nitrogen to the larva. Precluding larval access to leaf protein was seen as a difficult obstacle for larval counter-adaptation. However, caution has been recommended in regard to evolutionary and ecological generalizations based on apparent tannin/insect relationships with the reminder that tannins may enhance growth and survival of some insects [2.12, and references therein]. Given the content of this and other general reviews, it is appropriate to focus here on the investigation of the proposed physiological effect of tannins i.e. protein precipitation.

Experimental scrutiny employed spinach ribulose-1,5-bisphosphate carboxylase/oxygenase (RuBPC) the most abundant protein in photosynthetic tissues, representing up to 50% of soluble proteins and 25% of total proteins present. Tannic acid and quebracho precipitated many times their own weight of RuBPC and furthermore, mature foliage of pin, bar and black oak contained sufficient tannins to precipitate all RuBPC present. Yet pin oak tannins had 20-fold greater protein-precipitating potential than had black oak. So there was no basis for a claim that the more tannin-rich oak foliage was less digestible than that of black oak. This challenged the assumption that differences in tannin levels indicate corresponding differences in either nutrient availability or defence against herbivores. Precipitation of RuBPC was maximal near pH 4.0, extensive between pH 5.6 and 7.0, and minimal above 7.5. As midgut pH values larger than 9.0 are not uncommon among Lepidoptera it would seem that protein precipitation of tannins would be relatively uncommon in the insect gut [2.40].

Insoluble tannin-protein complexes are readily solubilized by detergents. Thus, 0.3g of tannic acid in gut fluid of the tobacco hornworm *Manduca sexta*, adjusted to pH 6.5, precipitated only 0.3mg RuBPC, as compared with precipitation of all 2.0mg RuBPC in buffer at the same pH. The surface tension of the *Manduca* gut fluid indicated the presence of surfactants, although these were not identified. However, lysoleathin and linolyl glycine, two surfactants occurring in insect gut, were as effective as gut fluid in preventing RuBPC precipitation of tannic acid. Accordingly, there was no further justification for assigning to tannins a role in meaningfully decreasing digestibility of plant tissue for insect herbivores. Subsequent data indicated that detergency was far more effective than alkalinity in decreasing protein-precipitation of tannins, even when the action of K, Mg, and Ca ions was taken into account [2.41, 2.42].

Ingested phenols are readily modified in the insect gut to highly reactive and toxic quinones. Thus, oxidation of phenols decreases assimilation of proteins and amino acids in the larva of the tomato fruitworm *Helicoverpa zea*. In contrast, ingested phenols may decrease susceptibility of this species to pathogens such as the nuclear polyhedrosis virus [2.27].

Recent data emphasize gut redox conditions in the insect herbivores' ability to provide a variable response to, and thus cope with, such allelochemicals. In a redox reaction there is a transfer of electrons, with or without an accompanying transfer of protons. The redox couple will exist in a reduced or oxidized form depending on the availability of protons and electrons in the system, and high and low availability favour reduction and oxidation respectively. The midguts of *M. sexta* and the *Polia latex* provide a milieu comparable to that of a waterlogged organic soil, reducing but not very aerobic. Interestingly, the foreguts and hindguts of these species are mildly oxidizing and transitional between anaerobic and aerobic. In contrast, oxidizing conditions predominate, through the guts of the gypsy moth *Lymantria dispar*, the common monarch butterfly *Danaus plexippus*, and the tiger swallowtail butterfly *Papilio glaucus*; soil under these conditions would be rated as aerobic. Given that the midgut is the location of food digestion and absorption in the Lepidoptera, then these significant interspecific differences might reflect their evolved strategies for coping with ingested allelochemicals such as phenols [2.3].

Terpenoids

This is the largest and most diverse group of plant organic compounds amounting to at least 15,000 chemicals known from plants. Exhibiting a remarkable range of structural diversity, and as yet with no known function in plant growth and development, they are firmly in the cohort of secondary plant chemicals serving as attractants for pollinators, and as toxins against herbivores [2.36, and references therein].

Acetyl-coenzyme A or acetate is the starting unit for terpenoid production, and mevalonic acid is the point of departure for production of non-terpenoids. Despite the remarkable structural complexity of terpenoids, their common ancestry is discernible by subdividing into repeating C_5 units, each having the C skeleton of isoprene and usually joined in a head-to-tail pattern (Fig. 11) [2.36, and references therein].

Distinct stages in the biosynthesis of monoterpenes (C_{10}), sesquiterpenes (C_{15}), and diterpenes (C_{20}) include: production of isopentenyl pyrophosphate (or diphosphate) (IPP) followed by isomerization to form dimethylallyl pyrophosphate (DMAPP), and subsequent condensation of such C_5 units to geranyl pyrophosphate (GPP); monoterpenes are derived in the cytoplasm from the latter, and 1'-4 additions to it of IDP units produce farnesyl diphosphate (FDP) that leads to sesquiterpenes, or geranylgeranyl diphosphate (GGDP) that leads to diterpenes (Fig. 11) [2.15].

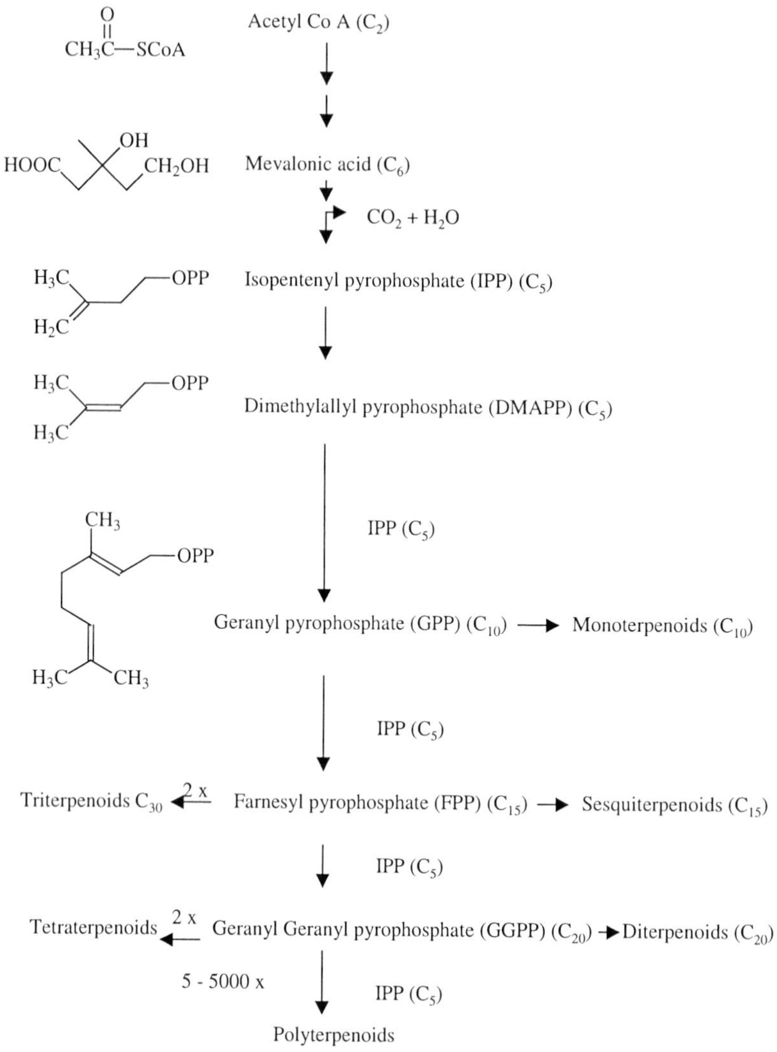

Figure 11. Schematic outline of terpenoid biosynthesis. Redrawn after Langenheim [2.36].

Wounding of grand fir *Abies grandis* induces biosynthesis of mono-, sesqui-, and diterpenes through gene activation, and enhanced activities of appropriate terpene synthases that account for the remarkable diversity of this group. Specifically, the synthases determine the structure and numbers of the ring systems, regulate electrophilic cyclizations, and protect intermediates. Such similarities in function undoubtedly derive from observed similarities in primary structure of the synthases studied so far. Some 40 synthases of plant origin have been cloned and all belong to one of three subfamilies of the *Tps* gene family [2.15, and references therein].

Monoterpenes

Monoterpenes (C_{10}) are among the most conspicuous constituents of the green plants examined by entomologists for chemicals affecting insect behaviour. Volatile and fragrant, they collectively account for much of the odour of plants as perceived by humans. As similar checklists of specific monoterpenes are reported from many different plant species, variations in their odours might reflect variations in their blends. Monoterpenes elicit a rich repertoire of responses from various insect pests as exemplified by conifer resin implicated in tree defence against bark beetles (Scolytidae). Specifically, concentrations of myrcene, α-pinene, and Δ^3-carene are elevated in *A. grandis* attacked by the fir engraver *Scolytus* spp. [2.45]. Myrcene, limonene and Δ^3-carene are toxic to western pine beetle *Dendroctonus brevicomis*, the latter two compounds amounting to almost 30% of the volatiles of resistant Ponderosa pine trees *Pinus ponderosa*. α-Pinene, limonene, camphene, geraniol and α-terpineol are attractants for bark beetles (the role of monoterpenes as pheromone precursors is considered in Chapter 3). Limonene is also an attractant for *A. grandis*, but kills the southern pine beetle *D. frontalis*. For the housefly *Musca domestica*, (-)-limonene is an attractant and (+)-limonene is a deterrent. The aldehyde *trans*-2-nonenal in carrot and its analogue *trans*-2, *cis*-6-nonadienal repel and kill the carrot fly larva *Psila rosae*, and a range of nonadapted pests. Attractants for *P. rosae* larva are: bornyl acetate, biphenyl, α-ionone, β-ionone and 2, 4-dimethyl styrene, concentrations of which are elevated in resistant cultivars. For completeness, it is relevant that α-pinene, limonene and terpineol are mildly toxic to mammals such as the rat, having dose lethal to 50% (LD_{50}) values of several g/kg. Accordingly, it is reasonable to view the original effect of these compounds as defensive and the attractant role of limonene, for example, as arising by insect adaptation [2.49, and references therein]. As with phenolics, the application of biochemical techniques to explore the mode of action has proved fruitful.

There are substantive similarities between molecular structure of many monoterpenes and that of acetylcholine (ACh), the natural ligand of acetylcholinesterase (AChE) (Fig. 12). Obviously, the anionic moiety of ACh has no counterpart in monoterpenes but the hydrophobic and carbonyl moieties are represented. Hydrophobic binding is established as the mechanism by which many organic compounds inactivate proteins, and the extent of hydrophobic binding to AChE by methyl carbamates is closely correlated with their *in vitro* potencies as insecticides. Compounds that inhibit or inactivate AChE cause ACh to accumulate at the cholinergic site. This produces continuous stimulation of cholinergic nerve fibres throughout the central and peripheral nervous system, followed by paralysis and death.

Five monoterpenes, representing a range of functional groups, have been investigated: citral (aldehyde), pulegone (ketone), linalool (alcohol), (-)-bornyl acetate (ester), and cineole (ether). As these monoterpenes are characteristic constituents of plant leaves, the test insect, the red flour beetle *Tribolium castaneum* a pest of stored grain, may be viewed as a nonadapted species. All five monoterpenes killed *T. castaneum* dose-responsively and beetles became paralyzed before dying. All were reversible, competitive inhibitors of electric eel AChE [2.49]. Subsequent data confirmed these effects with AChE purified from insect brain giving, for each inhibitor, a value for the inhibition constant (K_i) within the same order of magnitude as that for electric eel AChE [2.35].

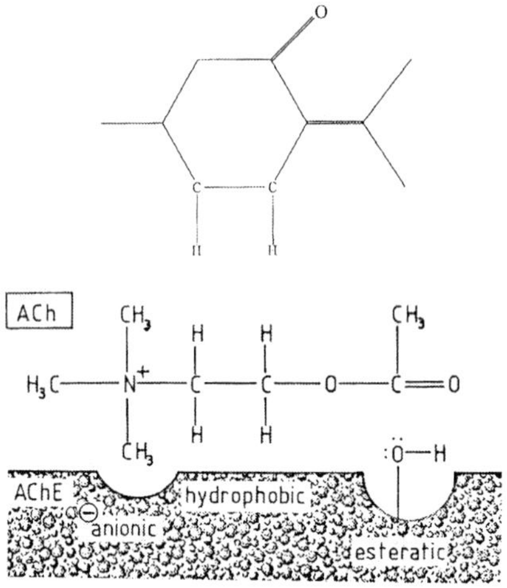

Figure 12. Diagrammatic representation of the hydrophobic and esteratic binding sites for monoterpenes in the active centre of acetylcholinesterase. Redrawn after Ryan and Byrne [2.49].

Although these particular terpenoids inhibit cholinesterase and kill a nonadapted insect, they are attractive to others. Specifically, linalool, a constituent of, for example, pine is an attractant for the silk worm *Bombyx mori.* (-)-Bornyl acetate is a host-location cue for *P. rosae* larva, and it seems particularly relevant that strong and statistically significant attraction was elicited by a concentration associated with 97% mortality of *Tribolium.* Thus, the mortality data are not assignable to concentration effects. (+)-Bornyl acetate is a mimic of the sex pheromone of the American cockroach *Periplaneta americana.* Citral, a constituent of citrus fruits, is also an alarm and defence pheromone of ants but an attractant and assembly pheromone for the honeybee worker [2.49].

The process by which insects adapt to new host plants was almost certainly catenary but there is debate on the sequence of steps. Detoxification of chemical defences might take place first, followed by behavioural adaptations. The contrasting view is that behavioural changes came first, followed by biochemical adaptations. The sequence could vary from one situation to the next and, indeed, that the sequence need not be significant [2.9, 2.10, and references in both]. In either event, the mechanisms of adaptation are of fundamental significance not least in the context of the spread of rain forests. Those of Central Africa are believed to have originated from small, isolated, patches of forest termed refugia. Insects are a major factor in pollination, but to serve such a function they would need to overcome or evade plant chemical defences, including monoterpenes.

Detoxifying enzymes of insects belong to the general categories of oxidases, hydrolases, transferases, and reductases. Insect success in detoxifying cholinergic inhibitors, as evidenced by strains with developed resistance to organophosphate and carbamate insecticides, is assigned to specific enzymes. One is a modified carboxyesterase that apparently lost most of its esterase activity, acquiring phosphatase activity instead; another is gluthathione-S-transferase. In addition, AChE has mutated in resistant strains to be more slowly inhibited than in susceptible ones: 1.2- to 6.0-fold less in organophosphate-resistant strains and 17- to 1570-fold less in carbamate-resistant strains. Accordingly, this enzyme should be considered where insect coevolution with, or adaptation to, plant hosts involves an interaction with monoterpenes [2.49, 2.35].

Sesquiterpenes
These C_{15} compounds formed from three isoprenoid units constitute the largest grouping of terpenoids. Conspicuous for their antifeeding activity are those with adjacent aldehydes, such as polygodial and warburganal, (see Chapter 7). The most numerous (ca 3,500) are sesquiterpene lactones occurring mainly in Compositae (Asteraceae) especially in hairs and latex ducts. Serving as toxins and relatedly as feeding deterrents, their activity seems dependent on the exocyclic α-methylene group which apparently acts through alkylation of nucleophiles. Particular targets would be the amino and sulphydryl groups of proteins with responses that include production of lesions. The sesquiterpene dimer gossypol in cotton glands decreases feeding, growth, development, and survival of cotton pests. This compound also reacts with acyclic monoterpenes such as myrcene and the sesquiterpene caryophyllene oxide to produce C_{25} compounds called heliocides. Toxic in their own right they also synergize the effects of gossypol. Nevertheless, pests of cotton such as the cotton leaf worm *Alabama argillacea* and *A. grandis* are insensitive to the defensive effects of gossypol, the latter species exploiting it as a feeding stimulant. It also binds to food proteins or/and digestive enzymes at a rate of up to four amino groups/gossypol molecule [2.28, and references therein].

Diterpenes
These C_{20} compounds are well established as feeding deterrents especially in interactions between conifer resins and sawflies. First-year growth needles, with 10-fold enhanced levels of diterpenes as compared with older growth needles, are avoided by young larvae in favour of the latter. Such diterpenes decrease larval growth, development, and survival. However, final instar larvae are unaffected by such high concentrations to the extent that they actively seek them, sequester them in

pouch glands of the foregut, and discharge them to deter predators. Diterpenes from the Euphorbiaceae and Thymeleaceae are toxic to insects and elicit skin blistering in mammals; the milky sap of the Euphorbiaceae is used both in folk medicine and in the study of cellular development [2.28].

Clerodane diterpenes, mainly from the Lamiaceae, Verbenaceae, and Asteraceae, are potent feeding deterrents for lepidopteran larvae. Activity is attributed to oxygenated functional groups on the 10-membered ring, and to the four furan group. Specifically, hydrolysis of the latter to an ene-dialdehyde group corresponding to that of drimane dialdehyde sesquiterpenes is invoked as the source of their activity [2.28].

Triterpenes
The largest terpenoid molecules, their basic configuration of six isoprene units may be modified such that 30 carbon atoms are no longer present as in cucurbitacins, tetracyclic structures derived from Cucurbitaceae, serving as toxins and feeding deterrents against insects and mammals. Beetles of the genus *Epilachna* (Coccinellidae) employ a behavioural stratagem to avoid the effects of cucurbitacins. By excavating a circular trench in the squash leaf such that the circumscribed zone is only narrowly connected to the rest of the leaf, they avoid the elevated cucurbitacin concentrations that are elicited by feeding.

The related families Rutaceae, Meliaceae, and Cneoraceae, produce limonoids, a large group of strongly oxygenated compounds having a fundamental configuration of 26 carbons. The best known limonoid with a plant-defensive role is azadirachtin, occurring in the neem *Azadirachta indica* and chinaberry *Melia azedarach* trees (for the potential of this compound in regard to pest control see Chapter 7).

Saponins are a large assemblage of widely distributed triterpenoid glycosides with soapy properties derived from their hydrophobic (terpenoid) and hydrophilic (sugar) moieties, subdivided into steroidal (C_{26}) and triterpene (C_{30}) saponins. Serving as feeding deterrents and toxins, saponins apparently take effect by binding in the insect gut to the free sterols that serve as precursors for moulting hormones. Saponins readily perforate cell membranes, and serve as significant molluscides [2.28, and references therein].

Alkaloids

This massive group of more than 12,000 compounds constitutes one of the most diverse assemblages of plant defence chemicals and is associated with some of the most potent known biological effects. Alkaloids have been variously defined as: cyclic compounds bearing nitrogen in a negative oxidation state and with limited distribution in plants; and more recently as nitrogenous compounds having at least one nitrogen atom derived from an amino acid and also with limited distribution

[2.62]. The latter definition incorporates nonbasic and noncyclic structures, as for example simple amines and amides, that otherwise would be termed protoalkaloids. It would also exclude nitrogen-containing diterpenoids as pseudoalkaloids. Occurring in almost all categories of microorganisms and especially in marine life forms, alkaloids are associated with more than 20 % of flowering species, and 60-70 % of species in the Solanaceae and Apocynaceae produce alkaloids. Their nitrogenous structure enables competition with a range of neurotransmitters thus promoting their use either naturally occurring or modified as: muscle relaxants; local anaesthetics; analgesics; antihypertensive; antimicrobial and antiarrythmic agents; as well as antimalarial and anticancer agents. Accordingly, 10 of the 12 principal plant-derived pharmaceutical agents are alkaloids. Within plants, alkaloids are preferentially distributed in regions subject to attack by herbivores such as peripheral cell layers, young tissues and inflorescences, such that alkaloid composition of the whole plant may be as low as 0.1-2 %, but 5 % of *Rauwolfia* root bark and 25 % of dried latex from poppy [2.67, and references therein].

Biosynthesis of alkaloids essentially entails Schiff base formation and Mannich condensation. Alkaloids are readily characterized in terms of their biogenic origin and particularly in terms of the nitrogen source. Thus, tyrosine and phenylalanine give rise to the 1-benzyltetrahydroisoquinoline group with the common skeleton C_6-C_2-N-C_2-C_6 (type A) and also C_6-C_2-N-C_1-C_6 (type B): isoprene or acetate units donate the carbonyl to the Mannich condensation to generate simple isoquinoline alkaloids C_6-C_2-N-C_n (type C); and a substantially modified derivative of dopa, betalamic acid (type D) serves as a building block for the betalains (Fig. 13). Tyrosine and tryptophan are nitrogen sources for emetine and indole-loganin alkaloids respectively, while ornithine is the source for both tropane and pyrrolizidine types, and lysine for quinolizidines [2.62, and references therein].

Sequestration

Although lipophilic compounds such as terpenoids may be stored in oil containers, resin ducts and trichomes, hydrophilic compounds such as alkaloids and glycosides may be confined to idioblasts, or to vacuoles termed defence and signal compartments. Weakly basic alkaloids may enter the vacuole by simple diffusion but the tonoplast represents a penetration barrier to polar alkaloids and glycosides, if they are charged in physiological pH. Accordingly, a carrier mechanism is required which may entail an alkaloidal channel, or a transporter protein that may also serve primary metabolites. The higher concentration of alkaloids inside as compared with outside the vacuole necessitates a driving force to overcome the concentration gradient. Prime candidates include H^+-ATPase that in conjunction with pyrophosphatase constitutes a proton motive force, specific ATPases, and glutathione –S-transferase. A supplementary process is "trapping" by which molecules already in the vacuole are modified such that they no longer influence the concentration gradient. Specific processes include: "ion-trapping" following the conversion of alkaloids to a more polar form; binding to the internal membrane of the tonoplast; conformational changes; complexing with compounds such as chelidonic and meconic acid, malic acid and malate, and chlorogenic acid [2.69, and references therein].

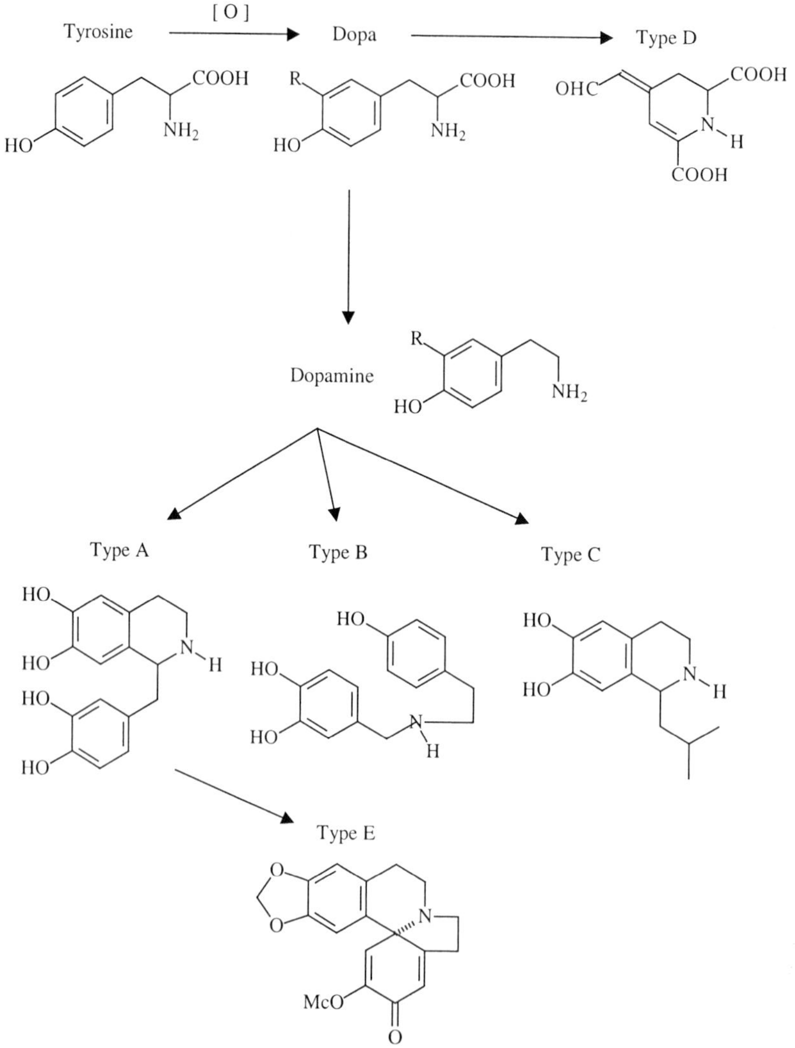

Figure 13. Biosynthetic pathways for four categories of alkaloids from tyrosine (details in text). Redrawn after Waterman [2.62].

Effects on insects

The most obvious effects include feeding deterrency and toxicity. A checklist of 55 plant-derived alkaloids indicated 54 as eliciting feeding deterrency, 18 as toxic, and

15 as active in both modalities [2.67]. As mentioned previously, the presence of a nitrogen atom and other structures may facilitate: competition with neurotransmitters for the corresponding receptor; inhibition of the deactivating enzyme in the cleft; transmitter biosynthesis and uptake. Specific neurotransmitters affected include acetylcholine, γ aminobutytric acid (GABA), dopamine, and serotonin. It is hardly coincidental that the first two are especially significant for neurotransmission in insects. Other modes of action include: voltage-gated Na, K, and Ca ion channels; Na^{+}/K^{+}-ATPases; enzymes involved in phosphorylation; DNA/RNA as affected directly by binding (intercalation), and indirectly by inhibition of DNA and RNA polymerases [2.68, and references therein].

Polyacetylenes and other compounds

The rather motley assemblage of "other compounds" includes several groups that are phototoxic and seem biosynthetically unrelated, so phototoxicity may have evolved independently and repeatedly, presumably because its effects are so potent. Among such compounds are polyacetylenes (polyyines), and sulphur derivatives including the nematicide thiophene from marigold *Tagetes* spp. Other phototoxic groups include extended quinones (hypericin), furanocoumarins (psoralen and xanthotoxin), beta-carbolines (harmane), furoquinolines (dictamnine) and furanochromones (khellin) [2.2].

Phototoxins take effect either through photogenotoxicity or photosensitized oxidations. Eliciting the former effect are the furanocoumarins, furanoquinolines, and furanochromones, the planar structure of which facilitates intercalation with RNA and DNA, and specifically with bases such as adenine (A) and thymine (T), that are associated with regulatory bases of TATA. Absorption of photons elicits formation of mono- and difuractional adducts that attack and severely damage chromosomes. Photosensitized oxidations occur only in the presence of oxygen and are associated with the polyyines, thiophenes, and quinones. The mode of action involves activation of the sensitizer by a photon firstly, to an excited state and then to a longer-lived triplet state. This in turn reacts with oxygen to catalytically produce millions of aggressive toxic, singlet oxygens. Biological molecules affected include unsaturated fatty acids and protein histidines.

Phototoxic effects on insects are best documented for the photoxidants that elicit black cuticular lesions that are fatal when moulting fails to shed the exuviae. Sublethal effects include: feeding deterrency; impaired assimilation of food as a result of punctured goblet cells in the midgut; delayed larval development; and pupal teratogenicities. The fact that several phototoxins co-occur with synergists that inhibit various detoxifying enzymes such as polysubstrate monooxygenases (PSMOs) is interpreted as consistent with evolution of plant defensive chemistry [2.2, 2.10, and references in both].

ALLELOCHEMICALS: FUNCTIONAL CATEGORIES

In 1971, the introduction of the term allelochemic or allelochemical was accompanied by a valuable and extended checklist of plant animal interactions mediated by allelochemics, and an account of biosynthetic pathways was included [2.66]. Two classes of allelochemics were particularly relevant to plant-insect interactions: (1) allomones, conferring an adaptive advantage on the producing organism (host plant); (2) kairomones, giving an adaptive advantage to the receiving organism (e.g. phytophagous insect). Subsequently, two further classes were distinguished: (3) synomones, benefiting both sender and receiver; and (4) antimones, maladaptive to both sender and receiver [2.64].

Allomones

The existence in plants of chemicals representing the first category, i.e., providing defence against herbivores, was long established and general identities such as phenolics and glycosides were well known. For example, plants of the Cucurbitaceae (including cucumber, watermelon and squash), synthesize cucurbitacins (hereafter Cucs) that rank as the bitterest chemicals known, and are perceived by humans in concentrations as low as 1 ppb (Fig. 14). Trace amounts elicit mouth and lip partial paralysis. Cucs are also toxic to mammals with LD_{50} values in the ranges of 1.2-6.8 mg/kg for mice injected intraperitoneally, and 5.0-40.0 orally. Cattle, sheep and humans may be poisoned following ingestion, especially where fruit has undergone a mutation to the allele for maximum bitterness. Cucs deter feeding by a broad range of insect herbivores. They also mediate tritrophic interactions as 24%-72% of *Diabrotica* species that fed on the fruit were rejected by praying mantids; none fed a control diet were rejected [2.43, 2.44] (for tritrophic interactions see below).

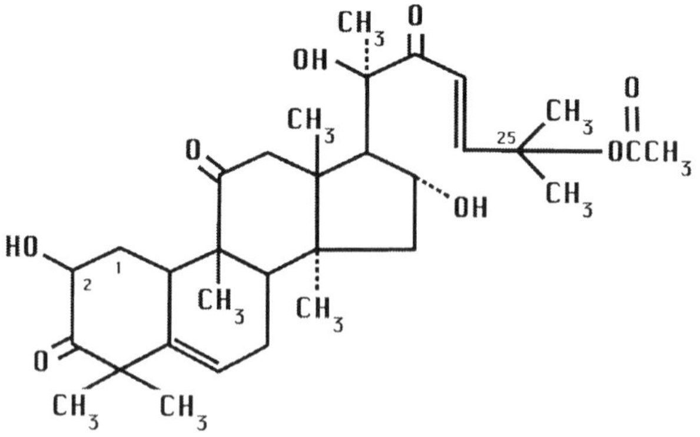

Figure 14. Structural formula of Cucurbitacin B occurring in 91% of all species characterized; Cuc D (69%) is C_{25}-OH; Cuc G (47%) is C_{24}-OH,C_{25}-OH; Cuc E (42%) is C_1=C_2; and Cuc I (22%) is C_1=C_2,C_{25}-OH. Compiled from Metcalf and Lampman [2.43].

Many plant species avoid the metabolic expense of investment in permanent chemical defence by producing allomones only when necessary, and as regulated by signal transduction pathways. Thus, insect attack on leaves of tobacco plants *Nicotiana tabacum* elicits extra production of jasmonic acid that migrates to the root system to evoke manufacture of the principal allomone, nicotine. This is conveyed to the leaf where it paralyzes the insect herbivore. Increased levels of jasmonic acid occur five minutes from attack, reach the root within two hours, and nicotine pulses are recorded at the leaf five hours later; this is not too slow as it may take several days for a caterpillar to eat a leaf. Once triggered, a plant's response is faster subsequently, as three doses of jasmonic acid (JA) applied six days apart elicited a much faster response than the first two [2.4]. This is an example of direct defence induced in a plant by an insect, and so serves as a line of defence additional to the presence of a constitutive toxin.

Plants produce a range of volatiles, derived from oxidative degradation of leaf lipids, known as green leaf volatiles (GLVs). These include such 6-carbon alcohols, aldehydes and derivative esters as 1-hexanol, hexanal and (*Z*)-hexenyl acetate. Intact plants emit low GLV levels that are substantially elevated in plants damaged physically, or by insect feeding. Soon after cotton plants are damaged by beet armyworms *Spodoptera exigua*, the emitted volatiles are characterized by elevated levels of lipoxygenase-derived volatiles such as (*Z*)-3-hexenol, (*Z*)-3-hexenyl acetate, and by a mixture of terpenes, cyclic (α-pinene, caryophyllene) and acyclic (myrcene) respectively. Exposing cotton plants to radiolabelled carbon ($^{13}CO_2$) for 36 hr, from 2 hr after the initiation of feeding, elicited significant ^{13}C incorporation as follows: (*E*)-β-ocimene, 92 %; linalool, 86 %; (*E,E*)-α-farnesene, 80 %; (*E*)-β-farnesene, 99 %; and the homoterpenes (*E*)-4,8-dimethyl-1,3,7-nonatriene (DMNT) and (*E,E*)-4,8,12-trimethyl-1,3,7,11-tridecatetraene (TMTT), 93 % and 54 % respectively. Accordingly, these compounds were synthesized in response to insect attack [2.58] and given the established mode of action of linalool, for example, as an AChE inhibitor, there is little doubt that they are a direct defence against herbivory. Intermediate levels of incorporation were exhibited by: myrcene, 35 % and limonene, 20 %; low levels were exhibited by: α-pinene, 5 %; β-pinene, 5 %; (*E*)-β-caryophyllene, 5 %; and α-humulene, 5 %. These latter compounds were derived from storage or were biosynthesized by a separate pathway [2.58]. Several biosynthetic pathways are involved as the compounds released include indole, together with various terpenoids including mono- and sesquiterpenes, the homoterpenes mentioned above, and GLVs (Fig. 15) [2.58, and references therein].

Levels of monoterpenes and other volatiles are much higher in response to insect attack than to mechanical damage alone, which is attributed to the oral secretion or regurgitant of the herbivore. The regurgitant of the cabbage white butterfly *Pieris brassicae* contains β-glucosidase that cleaves glycosides to release volatiles, and leaves treated with commercially available β-glucosidase released a blend of volatiles similar to that elicited by regurgitant. The principal elicitor in the oral secretion of *S. exigua* larva is volicitin, N-(17)-hydroxylinolenoyl)-L-glutamine (Fig. 16). The biosynthetic pathway entails adding a hydroxyl group and glutamine by the insect to linolenic acid obtained from the plant. Linolenic acid is an essential fatty acid for Lepidoptera and induces feeding in *M. sexta*, so this direct, induced interaction comprises the plant supplying the substrate for the herbivores oral secretion that enhances the production of allomones. Jasmonic acid is also produced from linolenic acid [2.58, and references therein].

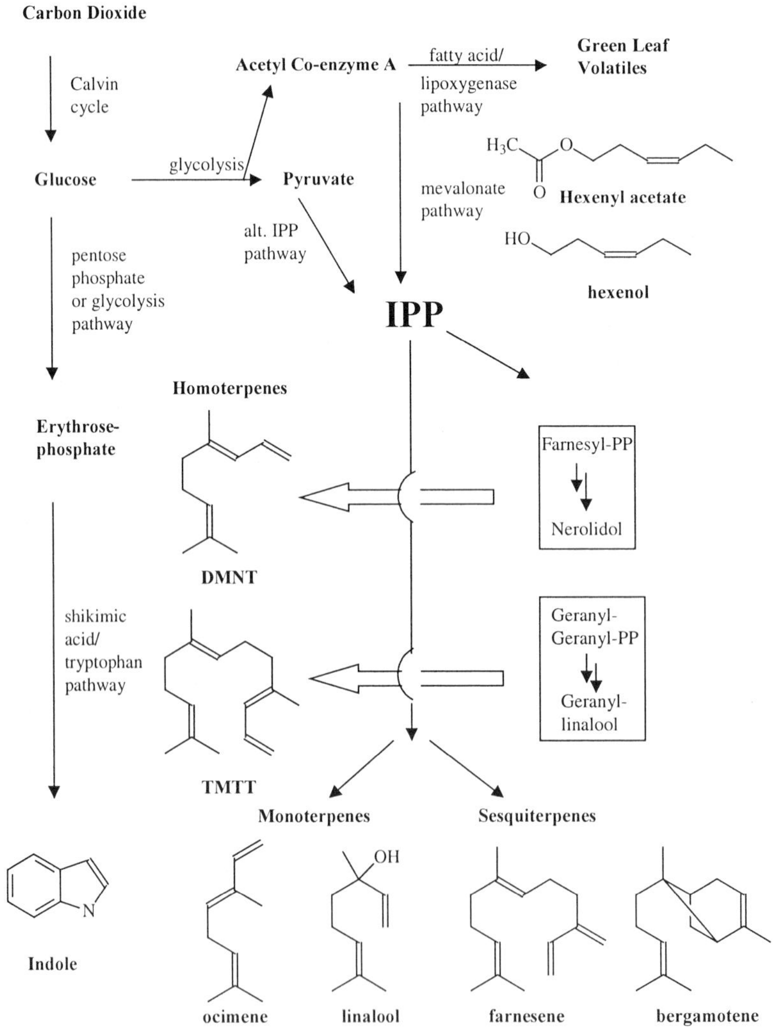

Figure 15. Biosynthetic pathways for the production by damaged plants of significant volatiles (indole, homoterpenes and various other terpenoids, and green leaf volatiles) Redrawn after Tumlinson et al. *[2.58].*

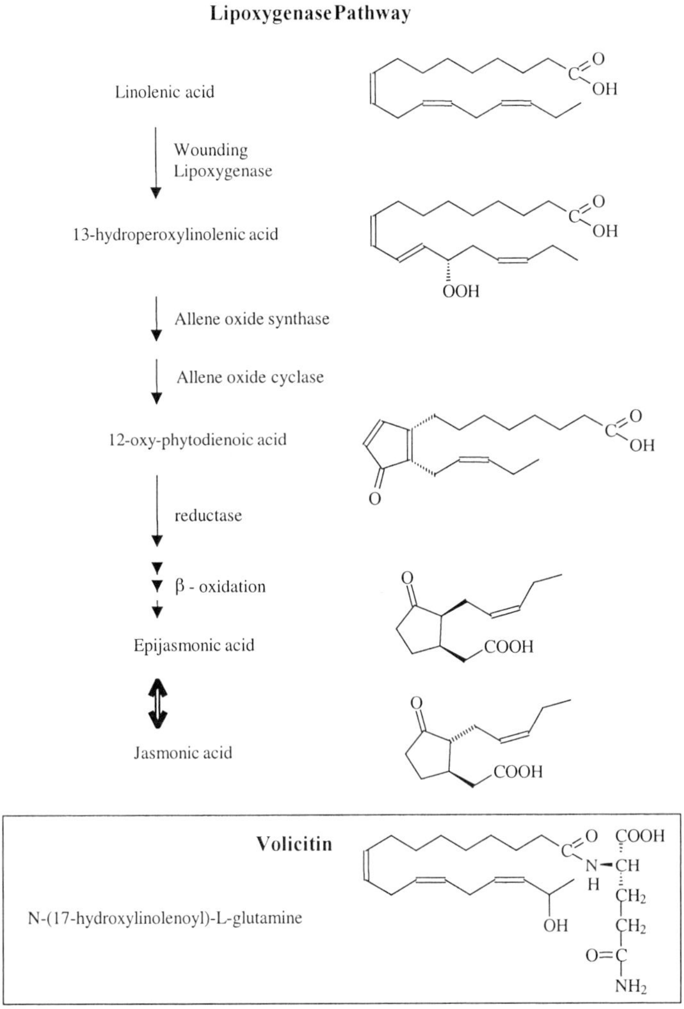

Figure 16. Schematic representation of the lipoxygenase pathway leading to jasmonic acid, together with the structure of volicitin. Redrawn after Tumlinson et al. *[2.58].*

Jasmonic acid

The evidence for the significance of jasmonic acid (JA) derives from a range of investigations. Firstly, the tomato mutant *defenceless* (*def* 1) that is incapable of elevating its JA level after attack, is less resistant to herbivores. Secondly, the triple mutant of *Arabidopsis thaliana* (*fad 3-2 fad 7-2 fad 8*) deficient in α-linolenic acid also fails to increase JA levels and is vulnerable to insect herbivores. Thirdly, another *A. thaliana* mutant (*coi*1), insensitive to JA, is also less resistant. Although JA translocates from shoot to root in tobacco, it serves a local defence factor in

A. thaliana as, in common with other octadecanoids, it does not leave the attacked leaf. It seems likely that leaf damage entails hydrolysis of lipids that releases fatty acids for modification to JA. However, neither the action of JA nor that of its methyl ester, methyl jasmonate (MeJa) entirely explain the in planta signalling process against herbivory. Other octadecanoids including the C_{18} metabolite 12-oxophytodienoic acid (OPDA) contribute, and perhaps other as yet unidentified compounds [2.63, and references therein].

Octadecanoids seem to mediate at the wound site of the plant perhaps through decompartmentation that, after hydrolysis of lipids, releases fatty acids. These serve as substrates for biosynthesis of JA. Alternatively, in chloroplasts, the free α-linolenic acid may be derived from the endoplasmic reticulum through the cleavage by lipases or acylesterases of membrane lipids. Consistent with this are elevated levels of α-linolenic acid associated with wounding [2.63, and references therein].

A further expression of the role of JA in direct defence against herbivory comes from the results of endogenous and exogenous application of JA. In the laboratory, the latter stimulated the response to herbivory and enhanced plant resistance probably through an induced response, as jasmonates in an artificial diet elicited no toxic effects. Endogenous treatment with JA of field-grown tomatoes elicited decreased populations densities of aphids, thrips, flea beetles and noctuid larvae. Specifically, the latter were killed as early instars, treated plants were avoided by flea beetles, but the mechanisms affecting aphids and thrips were unknown. Overall, treated plants sustained 60 % less damage than controls. Induced direct resistance against herbivores may have evolved because it is less costly to plants. Resources allotted to constitutive defence may be quite costly and impinge on life history requirements, in addition to increasing the plant susceptibility to pathogens and to herbivores that evolve to detoxify and exploit constitutive toxins [2.34, and references therein].

Insects may exploit allomones against plants as when using them to elicit galls. A fruitful and secure habitat for the herbivore, the gall may be rendered more secure by elaboration of protective spines and spicules. In a further escalation, allelochemicals within the gall may deter predators and parasitoids [2.64].

Proteinase inhibitors

In addition to plant defence by toxins already *in situ* is defence by proteinase inhibitors *in situ*. They are subject to development regulation as their activity is preferentially distributed in descending order in young, mature, and senescent foliage. In cabbage, the phenology of inhibitor expression matches the timing of invasion by the imported cabbage worm *Pieris rapae*, and the cabbage looper *Trichoplusia ni*. Furthermore, increased inhibitor levels were negatively and significantly correlated with larval growth of both species. Interestingly, their preferred feeding sites varied, as *P. rapae* larvae fed on the youngest foliage of the mature plant thus encountering higher levels of the inhibitors, eliciting a 56 % decrease in larval size. In contrast, *T. ni* larvae fed on the underside of the older leaves, thus encountering the smallest levels of inhibitors with no effect on growth [2.17, and references therein].

Induction of proteinase inhibitors has been extensively investigated in tomato and potato by C. A. Ryan and associates. Potato leaves have two inhibitors of which inhibitor I is 41 kilodaltons (kDa) in size and a pentamer of 8 kDa subunits, and inhibitor II is about 23 kDa. Developmental factors control physiological signals regulating the presence or absence of such inhibitors, but their quantity is governed by the proteinase-inhibitor inducing factor (PIIF). Released near the wound by degradation of the plant cell wall, this is conveyed in phloem to intact especially younger leaves, within 2-3 hr, to react with specific receptors; biosynthesis of proteinase inhibitors is triggered. Target proteinases are trypsin and chymotrypsin, serine proteinases crucial for food digestion by insects [2.17, 2.48].

PIIF elicits the rapid synthesis of proteinase inhibitors in aerial regions of the plant. A comparison of total levels of foliar phenols, taken as the sum of levels of phenol, protein, and proteinase inhibitors, in intact and fed upon tomato leaf tissue indicated no significant differences between levels of the first two. However, levels of proteinase inhibitors I and II were significantly increased by feeding, and the use of diets, comprising reconstituted tomato leaf, indicated that wounded leaves served as a poorer food source that intact ones. Larvae were significantly smaller on wounded tissue apparently as a result of chronic exposure inducing hyperproduction of gut proteases [2.48].

An 18-amino acid polypeptide designated systemin is the most probable signal for production of the two proteinase inhibitors. In femtomolar concentrations it actively induces synthesis of the inhibitors and it is readily conveyed through phloem. Systemin arises by proteolytic cleavage from a precursor, designated prosystemin that comprises 200 amino acid residues, the gene for which has been isolated and characterized. Its overexpression in tomato plants, in an antisense configuration, elicits decreased expression of the inhibitors in response to herbivory. Conversely, overexpression of this gene in the natural orientation elicits constitutive expression of the two inhibitor genes associated only with wounding, and evinced the inhibitors in concentrations > 1 mg/ml in leaf sap [2.51].

Overexpression of the prosystemin gene and consequent production of systemin is also associated with production of polyphenol oxidase (PPO) especially in response to wounding. This inducible enzyme oxidizes a broad spectrum of plant phenolics and is likely involved in plant defence against both pathogens and herbivores. PPO is sequestered in the chloroplast to preclude inappropriate interaction with phenolic substrates, but is released by wounding. The derived quinones alkylate proteins thus decreasing their nutritive properties. Specifically, amounts of PPO and of the phenolic derivative chlorogenic acid, bound to plant protein following digestion by *H. zea*, were positively correlated with decreased growth of this insect [2.51].

Furthermore, MeJa, another significant inducer of proteinase inhibitors in tomato, rapidly induced PPO activity. As MeJa is established as a constituent of the octadecanoid signal transduction system of tomato, this indicates that PPO genes are governed by this system. MeJa and/or wounding also regulate genes for some proteinase inhibitors, an aminopeptidase, and a threonine deaminase. Such associations led to the proposal that in tomato, and indeed in other plant species, systemin induces a broad spectrum of defensive genes as a response to damage by pathogens and herbivores [2.51]. Enhanced systemin expression must represent one of the truly exciting and feasible elements in plant-sourced defence against insects (technical details of gene manipulation are in Chapter 10).

Kairomones

There are many examples of plant-produced allelochemicals functioning as allomones for some organisms but as kairomones for others. For example: juglone (5-hydroxy-1,4-naphthoquinone) is an allomone in allelopathic interactions of walnut *Juglans* sp., hickory *Carya* sp., and perhaps other trees, by acting as a feeding deterrent to the smaller European elm bark beetle *Scolytus multistriatus*, but not to the closely related hickory bark beetle *S. quadrispinosus* [2.45]. Sinigrin stimulates feeding by *Brevicoryne brassicae*, but deters feeding by the pea aphid *Acyrthosiphon pisum*. Similarly phloridzin, a phenolic compound found in leaves of apple *Malus* sp., deters probing attempts by non-apple-feeding aphids *Myzus persicae* and *Amphorophora agathonica*, but not by the apple aphid *Aphis pomi*. Collectively, such early examples indicated that insects could adapt to chemical defences [2.64, and references therein].

Furthermore, hydrolysis of sinigrin when cruciferous leaf tissues are damaged releases isothiocyanate, a volatile mustard oil, that irritates and damages animal tissue. The larva of the black swallowtail butterfly *Papilio polyxenes*, does not feed on Cruciferae but is restricted to the Umbelliferae. Experimentally perfusing umbelliferous leaves with sinigrin not only rendered them somewhat repellent to *Papilio* larvae but inhibited larval growth and development; ultimately, there was 100% larval mortality at concentrations naturally occurring in Cruciferae. The mode of action is decreased food assimilation from the gut. Nevertheless, Cruciferae are attacked by many insect species that obviously have evolved appropriate detoxification mechanisms. Further, *Pieris rapae* and *P. brassicae* are attracted by sinigrin [2.50, and references therein]. This was a convincing early illustration of an allomone also serving as a kairomone.

Returning to a consideration of Cucs (Fig. 14): there are specific associations between Cucurbitaceae and some Diabroticite beetles (Luperini) including the plain pumpkin beetle, the red pumpkin beetle, the pumpkin beetle, the banded cucumber beetle, the spotted cucumber beetle, the cucurbit leaf beetle, the cucurbit beetle, and the striped cucumber beetles. A plausible sequence of events to explain these associations is: (1) primordial Cucurbitaceae having *Bi* genes for Cucs are readily exploited by herbivores; (2) mutation of the *Bi* genes produces bitter and toxic Cucs providing plant defence; (3) such mutated plants thrive in the absence of herbivorous attack; (4) mutant Luperini beetles detoxify the Cucs; and (5) in a competition-free milieu they readily exploit this resource; (6) such beetles now evolve specific receptors for the presence of Cucs as host-finding cues; (7) they sequester Cucs gaining significant defence from predators; (8) despite domestication of Cucurbitaceae for sub-threshold concentrations of Cucs, their presence in foliage and flowers and their ready induction preserves the association with Diabroticite beetles [2.43].

Although many plant-insect relationships reached stability over some hundred million years of association, adaptation may be observed in action. Defensive froth emitted by the grasshopper *Romalea microptera* contains several odorous compounds (phenols, terpenes, benzoquinone), including a chlorinated aromatic compound, 2,5-dichlorophenol. This is repellent to ants and, therefore, defensively useful to the grasshopper. It probably stems from a herbicide or herbicide-derivative ingested by the grasshopper as insects seem incapable of synthesizing chlorinated compounds. 2,4-Dichlorophenoxyacetic acid is a much-used herbicide in Florida where the grasshoppers were collected. In wild areas, where no herbicide was applied, hoppers lacked the 2,5- dichlorophenol constituent in their defensive froth. Apparently, the grasshopper's chemical defence system evolved to incorporate this derivative of a man-made product designed to protect plants [2.26].

Synomones

Some compounds, such as pine tree terpenoids, may act as allomones, kairomones, and synomones: allomones, in that they deter herbivores; kairomones, in that they attract adapted bark beetles; and synomones by attracting bark beetle predators [2.61, and references therein]. The best known and widely distributed plant-insect relationship mediated by synomones involves flowering plants and their pollinators: volatiles attract insects that pollinate the plant, and insects in turn feed on nectar and locate a mate. Flowers vary in terms of floral rewards and attractiveness that may determine visitation rate of not only pollinators but also entomophages of herbivores. It also seems likely that plants have evolved chemical and nectar profiles that confer protection on adapted insect pollinators. Thus, iridoid glycosides sequestered from the catalpa tree by the catalpa sphinx *Ceratomia catalpae*, render it unpalatable to birds; the adult and pollinating moth is tolerant of them but, opportunistic, or non-pollinating species are incapacitated [2.16].

Synomones may also influence tritrophic level interactions. Some plant species may respond to tissue damage induced by herbivory by releasing synomones that attract the herbivores' natural enemies. Thus, the spined soldier bug *Podisus maculiventris* orients to soybean plants damaged by *Trichoplusia ni*. The joint action of tree volatiles and the bark beetle pheromone frontalin, are necessary to attract the dolichopodid entomophage *Medetera bistriata*, to trees attacked by bark beetles. Monoterpenes released by bark beetles attract a range of beetle predators but the herbivore may counter this by entirely severing leaves. Crucifers damaged by insect feeding convert, by enzymatic action, inactive mustard oils into attractants for parasitoids. Practical implications of this general effect involve possible biological control of pests by use of interplanted crops, or other plants, the odours of which attract parasitoids of the main crop's pests [2.64].

Antimones

Evidence for antimones comes from the examples of substances released by a plant that damage both producer and receiver. Exudates of trichomes (see Chapter 8) obviously deter and impede herbivores but when the effect extends to enemies of the herbivore, then the plant sustains a disadvantage. Whether this is sufficient to constitute damage to the plant may be questionable. More convincing are the examples where honeybees are killed, either individually or as a colony, following return of individual bees bearing nectar containing plant defensive toxins [2.64, and references therein].

TRITROPHIC INTERACTIONS

As indicated, the influence of plant chemicals extends to the tritrophic level by mediating the behaviour of predators and parasitoids. By decreasing the development rate and longevity of a herbivore, allomones render it more susceptible to predators and parasitoids. Specifically, Mexican bean beetles *Epilachna varivestis* fed on a high tannin diet were more susceptible to predation by a pentatomid. An extended generation time would also adversely affect herbivorous species that depend on host plants with a narrow temporal definition [2.64].

Topically applied tannins promoted parasitism by hymenopterans on the larva of the leaf miner *Cameraria* sp., feeding on leaves of *Quercus emoryi*. However, this was balanced by decreased mortality from other causes especially bacteria, viruses, and abiotic factors to which this herbivore is especially susceptible, perhaps because of constant 100% relative humidity within mines. Pupal weight was significantly decreased in tannin-painted leaves with presumed knock on effects on fecundity [2.27].

However, the most elegant examples come from interactions with parasitoids, involving induced indirect defence. For example, female parasitoids from two different wasp families, *Microplitis croceipes* (Braconidae) and *Netelia heroica* (Ichneumonidae), oriented in a wind tunnel to artificially-damaged leaves of cotton

Gossypium hirsutum and cowpea *Vigna unguiculata*, and to leaves of cowpea damaged by feeding larvae of *H. zea*; male wasps evinced little interest. Accordingly, GLVs guide parasitoids to their caterpillar hosts and thus function as synomones. This indicates that herbivore-damaged plants communicate with insects at the tritrophic level. Such effects could be supplemented by GLVs emanating from caterpillar frass. Given that parasitic Hymenoptera evolved from phytophagous ones, and that the latter respond and orient to GLVs, it seems that retention of this property reinforced by superior host-detection rates could explain the present phenomenon [2.65, and references therein].

Corresponding results were obtained from another tritrophic systems, involving *Z. mays*, the larva of *S. exigua*, and the generalist parasitic wasp *Cotesia marginiventris*. This system showed additionally that release of volatiles by the plant was enhanced by oral regurgitant from the caterpillar. It also distinguished between GLVs emanating only from plant sites being fed on, and larger terpenoids including (*E*)-β-farnesene, α-*trans* bergamotene, and (*E*)-nerolidol that were released several hours after the attack. The plants response proved to be systemic as volatiles were released by the entire plant, presenting it as a beacon readily distinguishable from its neighbours [2.60, and references therein]. In a separate study, regurgitate from a range of caterpillar species (*S. frugiperda*, *H. zea*, *T. ni*, and the velvetbean caterpillar *Anticarsia gemmatalis*) and from the American locust *Schistocerca americana*, all enhanced volatile release from maize seedlings, without physical damage to the plant surface. Each regurgitate elicited the same eight compounds: (*Z*)-3-hexen-1-yl acetate, linalool, (*E*)-4,8-dimethyl-1,3,7-nonatriene (DMNT), indole, α-*trans*-bergamotene, (*E*)-β-farnesene, (*E*)-nerolidol, and (*E,E*)-4,8,12-trimethyl-1,3,7,11-tridecatetraene (TMTT) in the same ratio, although total amounts varied. *C. marginiventris*, and *M. croceipes* which parasitizes only *Helicoverpa* and *Heliothis* species, landed significantly more often but indiscriminately on plants treated with regurgitate; *M. croceipes* responded positively to maize treated with a nonhost regurgitate [2.59].

Evolutionary aspects

Exploration of tritrophic relationships favourable to the plant, or indirect defence, has led to a direct consideration of plant-carnivore interactions with special emphasis on: how such mutualism evolved; the extent to which such signalling constitutes an 'evolutionary stable strategy' (ESS); and their role in the wider context of other interacting species [2.61]. It is acknowledged that signalling through terpenoids would have evolved only after their primary function of direct defence through toxicity to herbivores was established. Once such volatiles were consistently released after damage by feeding, associative learning would have enabled herbivore detection and recognition by the carnivore. With a net benefit established for plant and carnivore, natural selection would elicit plants that emit clearly detectable chemical signals. This has led to the use of the term conspiracy to describe the plant-carnivore relationship, although it may be a trifle anthropomorphic.

There is a spectrum of plant chemical responses to damage from mechanical sources as distinct from herbivory. At the lower end, there is no qualitative difference in the chemical mixture released, as for example by brassicas, although both are attractive to carnivores and quantitative differences are elicited by different herbivores. Such differences can be discriminated by the carnivore through learning. Adjacent to this on the spectrum is the release of a novel and attractive signal in response to herbivore attack, e.g. maize damaged by *Spodoptera* spp., as distinct from an unattractive blend released after mechanical damage. This signal is also discriminated through learning. At the upper end is the production of qualitatively novel blends in response to attack by different herbivores as exemplified by broad bean/aphid interactions. These signals are discriminated without learning. Accordingly, the signal-to-noise ratio improves considerably from lower to upper regions of the spectrum with concomitant increases in foraging efficiency [2.61] (Fig. 17). A significant contribution to this signal derives from insect regurgitant, to the extent that plants treated with regurgitant from instars suitable or unsuitable for infestation are attractive and unattractive respectively to the carnivore (as indicated above, the known actives from regurgitant are β-glucosidase and volicitin from *P. brassicae* and *S. exigua* respectively). An intriguing and as yet unanswered question is whether discrimination by carnivores is based on differences between bouquets taken as a whole, or between a few indicator compounds. Nor may feedback responses by the herbivore be overlooked, as a *Pieris* sp. will oviposit on plants that are of inferior quality for larval development but are also less frequently searched by a parasitoid [2.59, and references therein].

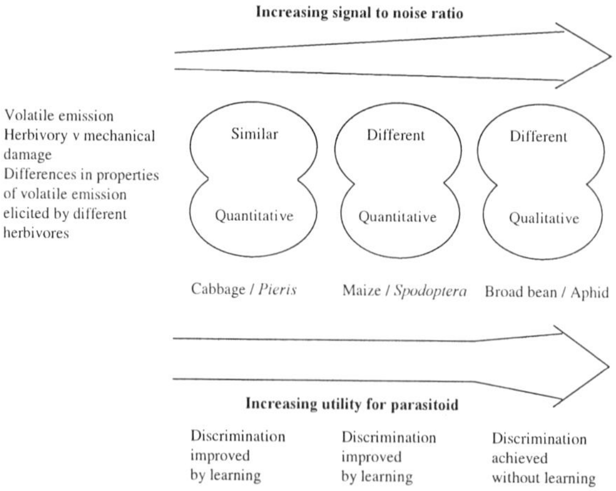

Figure 17. A spectrum of signal-to-noise ratios in terms of volatile emissions associated with plant damage. Redrawn after Vet [2.59].

Blend composition

A significant factor in clarifying this issue is the improved sensitivity of analytical techniques. Thus, original estimates (ca 1990) of the odorous composition released by herbivore-infested plants were in the region of 10-20 compounds. For example, 17 were reported from lima bean *Phaseolus lunatus* infested by spider mite *Tetranychus urticae*. Within 10 yr the number of compounds detected had risen to more than 60, and numbers from other systems amount to more than 100, many in trace amounts. Such composition is influenced by plant genotype, time of day, leaf age, light intensity, water stress, and species of infesting herbivore [2.23, and references therein]. The fact that higher light intensities elicit increased amounts of induced volatiles in cotton, maize, and lima beans was an indication that their production is *de novo*, and their release is governed by photosynthesis [2.38, 2.58, and references in both].

In respect of blend discrimination by carnivores, it is clear that they may distinguish between plants of different species that are sustaining herbivores of the same species. In assessing the significance of minor constituents of the blend, as revealed by increasingly sophisticated instrumentation, it is appropriate to employ behavioural assays that take account of various factors that influence choice by a carnivore. Such factors include prevailing environmental conditions, especially the presence of a predator or of a rival parasitoid that is avoided. Physiological pre-history is relevant in that starved parasitoids are less selective, especially when a specific dietary requirement must be obtained. Previous experiences and associative learning are very significant in that a prior encounter with a host may enhance responsiveness to the host's volatiles, especially in generalist carnivores. The reason for lesser discriminatory powers shown by specialist carnivores is unclear. Interestingly, unrewarding experiences may be as decisive as rewarding ones in that a combination of each led to enhanced discrimination between blends exhibiting merely quantitative differences. In terms of discrimination by carnivores between volatiles from the same plant species sustaining different species of herbivores, the record is varied; some but not all carnivores do so [2.23, and references therein].

Comparison of direct and indirect defence

This is constrained by the fact that studies of induced indirect defence are confined to two species of plants, maize and lima bean, that differ significantly in their responses expressed as volatiles to mechanical damage and herbivory. In contrast, studies of induced direct defence have employed tomato, potato, and cabbage that exhibit similar responses through volatiles to such stimuli. A true comparison will require the use of a single plant species. Nevertheless, a first approximation is possible. Firstly, the two patterns may exhibit incompatibility as, for example, when trichomes as a direct line of defence kill carnivores. Secondly, it seems that solanaceous plants such as tomato and potato, and crucifers such as cabbage, all equipped with potent allomones, release similar volatiles in response to mechanical damage and herbivory. In contrast, maize and bean that are not as conspicuously well defended and are exploited by a range of herbivores release novel constituents that attract carnivores [2.23, and references therein]. This juxtaposition is open to various interpretations, but it could be argued that it indicates complementarity between constitutive or direct, and indirect defence. In regard to monoterpenes, their role in direct and indirect defence may be viewed as a continuum from inhibition of the herbivores AChE to elicited death by carnivory.

MICROBIAL INTERACTIONS

Given the diversity and abundance of microorganisms, it would be surprising if microbe-insect interactions were trivial. In general, these work to the detriment of insects and over 100 insect diseases are elicited by protozoa rickettsias, viruses, bacteria and fungi [2.7, and references therein].

Allelochemicals may significantly affect the outcome of insect-microbe interactions. Tannins fed to the second and fifth instars of *P. polyxenes* a non-adapted species, elicited gut lesions 20-40 cells wide, exposing the basement cell membrane. Cells associated with enzyme secretion formed cytoplasmic extrusions with degeneration of cell functions. Such larvae evinced high mortality with the baggy appearance and unpleasant odour typical of bacterial septicaemia. Apparently, gut lesions facilitated bacterial migration into the haemocoel with fatal results. When resorcinol and gallic acid, phenolics occurring in most cotton cultivars resistant to insect attack, were coincluded with *Bacillus thuringiensis* var. *galleriae* in an artificial diet of the old world bollworm (=African bollworm) *Helicoverpa armigera*, larval feeding and weight decreased, and toxin efficacy was enhanced [2.53].

In contrast are beneficial effects association with extracellular symbionts, the presence of which is usually associated with an inadequate herbivorous diet. They may: fix atmospheric nitrogen and retrieve and recycle it from diet; produce essential amino acids; metabolize plant urates for insects lacking Malpighian tubules (e.g. aphids); contribute to synthesis and metabolism of fatty acids; hydrolyze plant matrix polysaccharides; synthesize enzymes, and vitamins especially those of the B-complex; contribute cholesterol; and metabolize plant allelochemicals, fungal toxins, and insecticides. One test of symbiont significance is to assess the insect response to their elimination by disinfectants and antibiotics. Typical responses include decreased size, survival, increased development time, physical deformities, and increased susceptibility to insecticides. Allelochemicals from some species of magnolia, oak and poplar exterminated the protozoan population of termite guts, causing death [2.7].

Intracellular symbiotes (endosymbiotes) in insect cells may aggregate to form a mycetome. Originally associated with insects having a narrow host range, the herbivores from which they are best known are the Homoptera and Coleoptera.

Among the former, aphids, leafhoppers, planthoppers, and scale insects have been the subjects of numerous studies, frequently employing artificial diets treated with antibiotics. In addition to establishing the transovarial transmission of endosymbiotes from the adult to the next generation, a role was indicated for them in the synthesis of amino acids, sterols, vitamins, antibiotics, and pigments [2.19, and references therein].

Ectosymbiosis benefits bark beetles as, for example, when fungal spores attached to the bodies of pioneer beetles feed the population. Caryophyllene epoxide associated with leaves of *Hymenaea courbaril*, a tree rarely attacked by attine ants, repels them and kills the attine fungus. Allelochemical concentrations in this and in other tree species declined significantly at the onset of the dry season when risk of fungal attack is lessened, raising the possibility that the compounds served primarily as fungicides and secondarily as insect repellents. More analytically, modification by microbes of plants as a resource was proposed as a test of their significance, leading in turn to the proposal of a net effects perspective (Fig. 18) [2.33].

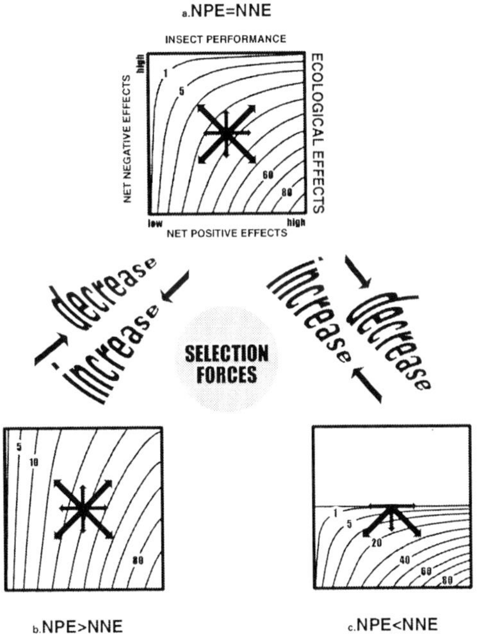

Figure 18. Schematic representation of the relationship between net positive effects (NPE) and net negative effects (NNE) as components of a net effects perspective. Redrawn after Jones [2.33].

This entails net positive effects (NPE) and net negative effects (NNE) with the overall insect response represented as the Z-axis or vector of a three-dimensional surface response, reflecting interactions between the X-axis (NPE) and the Y-axis (NNE), as visualized in a net effects diagram (NED). Selection or evolutionary forces are quantifiable from shifts in the relative dominance of NPE or NNE between NEDs. It is emphasized throughout that interactions between plant, microbial and insect attributes, rather than attributes per se, affect the outcome as expressed in insect performance. This will find quantifiable expression in such vital activities as feeding, food assimilation, growth, development, fecundity, and fertility.

A particular feature of the NED is the possibility of predicting outcomes of the effects of various factors such as chemicals including nitrogen, and microorganisms. Thus, an attack by plant pathogens that induces secondary metabolites with allelochemical properties should decrease insect fitness; conversely, compensatory plant growth eliciting increased nitrogen levels could increase insect growth. Both effects occurring contemporaneously will change the resultant to a new coordinate. If the above negative effects of induced allelochemicals are accompanied by decreased nutritional uptake elicited by vascular plant pathogens, then the resultant will move to a coordinate representing decreased insect growth. In a survey of a range of nutritional and allelochemical interactions, NEDs summarize the net effects. Ectomycorrhizae (fungi affecting nutrient and water relations of plants, and phytotoxicity) were predicted to affect above ground resistance of saplings relying on C-based defence. This followed from the fact that in nutrient-poor soils ectomycorrhizal plants are more efficient than non-ectomycorrhizal plants in nutrient sequestration, as less C is diverted below ground/unit nutrient sequestered. Accordingly, more C is available for shoot production and for protection.

Nitrogen-fixers

These endosymbiotic bacteria provide more N to the plant, at the expense of below ground C, leading to enhancement of insect growth but also of N-based plant defence. Insects are most adversely affected by a combination of enhanced N-based defence with decreased N-quality, although these could elicit detoxification adaptations.

Endophytes

These fungi are associated with the relatively uncomplicated effect of plant sequestration of N-based defences accompanied by additional vegetative growth and reproduction. Decreased resources for sexual reproduction due to endophytic infection are presumed to be offset by decreased damage by the insect.

Phylloplane microflora

The diversity of this grouping of fungi, bacteria and viruses is associated with a corresponding diversity of effects on the plant. These include the generation of secondary metabolites with either positive or negative effects. Among the former are generation and induction of secondary metabolites toxic to insects, and among the latter are: generation of host location cues; serving as a food resource, or facilitating food uptake by the insect; detoxification of allelochemicals; and enhancement of sex pheromone production.

Plant pathogens

The extraordinary diversity and multiplicity of effects of this grouping makes very difficult a prediction of the net effect in any given plant-insect interaction. A checklist of major interactions in terms of positive and negative effects for the plant could include as headings, alterations in: host-plant attractancy; nutritional content and physical defences; composition and quantity of allelochemicals, each capable of either decreasing or enhancing insect performance. In the event, the latter would be increased by three vectors, decreased by five, and would be unaffected by two, equivalent to seven to three in favour of either decreased performance or no effect.

Insect mutualists

Although another very diverse grouping, their overall effects seem beneficial to insects by supplying and facilitating exploitation of nutrients, enhancing pheromone production and thus mating, and detoxifying plant allelochemicals. Such effects could be offset by plant defences against such mutualists.

Insect Pathogens

In addition to direct mortality elicited by these, are sublethal effects such as decreased detoxification by insects of plant allelochemicals and decreased exploitation of nutrients. Nevertheless, the principal consideration is the interaction between plant metabolites and the pathogen or its toxin. This may either inhibit the pathogen or enhance its toxicity. A NNE on the plant will result from broad-spectrum allelochemicals that directly impair insect performance but also inhibit their pathogens; this is in contrast to the effect of such allelochemicals on mutualists.

Overall, it would seem that the effect of microorganisms is to alter the outcome otherwise expected from the interaction of the plant and insect. The categories decreasing insect performance outnumber those increasing it, but a corresponding outcome requires that the incidence of the effects be evenly distributed between categories [2.33, and references therein]. Of the foregoing interactions, those involving endophytes seem to merit special attention. Endophytic fungi are defined as those forming inconspicuous infections within the tissue of healthy plants for all or most of their life cycle and as such, the interaction may be mutualistic. Furthermore, most plant-fungal-herbivore interactions lessen plant susceptibility to herbivores, primarily by fungal production of alkaloids [2.22, and references therein].

THE CONCEPT OF COEVOLUTION

The importance of Fraenkel's publications in establishing the significance of secondary plant compounds has been considered in Chapter 1. A crucial intervening paper by Ehrlich and Raven examined interactions between butterflies and plants, and coined the term coevolution [2.25]. This described the community evolution that was influenced by reciprocity between major groups of organisms with a close and evident relationship as between plants and herbivores. Addressing the role of secondary plant substances, Ehrlich and Raven recognized both their defensive role and the possibility of a genetic mutation that would allow insects to feed on hitherto protected plants. This would promote insect success in the absence of competition from phytophagous but non-adapted animals. They specifically visualized that, for an insect group restricted to a narrow range of food plants, compounds formerly repellent could become attractants.

Although Ehrlich and Raven's paper was the first to invoke coevolution in the context of plant-insect relationships, it was by no means the first to recognize the concept, as already indicated by the quotation from Darwin (see Chapter 1). Three subsequent approaches were significant. First, Flor's approach involving gene-for-gene interactions underlying the evolution of resistance patterns between plants and pathogens. This invokes complementary *loci* such that each gene eliciting virulence in the parasite is matched by one governing resistance in the host. However significant this may be for plant-pathogen interactions, such a relationship is unlikely to explain behaviourally more complex, and polygenic, plant-insect interactions. Second, mathematically modelling of coevolution that stems from an analysis of coevolution between obligate parasites and their hosts. Third, the concept of evolution involving reciprocal adaptations in interacting populations as developed by Pimentel. Reciprocity was a key element in Ehrlich and Raven's thesis that extended to a connection between adaptation and speciation. In its most explicit form, coevolution has been defined as:

> An evolutionary change in a trait of the individuals of one population, elicited by selective pressure from a second population, followed by an evolutionary response by the second population to the elicited change in the first [2.30].

So, both specificity and reciprocity are involved and, in a further requirement, these must evolve simultaneously. By this definition, the evolution of each trait is due to the evolution of the others and plant-insect interactions are viewed as one to one. Although such requirements are rather difficult to establish, the term was applied to cover a broad range of interactions not involving reciprocal change, suggesting that this strict definition was not universally accepted. Relatively quickly coevolution was invoked in many forms and in a rich array of contexts. Four mechanisms of coevolution have been identified as: specific coevolution; diffuse or guild coevolution; escape and radiation coevolution; and diversifying coevolution [2.57, and references therein].

Specific reciprocal coevolution

This involves a coadaptation confined to two species, not involving a gene-for-gene interaction, with a range of results from escalation of attack and defence mechanisms to trait convergence as expressed in mutualism. A significant question awaiting elucidation is: what factors favour which outcome? However, truly specific and reciprocal coevolution must be rather rare in plant-insect relationships given the substantial evolution of microbes including fungal pathogens, and that plant chemical defences may have evolved in the cretaceous period but the feeding habits of insects evolved more recently. Many plant chemical constituents such as monoterpenes are hydrophobic and potentially fungicidal, which would reinforce their continued production as a defence against insects. This function lies outside chemically-mediated, specific, and reciprocal plant-insect coevolution. Overall, the presence of fine-tuned adjustments between species is insufficient proof for coevolution.

Diffuse coevolution

Less specific or diffuse coevolution is where a trait evolves in one or more species in response to a trait or suite of traits in several other species. Specifically, this provides for plants evolving defences against a diverse group of insects, and also for many insects acquiring the ability to detoxify a wide range of plant chemicals. As most plants are attacked by a range of insect species, a specific single change in plant chemistry might not improve defence against them all. Accordingly, selection pressures will favour a broad-spectrum effect.

If, however, relationships are very broad then it becomes difficult to distinguish coevolution from evolution or adaptation. For example, if plant defences that evolved against beetles are countered by moths but not by beetles, then evolution but not coevolution has been detected. However, the validity of the assessment obviously hinges on accuracy in assigning the plant's defences as originally evolved to deter beetles. This is not always attainable as the chemicals may have evolved in response to fungi but, by serendipity, are deleterious to insects as indicated in the previous section. Also known as guild coevolution, the concept of diffuse coevolution serves to focus attention on interacting groups of species, some of which may have arisen by speciation.

The concept of diffuse coevolution may be deficient in terms of insufficient recognition of the significance of geographical specialization. At one extreme, is the butterfly *Euphydryas editha*, that exhibits significant differences in ovipositional preferences both within and between populations. At the conserved extreme,

Californian populations of the butterfly *Papilio zelicaon* feed exclusively on introduced fennel, and laid significantly more eggs on this plant than did butterflies from Washington State; but the Californian butterflies did not discriminate between fennel and two native plant species. So, although capable of switching to a novel host, there was no genetic specialization for that host. Interspecific matings between *P. zelicaon* and *P. oregonius* indicated that ovipositional preferences are governed by at least one locus on the X chromosome. These and many other studies have enabled the geographic mosaic theory of coevolution [2.57, and references therein].

Escape and radiation coevolution

This explicitly provides for intervals during which there is no interaction between the rival taxa and during which the host escapes from its exploiter to radiate extensively before being again overtaken. Some relationships involving coumarin production by Umbelliferae seem relevant [2.8, and references therein].

Coumarins are derivatives of p-coumaric acid that are activated by ultraviolet light to form very reactive states that bind a range of biomolecules including DNA, and disrupt physiological processes. The simplest coumarins are hydroxycoumarins, established as having biocidal activities. Addition of a third ring gives rise to furanocoumarins of which there are two distinct classes associated with the Umbelliferae: linear furanocoumarins, enabled by an enzyme prenylating umbelliferone at the 7-position, and restricted to six tribes; and angular ones, enabled by an enzyme prenylating at the 8-position, and known from four.

Seasonal changes in wild parsnip *Pastinaca sativa* (Umbelliferae) indicated that reproductive structures, taken as indicators of fitness, contained 10-fold more nitrogen and furanocoumarins than vegetative structures such as senescent leaves. Furanocoumarin distribution was more closely associated with nitrogen, a growth limiting factor, than with biomass. The linear furanocoumarins, xanthotoxin, imperatorin, and bergapten, occurred in all above ground parts of the plant; the angular furanocoumarins, angelicin and sphondin, occurred in umbels [2.11].

Generalized insect herbivores feed preferentially on plant parts, such as foliage, low in furanocoumarin concentration, but parsnip specialists, such as parsnip webworm *Depressaria pastinacella*, feed exclusively on umbels. By tolerating the furanocoumarins, this herbivore gained access to significantly larger amounts of nitrogen with consequential likely gains in growth rate, fecundity, longevity, and survivorship. Thus, the factor that best predicts numbers of *D. pastinacella* on umbels is the presence of two minor constituents, angelicin and sphondin, biosynthetically distinct as angular coumarins from the others.

Apparently, escalation by Umbelliferae in the production of coumarins from hydroxycoumarins through linear to angular ones, with consequential escape after each escalation, was matched by reciprocal adaptation (coevolution) in lepidopterous species. However, a reanalysis of the data involving a different approach to the systematics of the Umbelliferae did not sustain this conclusion [2.57]. Nevertheless, almost 75% of variation in resistance of *P. sativa*, as measured by seed production, was assignable to four furanocoumarins, each with significant heritability. Thus, one principal requirement for coevolution is met, ability of herbivores to change the genetic and chemical constitution of a plant population [2.11].

Diversifying coevolution

A variant of escape-and-radiate coevolution is diversifying coevolution in which the coevolutionary process is again the product of interacting populations but without the escape intervals of the former. Examples would include speciation induced in invertebrates by maternally transmitted symbiotic microorganisms such as spiroplasmas and rickettsias. Such symbionts have been invoked to explain hybrid unfitness and reproductive isolation in Lepidoptera, Diptera and Coleoptera. If significantly different hybrid populations then colonize different environments, the result could be speciation.

Objections

The rather slick sequence of events presumed by coevolutionary theory is not without its opponents especially Jermy. A key objection claimed that there is little evidence to support this hypothesis, which is essentially based on a series of assumptions. The proposed alternative was sequential evolution in which: a constellation of factors such as climate, soil- and plant-plant interactions drove the evolution of flowering plants, the biochemical diversification of which provided a trophic base for the evolution of phytophagous insects i.e. the converse of conventional coevolutionary theory [2.31].

In a more detailed rebuttal [2.32], the same author argued for the existence of a continuum to described insect effects on plants. This would range from plant destruction through minor to beneficial effects, based on plant recruitment. The lack of a consistent fitness decreasing effect is seen as weakening the assumption of universal chemical defence by plants. Allied to this is the view that plants may be characterized as having a high diversity of mainly inactive compounds that may have been selected before the evolution of terrestrial plants, and thus long before the evolution of insects. The emphasis on the role of individual compounds is criticized on the basis that feeding insects are simultaneously exposed to a range of chemicals that may have opposing effects. Diffuse coevolution is denied on the basis that secondary plant chemicals deterring one insect species may be phagostimulants for another.

Instead of host-plant selection driven by plant chemicals, spontaneous mutations are visualized as occurring in the insect chemosensory system such that hitherto unpalatable or chemically-defended plants become palatable or exploitable. A feature of this construct is emphasis on a primary role for feeding deterrents as illustrated by small ermine moths (*Yponomeuta* spp.) for which tolerance of the antifeedant phloridzin enabled a shift from feeding on *Prunus* to *Malus* (*Y. malinellus*), and tolerance of the deterrent salicin promoted a corresponding shift from Rosaceae to Solicaceae (*Y. rorellus*) [2.32, and references therein].

In fact, reference to the paradigm represented by the relationship between Luperini beetles and the Cucurbitaceae reveals many points of similarity with the above schema. Specifically, both envisage as a key event the mutation of the herbivore such that a hitherto chemically-defended plant becomes exploitable. Given that protagonists of the concept of coevolution have moved significantly from claiming it as being both specific and reciprocal, then the gap between the two positions on the course of events after this mutation is not substantial. Nevertheless, a difficulty with the above schema is the heavy reliance on only feeding deterrents as determining the host plant range for oligophagous insects. This seems to overlook the role of insecticidal toxins, the category to which many plant defensive chemicals belong [2.10]. And very many studies on insect resistance of crop cultivars to oligophagous pests associate resistance and toxin concentrations (see Chapter 8). It, therefore, becomes difficult if not impossible to dismiss selection pressure of insects on chemical adaptation by plants.

The concept of coevolution has also been criticized for its lack of predictive properties at the level of specific taxa [2.21] although it is unclear that any evolutionary construct will meet the requirement. The broad concept of evolution will not do so and is not a theory, sensu strictu, but is the validity of the concept really in doubt as a result? Nevertheless, a key issue requiring explanation is the evolution of host shift [2.21].

Although this shift may entail adaptation to a new host with chemical similarity to the original ones, close ecological proximity may also be a relevant factor. As larvae of such adapted insects may still utilize the original host as food, the behavioural shift is attributable to the ovipositing female. She may be more discriminating in the early than in the latter stages of host selection, as this would minimize time wasted on wrong choices. This led to the development of a hierarchy-threshold model by which the balance of attractants and deterrents determined acceptability, a threshold that alters in response to extrinsic and intrinsic factors, such as flying time and egg load respectively. Thus, a high threshold will exclude all but a few plant species, but a lower one will render additional species as ovipositional hosts. A strength of the approach was that it did not call for a mutation in the insects chemosensory system for a host shift to occur in the first instar. The Hopkins Host Selection Principle (HHSP), by which larval experience of a specific food plant would enhance adult preference for it, would reinforce the host shift in the next generation and succeeding ones (for a critical analysis of HHSP see 2.56). Prolonged exposure over very many generations of the larval stage to chemicals of this plant might serve as a selection pressure for the development of appropriate receptors at the molecular level (see reference to Cucs above).

In any event, the ovipositional model was then assessed under four headings proposed by Tinbergen to explain a behavioural development: (1) proximate causes including biochemical responses to sensory cues; (2) ontogenetic causes including learning; (3) ultimate causes attributable to natural selection; (4) phylogenetic causes derived from evolutionary history.

Criterion (1) was deemed to be met as the model made no assumptions regarding a necessary relationship between host acceptability and suitability. In regard to criterion (2), the theory accommodates cross induction in that a lowered threshold of acceptability will access a range of hosts with no chemical similarity. Criterion (3), could be met if the chemoreceptors evolved as indicated above. Criterion (4) is more demanding and requires determining which host shifts are most probable and at which point they are most likely to occur in the host selection process. By any standards this is a tall order. Overall, the emphasis on ovipositon in a general concept of acceptability provides a useful focus for exploration of the evolution of host specificity [2.21, and references therein].

FATE IN INSECTS OF ALLELOCHEMICALS

There seem to be four principal procedures by which insects cope with ingested allelochemicals. They are: (1) evacuated unaltered; (2) metabolized; (3) sequestered for insect defence; (4) employed as precursors for pheromones; and combinations of the foregoing [2.1, 2.37, and references in both].

Evacuation

The *M. sexta* larva feeding on tobacco leaves containing up to 5% nicotine (dry weight) excretes it unaltered, except for sequestration of small amounts of nicotine and the metabolite continine [2.5]. Iridoid glycosides ingested by the larva of *Junonia coenia* and *C. catalpae* are eliminated intact with some sequestration [2.16]. More commonly, however, plant allomones are detoxified through metabolic pathways.

Metabolism

Hydrophilic compounds are usually metabolized such that constituent carbons and nitrogens are reused during metabolic activity of the insect. In contrast, hydrophobic allelochemicals that readily insert into hydrophobic tissues, especially membranes and the fat body where concentrations could reach life-threatening levels, are countered by enzymes operating in two phases. Firstly, the hydrophobic molecule is converted to a less toxic, lipophilic derivative by oxidases, reductases, hydroxylases, esterases, and transferases. Secondly, the derivative is conjugated by transferases to a very hydrophilic group such as a sugar, phosphate or sulphate prior to excretion. The commonest primary step is oxygenation enabled by cytochrome P-450 [2.1, 2.37].

Cytochrome P-450-dependent polysubstrate monooxygenases (PSMOs)
Mixed-function oxidases (MFO), termed cytochrome P-450-dependent polysubstrate monooxygenases (PSMOs), are the principal enzymes by which such transformations are accomplished. As the name implies, these enzymes insert into the substrate one atom from O_2; dioxygenases insert both oxygen atoms. Some 400 million years before present, a remarkable increase occurred in the abundance and diversity of cytochrome P-450 forms. This might reflect original interactions between plants and animals as it also marks the earliest known fossil records of neopterous insects, ancestors of present consumers of secondary metabolites. The usefulness of these enzymes reflects three properties: (1) the oxidation steps they catalyze yield more charged and thus more hydrophilic products that are more readily excreted; (2) they oxidize a broad range of substances; (3) they are readily induced by ingested allelochemicals [2.1, and references therein].

This system is frequently at maximum activity during the actively-feeding life stages of insects. Furthermore, activity levels are affected by specific allelochemicals and are elevated after ingestion of monoterpenes. Gut tissue and the fat body evince the highest rate of PSMO activity, with significant rates in Malpighian tubules, gastric caeca, and lower rates from hindgut and nerve tissue. Within-cell *loci* of PSMO activity include the endoplasmic reticulum and mitochondrial membranes. Essentially, the system comprises NADPH cytochrome P-450 reductase, the red pigment cytochrome P-450, and phosphatidylcholine. The essential reaction with a substrate R is given by: $RH+NADPH+H^{+}+O_2 \rightarrow ROH+NADP^{+}+H_2O$ in which electron-deficient heme of the cytochrome binds to and oxygenates the most electronegative site of the substrate. The cytochrome moiety is attached to the membrane of the endoplasmic reticulum where it receives electrons from NADPH by the action of the also attached oxidoreductase. Depending on the properties of the original substrate, the resultant monooxygenations may entail such varied reactions as dealkylation, epoxidation, desulphuration, and oxidative deamination and dehalogenation. The various molecular forms of this system enable the oxygenation of relatively hydrophilic compounds including the organophosphate insecticide parathion. Oxidation reactions may be counter productive, if fleetingly, when they oxygenate a double bond to an epoxide as this unstable group progressively attacks proteins, DNA and RNA; however, these reactions may be deactivated by conjugation [2.1].

In mammals, monoterpenes are usually oxidated followed by conjugation to glucuronic acid. For example, limonene is metabolized via PSMO to transidols involving epoxidation of either one of the two double bonds followed in turn by epoxide hydration. Limonene is also derivatized into various alcohols through the action of dehydrogenases. These compounds are non-toxic and are excreted after glucuronidation. Frequently, and for maximum effectiveness, the system must be induced to greater activity. Microsomes from midguts of control larvae of the southern armyworm *Spodoptera eridania* oxidize in vitro only trace amounts of pulegone. Following enzyme induction by either α-pinene or pentamethylbenzene, however, this compound is oxidized to 9-hydroxypulegone and 10-hydroxypulegone. Other monoterpenes acting as powerful inducers include α-pinene, myrcene, camphene, (-)-limonene, and γ-terpinene. Classical inducers include hormonal phytosteroids such as sitosterol, stigmasterol and ergosterol and the insect moulting hormones ecdysone and ecdysterone. It seems likely that this system is generally in an induced state in insect herbivores.

Following ingestion of various monoterpenes commonly occurring in green plants, larval midgut tissues of *S. eridania*, exhibited enhancement of: cytochrome P-450 levels 2.9-fold; NADPH cytochrome P-450 reductase 1.6-fold; NADPH-oxidation 3.8-fold; aldrin epoxidation 1.5-fold (maximum values). Evidently, insects do not rely on a single line of defence against a particular toxin. Plants containing photoxic, linear furanocoumarins do not adversely affect the growth and development of *D. pastinacella*, due to toxin-induced metabolism by PSMOs. Such metabolism is not perfect and furanocoumarins that escape could produce toxic oxyradicals. Adding the cytochrome P-450 inhibitor, piperonyl butoxide, to *D. pastinacella* larval diet decreased food utilization but enhanced superoxide dismutase activity. Apparently, cytochrome P-450s as primary detoxifiers may be supported by second-line antioxidant enzymes [2.9]. However, myristicin, an insecticide synergist naturally occurring in parsnip is a more potent synergist of the action of furanocoumarins. Other allelochemicals serving as synergists in plant defence include constituents of oils of sesame and of black pepper [2.1, and references therein].

Rates of toxin elimination by the hemimetabolous, migratory grasshopper *Melanoplus sanguinipes* (Acrididae), and the holometabolous variegated cutworm *Peridroma saucia* (Noctuidae) have been compared. This was prompted by the hypothesis that the large amounts of cuticle produced by Orthoptera, could protect against topical and ingested toxins. This in turn would necessitate less investment than by Lepidoptera in toxin metabolism. In the event, as the phototoxic linear furanocoumarin xanthotoxin and the cardiac glycoside digitoxin were similarly metabolized by both species, the hypothesis was not sustained [2.54].

Esterases and transferases

Second in importance only to the PSMOs are the carboxyl-esterases and transferases with the activity of the esterases up to three orders of magnitude greater than that of PSMOs, perhaps because the ester group enables tighter binding to the substrate. Their intrinsic role seems to be inactivation of juvenile hormone to allow insect development to proceed. They are effective metabolizers of a range of insecticides including organophosphates, carbamates, and pyrethroids.

Transferases mediate conjugations of allelochemicals some of which are ingested as conjugates, and are thus reconjugated following cleavage within the insect. A key enzyme is glutathione transferase, occurring as at least 10 different isozymes, that catalyses conjugation to the tripeptide glutathione (GSH). Confined to the cell cytosol, GSH transferases are most concentrated in midgut, fat body and Malpighian tubules of the caterpillar. They metabolize allelochemicals from a range of chemical categories including terpenoids and glucosides. Like PSMOs, they are induced by a wide range of compounds including monoterpenes, flavones, and coumarins. Overall, GSH transferases, by almost invariably conjugating only toxins, may be more significant and more truly detoxifying than PSMOs that produce toxic products, at least if their first stage activity entails epoxidation [2.1, and references therein].

In naturally occurring concentrations (1-2% fresh leaf weight), a suite of salicin-derived phenolic glycosides elicit midgut lesions and kill the larva of the eastern tiger swallowtail butterfly subspecies *Papilio glaucus glaucus*, whereas the larva of the conspecific northern tiger swallowtail butterfly *P. glaucus canadensis* is unaffected. The phenolic glycosides, salicortin and tremulacin were administered with and without the inhibitors appropriate to inhibit PMSOs, esterases, and glutathione transferases respectively. Deactivation of the esterases in *P. glaucus canadensis* increased larval mortality 4.6- and 4.1-fold when fed salicortin and tremulacin respectively (Fig. 19) [2.37, and references therein].

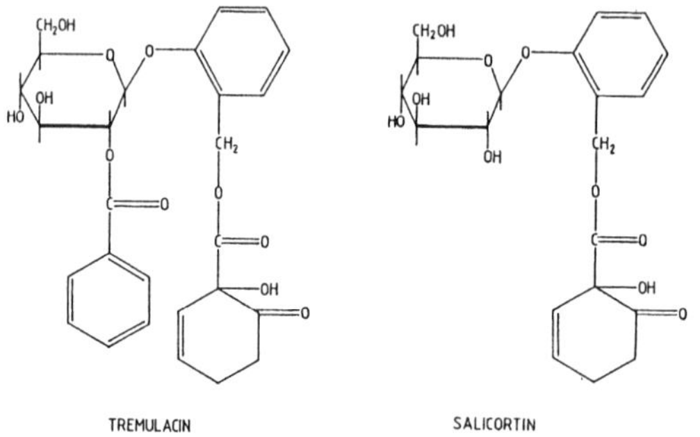

Figure 19. Structural formulae of the phenolic glycosides tremulacin and salicortin. Compiled from Lindroth [2.37].

Reductases

Flavins (FAD and FMN) reduced in the fat body of the Madagascar cockroach *Gromphadorhina portentosa* serve to non-enzymatically reduce nitrobenzene. Catalysis in insects of nitro- and azo-compounds is accomplished by reducing enzymes associated with the anaerobic flora of the gut. In contrast to the usual sequence of events, reduction of N-oxides produces less polar and potentially more toxic derivatives. This has led to the suggestion that N-oxides are a form in which allelochemicals may be translocated from plant roots as they are so readily N-oxidized. However, it is likely that gut microorganisms contribute significantly to reductions of allelochemicals [2.1, and references therein].

Sequestration

Allelochemicals may be sequestered either as ingested, or after modification to more polar forms in three phases: (1) absorption by gut; (2) transport from gut and in haemolymph; (3) deposition. The efficiency of each phase is affected by the hydrophobicity of the compound, its partition coefficient, molecular weight, and the energy requirement of each phase in relationship to available energy [2.24, and references therein].

Sequestration obviously requires that the allelochemical should not be denatured. Accordingly, generalist insect herbivores in encountering a broader array of allelochemicals than specialists, are expected to have more potent denaturation systems and, correspondingly, less sequestration. Equally, the insect requires some defence against the allelochemical present and one option is compartmentalization in the cuticle or in special glands. The pine sawfly larva *Neodiprion sertifer* sequesters pine monoterpenes and sesquiterpenes, storing them unaltered in foregut pouches. Their discharge deters ants and spiders from preying on the larva. Cardenolides sequestered by the milkweed moth *Oncopeltus fasciatus* are stored mainly in dorsolateral glands with very little in ventral ones; as such they are encountered earlier by a potential predator.

Iridoid glycosides are cyclopentanoids derived from monoterpenes and are bitter tasting constituents of over 50 plant families (Fig. 20). Their fate varies according to the species of butterfly larva that ingests them. Larvae of aposematic and unpalatable *Euphydryas phaeton* (Nymphalidae) use them as feeding cues, sequestering them through to the adult stage. The glycosides are selectively sequestered by the larva of *J. coenia* (Nymphalidae) and of *C. catalpae* (Sphingidae) but evacuated in the meconium at emergence, so the cryptic adults lack them. In contrast, the generalist feeder, the gypsy moth *L. dispar*, eliminates iridoid glycosides in the frass and does not store them although they may be metabolized. Accordingly, the fate of these compounds in four lepidopteran herbivores is related to strategies of feeding, and of defence against predators [2.13]. In an unusual reaction, the meconium was squirted a distance of two feet by an adult *C. catalpae*, indicating a defensive role for this structure previously regarded as a repository for pupal waste [2.15, and references therein].

Perhaps the most studied example of sequestration involves the milkweed genus *Asclepias* and the monarch butterfly *Danaus plexippus* [2.39]. This species breeds in regions of milkweed abundance, migrates to an overwintering site, then returns to the original breeding site. Essentially, the Rocky Mountains separate eastern and western populations of *D. plexippus*. More complete data are available for eastern populations that migrate in September and October to produce concentrated breeding populations in central Mexico, from where they migrate to the Gulf region of the U.S.A. Here they produce two spring generations that extend further north where three summer generations may be produced, thus completing the annual cycle. Western populations migrate to breed in central and southern California, from where they migrate to coastal mountain ranges and the Sierra Nevada. Cardenolides (cardiac-active, C_{23} steroid derivatives) sequestered from *Asclepias* spp. contribute to the defence of monarchs against predators, especially birds.

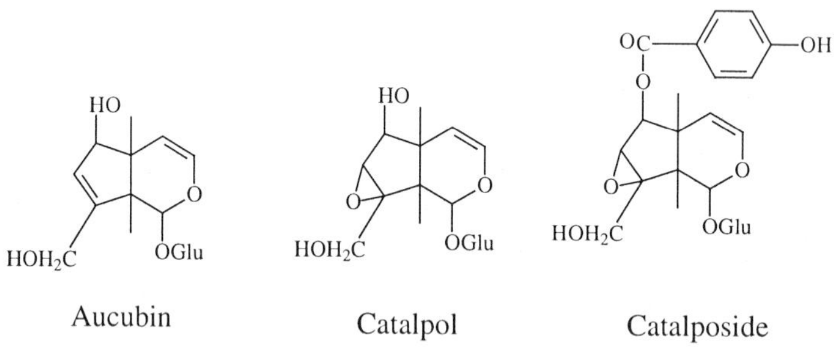

Figure 20. Structures of iridoid glycosides variously deployed in feeding and defence strategies. Compiled from Bowers and Puttick [2.16].

Monarchs feed on only 27 of 108 *Asclepias* species and they concentrate cardenolides most effectively from *Asclepias* species with intermediate or low rather than high cardenolide titres; *A. syriaca* and *A. californica*, species with intermediate titres (50 and 66 mg digitoxin equivalents/0.1 g dry weight), were most effectively exploited. Furthermore, chemical fingerprints from the plants enable the identification of the host species of a particular monarch, and as the distribution of such *Asclepias* species is known, then the monarch's origin can be inferred. Although 75 of the 108 species of *Asclepias* species occur north of Mexico, only 27 are fed on by monarch larvae. Furthermore, only three, *A. viridis*, *A. humistrata* and *A. syriaca*, comprise the diet of 94% of spring migrants east of the Rockies. Their distribution in geography and time is consistent with monarch migration, which is interpreted as an evolved response to the availability of host plants. High-cardenolide species are readily available in spring in the Gulf states, strengthening monarch defences before they migrate north, tracking *A. syriaca*. Thus, cardenolide sequestration increases the probability of survival during such migratory searches for food [2.39, and references therein].

Variation in the cardenolide concentration of host plants will lead to variability in unpalatability of the aposematic larvae, thus creating a spectrum of unpalatability. This is automimicry in the more unpalatable ones, and may be ubiquitous in unpalatable species (Fig. 21).

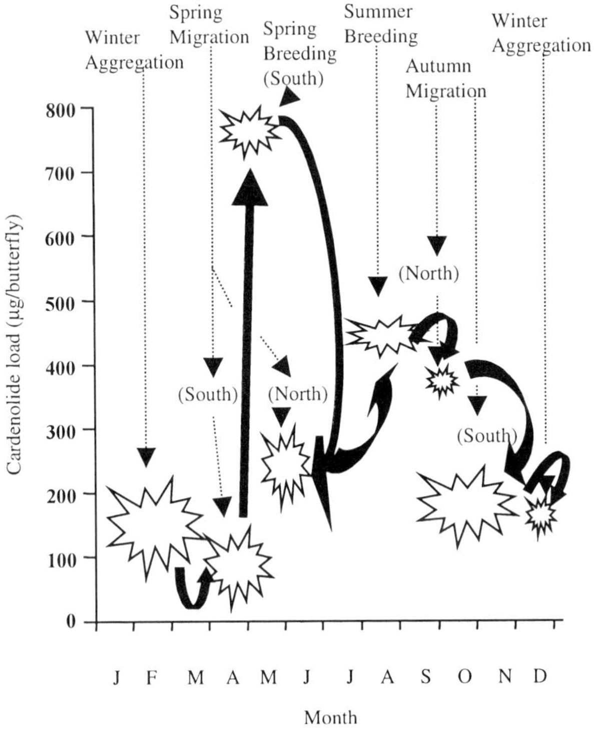

Figure 21. Diagrammatic representation of variations in cardenolide loads associated with activities (sequence in bold) of the monarch butterfly, Danaus plexippus. *Redrawn after Malcolm and Brower [2.39].*

Pheromone precursors

Pharmacophagy occurs where an insect is phago-stimulated by an allelochemical, in the absence of others, and where this serves a function independent of nutrition. More than 6,000 plant species produce in excess of 350 pyrrolizidine alkaloids (PAs) toxic to human liver and lung by forming pyrroles as reactive metabolites. PAs sequestered from host plants *Crotalaria* spp., by the larva of *Utetheisa ornatrix,* are derivatized to hydroxydanaidal, the adult sex pheromone. Larvae fed a precursor-free diet give rise to adults with inadequate amounts of pheromone and decreased mating success. PAs ingested by danaid butterflies, either as larvae or adults, from their host plants especially *Heliotropium* spp., regulate scent organ morphogenesis and serve as aphrodisiacal pheromones, as, for example, danaidone. PAs also occur in the spermatophore and incorporated into eggs provide them with some protection from predators. PA content of the pheromone secretion would, therefore, signal the extent of egg protection available from a particular male [2.52, and references therein].

REFERENCES

2.1 Ahmad, S., Brattsten, L.B., Mullin, C.A. and Yu, S.J. (1986) Enzymes involved in the metabolism of plant allelochemicals. In Brattsten, L.B. and Ahmad, S. (eds.) *Molecular Aspects of Insect-Plant Associations*. Plenum Press, New York, pp. 73-151.

2.2 Arnason, J.T., Philogène, B.J.R. and Towers, G.H.N. (1992) Phototoxins in plant-insect interactions. In Rosenthal, G.A. and Berenbaum, M.R. (eds.) *Herbivores: Their Interaction with Secondary Plant Metabolites*, vol II. Academic Press, San Diego, pp. 317-341.

2.3 Appel, H.M. and Martin, M.M. (1990) Gut redox conditions in herbivorous lepidopteran larvae. *J. Chem. Ecol.* **16**, 3277-3290.

2.4 Baldwin, I.T., Schmelz, E.A. and Ohnmeiss, T.E. (1994) Wound-induced changes in root and shoot jasmonic acid pools correlate with induced nicotine synthesis in *Nicotiana sylvestris* Spegazzini and Comes. *J. Chem. Ecol.* **20**, 2139-2157.

2.5 Barbosa, P. and Saunders, J.A. (1985) Plant allelochemicals: Linkages between herbivores and their natural enemies. In Cooper-Driver, G.A., Swain, T. and Conn, E.E. (eds.) *Chemically Mediated Interactions between Plants and Other Organisms*. Plenum Press, New York, pp. 107-137.

2.6 Barbosa, P., Krischik, V.A. and Jones, C.G. (eds.) (1991) *Microbial Mediation of Plant-Herbivore Interactions*. Wiley and Sons, New York.

2.7 Berenbaum, M.R. (1988) Allelochemicals in insect-microbe-plant interactions; Agents provocateurs in the coevolutionary arms race. In Barbosa, P. and Letourneau, D.K. (eds.) *Novel Aspects of Insect-Plant Interactions*. Wiley and Sons, New York, pp. 97-123.

2.8 Berenbaum, M.R. (1990) Evolution of specialization in insect-umbellifer associations. *Ann. Rev. Entomol.* **35**, 319-343.

2.9 Berenbaum, M.R. (1995). The chemistry of defense: Theory and practice. *Proc. Natl. Acad. Sci.* **92**, 2-8.

2.10 Berenbaum, M.R. and Seigler, D. (1992) Biochemicals: Engineering problems for natural selection. In Roitberg, B.D. and Isman, M.B. (eds.) *Insect Chemical Ecology. An Evolutionary Approach*. Chapman and Hall, New York, pp. 89-121.

2.11 Berenbaum, M.R., Zangerl, A.R. and Nitao, J.K. (1986) Constraints on chemical coevolution: Wild parsnips and the parsnip webworm. *Evolution* **40**, 1215-1228.

2.12 Bernays, E.A. (1978) Tannins: An alternative viewpoint. *Entomol. Exp. Appl.* **24**, 244-253.

2.13 Bernays, E.A. (ed.) (1989-1994) *Insect-Plant Interactions*, 5 vols. CRC Press, Boca Raton.

2.14 Bernays, E.A. and Chapman, R.F. (1994) *Host-Plant Selection by Phytophagous Insects*. Chapman and Hall, New York.

2.15 Bohlmann, J. and Croteau, R. (1999). Diversity and variability of terpenoid defences in conifers: molecular genetics, biochemistry and evolution of the terpene synthase gene family in grand fir (*Abies grandis*) In Chadwick, D.J. and Goode, J.A. (eds.) *Insect-Plant Interactions and Induced Plant Defence*. Novartis Foundation Symposium 223, Wiley and Sons, Chichester, pp. 132-149.

2.16 Bowers, M.D. and Puttick, G.M. (1986) Fate of ingested iridoid glycosides in lepidopteran herbivores. *J. Chem. Ecol.* **12**, 169-178.

2.17 Broadway, R.M. and Colvin, A.A (1992) Influence of cabbage proteinase inhibitors *in situ* on the growth of larval *Trichoplusia ni* and *Pieris rapae. J. Chem. Ecol.* **18**, 1009-1024.
2.18 Bryant, J.P., Chapin, F.S., III and Klein, D.R. (1983) Carbon/nutrient balance of boreal plants in relation to vertebrate herbivory. *Oikos* **40**, 357-368.
2.19 Campbell, B.C. (1989) On the role of microbial symbiotes in herbivorous insects. In Bernays, E.A. (ed.) *Insect-Plant Interactions*, vol I. CRC Press, Boca Raton, pp.1-44.
2.20 Chapin, F.S., III McKendrick, J.D. and Johnson, D.A. (1986) Seasonal changes in carbon fractions in Alaskan tundra plants of differing growth form: Implications for herbivory. *J. Ecol.* **74**, 707-731.
2.21 Courtney, S.P. and Kibota, T. (1992) Mother doesn't know best: Selection of hosts by ovipositing insects. In Bernays, E.A. (ed.) *Insect-Plant Interactions*, vol II. CRC Press, Boca Raton, pp. 161-188.
2.22 Dahlman, D.L., Eichenseer, H. and Siegel, M.R.. (1991) Chemical perspectives on endophyte-grass interactions and their implications to insect herbivory. In Barbosa, P., Krischik, V.A. and Jones, C.G. (eds.) *Microbial Mediation of Plant-Herbivore Interactions*. Wiley and Sons, New York, pp. 227-252.
2.23 Dicke, M. (1999) Specificity of herbivore-induced plant defences. In Chadwick, D.J. and Goode, J.A. (eds.) *Insect-Plant Interactions and Induced Plant Defence*. Novartis Foundation Symposium 223, Wiley and Sons, Chichester, pp. 43-59.
2.24 Duffey, S.S. (1980) Sequestration of plant natural products by insects. *Ann. Rev. Entomol.* **25**, 447-477.
2.25 Ehrlich, P.R. and Raven, P.H. (1964). Butterflies and plants: A study in coevolution. *Evolution* **18**, 586-608.
2.26 Eisner, T., Hendry, L. B., Peakall, D. B. and Meinwald, J. (1971) 2,5-Dichlorophenol (from ingested herbicide?) in defensive secretion of grasshoper. *Science* **172**, 277-278.
2.27 Faeth, S.H. and Bultman, T.L. (1986) Interacting effects of increased tannin levels on leaf-mining insects. *Entomol. Exp. Appl.* **40**, 297-300.
2.28 Gershenzon, J. and Croteau, R. (1991) Terpenoids. In Rosenthal, G.A. and Berenbaum, M.R. (eds.) *Herbivores: Their Interaction with Secondary Plant Metabolites,* vol I. Academic Press, San Diego, pp. 165-219.
2.29 Hedin, P.A., Williams, W.P., Davis, F.M. and Buckley, P.M. (1990) Roles of amino acids, protein, and fiber in leaf-feeding resistance of corn to the fall armyworm. *J. Chem. Ecol.* **16**, 1977-1995.
2.30 Janzen, D. (1980) When is it coevolution? *Evolution* **34**, 611-612.
2.31 Jermy, T. (1976) Insect-host-plant relationship – co-evolution or sequential evolution? *Symp. Biol. Hung.* **16**, 109-113.
2.32 Jermy, T. (1993) Evolution of insect-plant relationships - a devil's advocate approach. *Entomol. Exp. Appl.* **66**, 3-12.
2.33 Jones, C.G. (1991) Interactions among insects, plants and microorganisms: A net effects perspective on insect performance. In Barbosa, P., Krischik, V.A. and Jones, C.G. (eds.) *Microbial Mediation of Plant-Herbivore Interactions*. Wiley and Sons, New York, pp. 7-35.
2.34 Karban, R. (1999) Future use of plant signals in agricultural and industrial crops. In Chadwick, D.J. and Goode, J.A. (eds.) *Insect-Plant Interactions and Induced Plant Defence*. Novartis Foundation Symposium 223, Wiley and Sons, Chichester, pp. 223-238

2.35 Keane, S. and Ryan, M.F. (1999) Purification, characterisation, and inhibition by monoterpenes of acetylcholinesterase from the wax moth, *Galleria mellonella* (L.). *Ins. Biochem. Mol. Biol.* **29**, 1097-1104.
2.36 Langenheim, J.H. (1994) Higher plant terpenoids: A phytocentric overview of their ecological roles. *J. Chem. Ecol.* **20**, 1223-1280.
2.37 Lindroth, R.L. (1991) Differential toxicity of plant allelochemicals to insects: Roles of enzymatic detoxication systems. In Bernays, E.A. (ed.) *Insect-Plant Interactions*, vol III. CRC Press, Boca Raton, pp. 1-33.
2.38 Loughrin, J.H., Manukian, A., Heath, R.R., Turlings, T.C.J. and Tumlinson, J.H. (1994) Diurnal cycle of emission of induced volatile terpenoids by herbivore-injured cotton plants. *Proc. Natl. Acad. Sci.* **91**, 11836-11840.
2.39 Malcolm, S.B. and Brower, L.P. (1989) Evolutionary and ecological implications of cardenolide sequestration in the monarch butterfly. *Experientia* **45**, 284-295.
2.40 Martin, J.S. and Martin, M.M. (1983) Tannin assays in ecological studies. *J. Chem. Ecol.* **9**, 285-294.
2.41 Martin, M.M. and Martin, J.S. (1984) Surfactants: Their role in preventing the precipitation of protein by tannins in insect guts. *Oecologia* **61**, 342-345.
2.42 Martin, M.M., Rockholm, D.C. and Martin, J.S. (1985) Effects of surfactants, pH, and certain cations on precipitation of proteins by tannins. *J. Chem. Ecol.* **11**, 485-494.
2.43 Metcalf, R.L. and Lampman, R.L. (1989) The chemical ecology of Diabroticites and Cucurbitaceae. *Experientia* **45**, 240-247.
2.44 Metcalf, R.L. and Metcalf, E.R. (1992) *Plant Kairomones in Insect Ecology and Control. Contemporary Topics in Entomology 1*. Chapman and Hall, New York.
2.45 Norris, D.M. (1977) Role of repellents and deterrents in feeding of *Scolytus multistriatus*. In Hedin, P.A. *Host Plant Resistance to Pests*. ACS Symposium Series 62, American Chemical Society, Washington, DC, pp. 215-230.
2.46 Roitberg, B.D. and Isman, M.B. (eds.) (1992) *Insect Chemical Ecology. An Evolutionary Approach*. Chapman and Hall, New York.
2.47 Rosenthal, G.A. and Berenbaum, M.R. (eds.) (1992) *Herbivores: Their Interactions with Secondary Plant Metabolites,* vol. II. Academic Press, San Diego.
2.48 Ryan, C.A. (1990) Protease inhibitors in plants: genes for improving defenses against insects and pathogens. *Ann. Rev. Phytopathol.* **28**, 425-449.
2.49 Ryan, M.F. and Byrne, O. (1988) Plant-insect coevolution and inhibition of acetylcholinesterase. *J. Chem. Ecol.* **14**, 1965-1975.
2.50 Ryan, M.F., Guerin, P.M. and Behan, M. (1978) Possible roles of naturally occurring chemicals in the biological control of carrot fly. In Duggan, J.J. (ed.) *Biological Control*. Symp. Roy. Ir. Acad., pp. 130-143.
2.51 Schaller, A. and Ryan, C.A. (1996) Systemin – a polypeptide defence signal in plants. *BioEssays* **18**, 27-33.
2.52 Schneider, D. (1987) The strange fate of pyrrolizidine alkaloids. In Chapman, R.F., Bernays, E.A. and Stoffolano, J.G. (eds.) *Perspectives in Chemoreception and Behavior*. Springer, New York, pp. 123-142.
2.53 Sivamani, E., Rajendran, N., Senrayan, R., Ananthakrishnan, T.N. and Jayaraman, K. (1992) Influence of some plant phenolics on the activity of δ-endotoxin of *Bacillus thuringiensis* var. *galleriae* on *Heliothis armigera*. *Entomol. Exp. Appl.* **63**, 243-248.

2.54 Smirle, M.J. and Isman, M.B. (1992) Metabolism and elimination of ingested allelochemicals in a holometabolous and a hemimetabolous insect. *Entomol. Exp. Appl.* **62**, 183-190.
2.55 Southwood, T.R.E. (1973) The insect/plant relationship: An evolutionary perspective. *Symposia of the Royal Entomological Society of London* **6**, 3-30.
2.56 Szentesi, A. and Jermy, T. (1992) The role of experience in host plant choice by phytophagous insects. In Bernays, E.A. (ed.) *Insect-Plant Interactions*, vol II. CRC Press, Boca Raton, pp. 39-74.
2.57 Thompson, J.N. (1994) *The Coevolutionary Process*. The University of Chicago Press, Chicago.
2.58 Tumlinson, J.H., Paré, P.W. and Lewis, W.J. (1999). Plant production of volatile semiochemicals in response to insect-derived elicitors. In Chadwick, D.J. and Goode, J.A. (eds.) *Insect-Plant Interactions and Induced Plant Defence*. Novartis Foundation Symposium 223, Wiley and Sons, Chichester, pp. 95-109.
2.59 Turlings, T.C.J., McCall, P.J., Alborn, H.T. and Tumlinson, J.H. (1993) An elicitor in caterpillar oral secretions that induces corn seedlings to emit chemical signals attractive to parasitic wasps. *J. Chem. Ecol.* **19**, 411-425.
2.60 Turlings, T. C. J., Tumlinson, J. H. and Lewis, W. J. (1990) Exploitation of herbivore-induced plant odors by host-seeking parasitic wasps. *Science* **250**, 1251-1253.
2.61 Vet, L.E.M. (1999) Evolutionary aspects of plant-carnivore interactions. In Chadwick, D.J. and Goode, J.A. (eds.) *Insect-Plant Interactions and Induced Plant Defence*. Novartis Foundation Symposium 223, Wiley and Sons, Chichester, pp. 3-20.
2.62 Waterman, P.G. (1998) Chemical taxonomy of alkaloids. In Roberts, M.F. and Wink, M. (eds.) *Alkaloids. Biochemistry, Ecology and Medicinal Applications*. Plenum Press, New York, pp. 87-107.
2.63 Weiler, E.W., Laudert, D., Stelmach, B.A., Hennig, P., Biesgen, C. and Kubigsteltig, I. (1999) Octadecanoid and hexadecanoid signalling in plant defence. In Chadwick, D.J. and Goode, J.A. (eds.) *Insect-Plant Interactions and Induced Plant Defence*. Novartis Foundation Symposium 223, Wiley and Sons, Chichester, pp. 191-204.
2.64 Whitman, D.W. (1988) Allelochemical interactions among plants, herbivores, and their predators. In Barbosa, P. and Letourneau, D.K. (eds.) *Novel Aspects of Insect-Plant Interactions*. Wiley and Sons, New York, pp. 11-64.
2.65 Whitman, D.W. and Eller, F.J. (1990) Parasitic wasps orient to green leaf volatiles. *Chemoecology* **1**, 69-75.
2.66 Whittaker, R.H. and Feeny, P.P. (1971) Allelochemics: Chemical interactions between species. *Science* **171**, 757-770.
2.67 Wink, M. (1998) Chemical ecology of alkaloids. In Roberts, M.F. and Wink, M. (eds.) *Alkaloids. Biochemistry, Ecology and Medicinal Applications*. Plenum Press, New York, pp. 265-300.
2.68 Wink, M. (1998) Modes of action of alkaloids. In Roberts, M.F. and Wink, M. (eds.) *Alkaloids. Biochemistry, Ecology and Medicinal Applications*. Plenum Press, New York, pp. 301-326.
2.69 Wink, M. and Roberts, M.F. (1998) Compartmentation of alkaloid synthesis, transport, and storage. In Roberts, M.F. and Wink, M. (eds.) *Alkaloids. Biochemistry, Ecology and Medicinal Applications*. Plenum Press, New York, pp. 239-262.

CHAPTER 3

PHEROMONES

Although food plants and pheromones constitute separate resources for an insect, their relationship is closer than simply the use of the former to derive the latter. In the cabbage looper *Trichoplusia ni* increased emission of the male pheromone is stimulated by presence of the female pheromone, alone or with cotton bouquet. In addition, unmated males and females were more attracted to opposite sex conspecifics presented in the presence of cotton foliage or its volatiles. Enhanced attractiveness of the female was assigned to increased pheromone release in response to host odour; increased attractiveness of the male was ascribed to enhancement by host odour of sex attraction reactions of the female [3.30].

Pheromones are generally defined as chemicals secreted to the outside by one individual and perceived by another of the same species in which they release a specific behavioural or developmental process [3.26]. Understandably, considerable emphasis has been placed on the chemical identification of pheromones especially from pest species, but it seems appropriate to begin with a consideration of the glands that produce them especially as, by comparison with chemical studies, this is a somewhat neglected area of investigation.

PHEROMONE-PRODUCING GLANDS

Coleoptera

In the red flour beetle *Tribolium castaneum*, the male produces an aggregation pheromone from modified epidermal cells beneath a setiferous sex patch on the prothoracic femur [3.17]. Confined to the male, and apparently derived from epidermal stem cells, these secretory cells are arranged as clusters of goblet-shaped, lipid-secreting, gland cells with large nuclei and a basophilic, vacuolated cytoplasm. Each cell evacuates its contents as a waxy secretion to the leg exterior through an individual, lined, secretory duct (Fig. 22). Externally, there are foveae (secondary reservoirs) with cribiform plates through which the pheromone in wax probably flows, as secretory ducts within the plates are continuous with secretory cells. In contrast, in six species of *Trogoderma* (Coleoptera: Dermestidae), the sex-pheromone gland is constituted from the columnar or epithelial lining of the seventh abdominal sternite and opens through cuticular pores [3.34].

In the female tobacco beetle *Lasioderma serricorne*, the sex pheromone is emitted from a ducted gland 400 µm long, 250 µm wide and 135 µm deep, in the second abdominal segment. Situated vertically beneath the alimentary canal, this gland connects to its opening by an inverted V-shaped apodeme comprising two sclerotized rods ca 700 µm long and 20 µm diam. The apodeme is enveloped by a cuticular sheath, ventrally covered with denticles, directed toward the gland opening. Gland tissue is represented by many secretory cells with large nuclei, droplets of an opaque secretion, and contains secretion-loaded tubules that lead to the duct (Fig. 22). The gland was confirmed as pheromone-secreting when antennal receptor potentials were recorded from male beetles to extracts of the gland secretion from virgin females. 1×10^{-3} Gland equivalent was the threshold concentration for a response that was dose-responsive up to one gland equivalent [3.34].

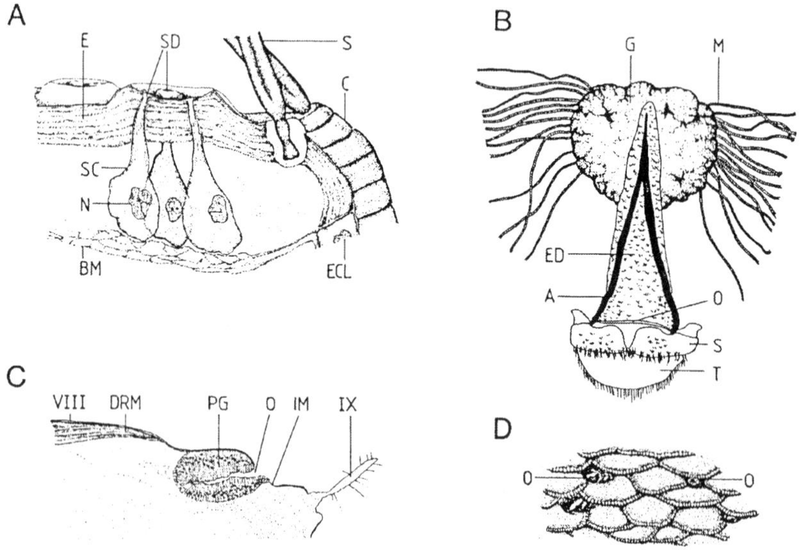

Figure 22. Structures and ultrastructure of some pheromone-producing glands. A) The aggregation pheromone-producing gland of the prothoracic femur of Tribolium castaneum *male; BM, basement membrane; C, cuticle; E, endocuticle; ECL, epidermal cell layer; N, nucleus; S, seta; SC, secretory cell; SD, secretory duct. Redrawn after Faustini* et al. *[3.17]. B) The sex-pheromone-producing gland, apodeme, and efferent duct in the abdomen of the female* Lasioderma serricorne*; A, apodeme; ED, efferent duct; G, glandular tissue; M, muscle fibres; O, orifice; S, eighth sternite; T, tenth tergite. Redrawn after Levinson* et al. *[3.34]. C) Sagittal section though the sex-pheromone-producing gland of the female nantucket pine moth* Rhyacionia frustrana, *between abdominal segments VIII and IX; DRM, dorsal retractor muscle; IM, intersegmental membrane; O. orifice; PG, pheromone gland. Redrawn after Baer* et al. *[3.3]. D) Cuticular surface of sex-pheromone-producing glands of the male cockroach* Leucophaea maderae, *on the fifth abdominal sternite; O, orifice. Redrawn after Sreng [3.57].*

Lepidoptera

In contrast, lepidopteran pheromones are usually produced by ductless glands frequently located in the intersegmental membrane between the eighth and ninth abdominal segments (Fig. 22). Modified cells of the intersegmental, pheromone-producing gland differ from unmodified cells in terms of size, shape and structure. Although similar to unmodified epithelium one to two days before adult emergence, one hour before emergence gland cells have assumed pheromone-secreting characteristics. Specifically, the basal region below the nucleus is rich in Golgi complexes, lipid droplets, glycogen deposits, and rough endoplasmic reticulum (RER). The apical region contains extensive smooth endoplasmic reticulum (SER), and microvilli of the apical surface contain a core of SER. In contrast, unmodified cells contain only RER and have trivial apical projections. The structure of the gland cells and especially the transition from RER to SER is indicative of involvement in lipid metabolism: mammalian SER contains steroidgenic enzymes and their formation corresponds to triglyceride absorption and resynthesis. In the white-marked tussock moth *Orgyia leucostigma*, columnar cells of the pheromone gland change to goblet-shaped cells the day before emergence. The day after emergence, RER and free ribosomes are conspicuous in the cytoplasm but thereafter, extensive SER and RER predominate. The SER resembles that reported from testicular interstitial cells of the guinea pig, a location for enzymes contributing to androgen biosynthesis [3.43, and references therein]. In the female *T. ni*, the basement membrane underlying pheromone-producing gland cells differs from that of unmodified cells. Both cell types share a thin amorphous layer but pheromone gland cells have a second layer apposing the haemocytes, and apparently formed from granules of granular haemocytes that also contain lipid spheres. Haemocytes beneath unmodified epidermal cells do not contain granules nor do they contribute to basement membrane formation [3.42].

More recent descriptions include those of the complex abdominal glands of the male cockroach *Nauphoeta cinerea*. In this species, and unlike other cockroaches, the male produces two successive chemical signals. The sex pheromone "seducin" from the sternal glands attracts the female; thereafter aphrodisiacs, when licked by the female from the tergal glands, elicit mating. The male produces both secretions simultaneously and the ultrastructure of both gland types is similar. Essentially, each gland is composed of four categories of cells, glandular units (a majority) comprising a secretory cell and a duct cell, basal cells and microtubule cells (Fig. 23). The secretory cells are 40 μm high and 10 μm long with a cytoplasm rich in vesicles and Golgi bodies. The plasma membrane contains many microvilli creating space for the receptor canal, a cuticular structure composed of fibrous material. Each duct cell is located at the apex of a secretory cell. Microtubule cells are distributed throughout the gland and the microtubules are located between apical microvilli. Basal cells are voluminous and characterized by extensive SER, elongated mitochondria, and clear vesicles (Fig. 23). These are probably oenocytes modified to secrete products likely to enter the haemolymph. The microtubule cells represent transformed, epidermal cells. Tergal glands secrete proteins and fatty acids and are comparable to those of other cockroaches. It seems probable that secretory cells produce the proteins, and basal cells rich in SER produce the fats [3.58, 3.59].

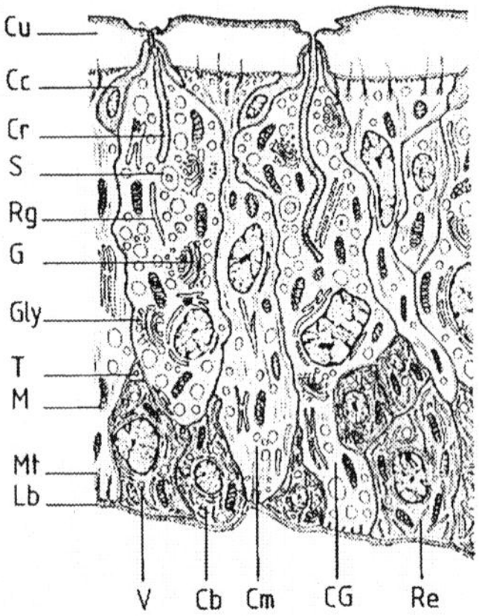

Figure 23. Schematic representation of the secretory unit (type three) from the tergal sex-pheromone-producing gland of the male cockroach, Nauphoeta cinerea, *comprising: BM, basement membrane; a glandular cell CG associated with a duct cell CC; Cb, type 2 basal cell; CM, microtubule cell; Cr, receptor canal; Cu, cuticle; G, Golgi apparatus; Gly, glycogen; M, mitochondrion; Mt, microtubules; Re, smooth endoplasmic reticulum; Rg, rough endoplasmic reticulum; S, secretory product; T, tracheole; V, clear vesicle. Redrawn after Sreng [3.59].*

The site of pheromone production in an archaic lepidopteran, *Eriocrania cicatricella* (Eriocraniidae) has been experimentally verified. Electrophysiological recordings from the male antenna (see Chapter 5), quantified the efficacy of the female whole abdomen, ovipositor, and the Vth abdominal segment. The response to the latter was as strong as to the whole abdomen, thus identifying glands in this segment as the pheromone source [3.66]. The phylogenetic significance is that a pair of glands opening on the abdominal segment could constitute an autapomorphy of the superorder Amphiesmenoptera (Trichoptera + Lepidoptera) i.e. a derived character state found in only one of two sister groups. Although usually present in both sexes of the Eriocranioidea, within the genus *Eriocrania* they occur in only the female. Female sex pheromones of the Trichoptera usually are released from the sternum V glands, but these are absent in the Lepidoptera. Their pheromones are usually released between abdominal segments VIII and XI. The ultrastructure of these glands is consistent within archaic Lepidoptera, as reservoir walls are covered with secretory cells of "type three". Nevertheless, as these glands occur independently in various lineages of both orders, they were viewed as a synapomorphy or the sharing of ancestral characters by different taxa.

There is an interesting association between ovarian development and numbers of sex-pheromone glands in the hind tibiae of oviparous aphid females; viviparous parthenogenetic females that do not copulate lack these structures. Apparently, juvenile hormone (JH) regulates morphogenesis of such plaques but it was not established if this takes effect through the endocrine system, or by a direct action on the plaques. Nevertheless, there is a significant positive correlation between numbers of plaques and vitellogenic eggs, indicating that JH affects glands not directly but through the ovary, and suggesting ovarian control of plaque morphogenesis [3.15].

CHEMICAL STRUCTURES AND NOMENCLATURE OF PHEROMONES

A relatively recent account [3.59] is a primary source for the following summary. Most pheromones comprise a hydrocarbon skeleton, as illustrated by decane (Fig. 24a, b) comprising ten carbons (dec-) with no multiple bonds (-ane), but with a terminal hydrogen substituted by a functional group. This may comprise a ketone (-CO); an alcohol (-OH) (Fig. 24c); ester (-OCOH) formed from combination of an alcohol with an organic acid with loss of water; aldehyde (-CHO) formed by oxidation of the alcohol; carboxylic acid (-COOH) formed by further oxidation of the aldehyde. Such functional groups take precedence in the naming of a compound by terminating the name with –one, -ol, -acetate, -al, and -oic acid respectively. The carbon to which the functional group is attached is designated –1–.

The presence of one or more double bonds, generated by the loss of hydrogens from adjacent carbons, determines the degree of unsaturation of the molecule and alters the designation as, for example, a hydrocarbon from –ane to –ene. The position of each double bond is represented by a numeral corresponding to that of the carbon from which it begins, with each carbon numbered from that attached to the functional group. Thus, Fig. 24c represents dodecan-1-ol. Most pheromones have chain lengths numbering 10 (deca-), 12 (dodeca-), 14 (tetradeca-), 16 (hexadeca-), or 18 (octadeca-) carbons long. The presence of a double bond has another effect. It precludes rotation of the molecule by fixing it in one of two possible configurations, each representing geometric isomers that are different molecules. These are designated either *E* (from the German word *Entgegen*, opposite) or *Z* (*Zusammen*, together), when the carbon chains are connected on the opposite (*trans*) or same (*cis*) side respectively of the double bond (Fig. 24d). Insects readily discriminate between the geometric isomers of a pheromone.

The presence of two and three double bonds is indicated by ending the name with –diene and –triene respectively, and the position of each is again indicated by reference to the carbon number from which it originates. Thus, Fig. 24e represents (*E,Z,Z*)-4,7,10-tridecatriene-1-ol acetate, a constituent of the pheromone of the potato tuberworm *Phthorimaea operculella*.

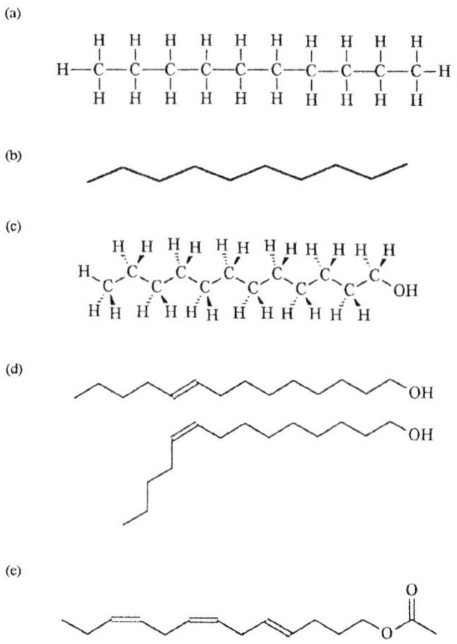

Figure 24. Structural aspects of insect pheromones. (a) Typical simple (straight chain) framework. (b) Simplified representation of same showing only the carbon-carbon bands. (c) Planar representation of framework, with continuous wedges indicating bands above the plane of the paper and discontinuous ones indicating those below, and with substitution of a hydrogen by an oxygen to create an alcohol functional group. (d) Distinction between geometric isomers on the basis of the attachment of the carbon chains on each side of the double bond; attachment on opposite sides is designated (E) (top), and on the same side (Z). (e) Exemplification of the use of the nomenclature, (E,Z,Z)-4,7,10-tridecatriene-1-ol acetate; carbons are numbered from the functional group carbon as –1-. Compiled from Stevens [3.60] and reprinted with kind permission from Kluwer Academic Publishers.

Two systems of nomenclature are in use, the first based on that of the International Union of Pure and Applied Chemistry (IUPAC). Essentially, the first two letters represent and are abstracted from the term describing chain length, with the addition of the terminal letters describing the functional group as: DDOL for dodecanol; DDA used for the acetate; and DDAL for the aldehyde. The presence of a double bond and of a geometric isomer is represented by prefixing with the appropriate symbol and numeral as in (*Z*)-8-DDOL. The presence of two or three double bonds is represented by the use of the prefixes D and T for –diene and –triene respectively. The potential ambiguity in using T for triene when it might equally represent tetra, and H for hexa- when it might equally represent hepta-, is avoided as no decanes with four double bonds are known, and odd-numbered chains are rare. Thus, (*E,Z,Z*)-4,7,10-tridecatriene-1-ol acetate (Fig. 24e) is abbreviated to (*E,Z,Z*)-4,7,10-TriDTA.

The second method accommodates functional groups additional to alcohol, acetate, and aldehyde (-OH, -Ac, and -Al, respectively) through representation of: ketone by –Kt; epoxide (oxygen bound to two adjacent carbons) by –epo; and hydrocarbons without functional groups containing oxygen by Hy. Double bonds and isomers are designated as before but chain length is represented by its number of carbons. (*E,Z,Z*)-4,7,10-Tridecatriene-1-ol acetate (Fig. 24e) or (*E,Z,Z*)-4,7,10-TriDTA by the first system is more simply designated (*E*)4(*Z*)7(*Z*)10-13:Ac with the advantage of immediate recognition of chain length. This is the system adopted in the authoritative *List of Sex Pheromones of Lepidoptera and Related Attractants* [3.2].

Pheromones of Female Lepidoptera

Presenting almost exclusively as straight chain compounds, their specificity is engineered from the interaction of five factors: chain length; functional group; position and number of double bonds; and stereochemistry. Although providing scope for a great many diverse compounds, there is repeated use of the same chain lengths and double bond position in many families. For example (*Z*)-9-tetradecenyl acetate has been identified from 223 species (108 genera); (*Z*)-11-tetradecenyl acetate from 214 species (106 genera); (*Z*)-11-hexadecenyl acetate from 189 species (107 genera); (*E*)-11-tetradecenyl acetate from 161 species (81 genera); (*Z*)-7-dodecenyl acetate from 147 species (78 genera) [*http://www-pherolist.slu.se*]. Such repeated usage of the same molecules may reflect their facile biosynthesis from fatty acids.

The even-numbered feature of chain lengths derives from the construction of fatty acids from acetic acid units of two carbons each (CH_3CO_2H). As an exception, compounds without an oxygen at C-1 are almost always odd-numbered, perhaps because of decarboxylation (removal of CO_2) of the precursor fatty acid.

Pheromones of Male Lepidoptera, and of other Orders

Many male Lepidoptera and non-lepidopteran females produce branched compounds, the simplest being a single carbon off the main chain. Branch position is designated by a numeral, and the number of carbons in the branch by modifying the appropriate hydrocarbon root to end in -yl as hexyl from hexane. Thus the *Trogoderma* spp. pheromone (Fig. 25a) is 14-methyl-*Z*-8-hexadecanal.

Asymmetry in a molecule produces optical isomers (=enantiomers) that cannot be superimposed on their mirror images (Fig. 25a, b) or do not have a plane or centre of symmetry, and arises when the same carbon is linked to four different groups as, for example, a hydrogen, a methyl, an ethyl, and the remainder of the chain (Fig. 25d). This carbon is the chiral centre or chiral carbon.

Optical isomers are so named because they rotate plane polarized light equally but in opposite directions. An equal or racemic mixture of two enantiomers is optically inactive. Most chiral pheromones exist in a single form i.e. are homochiral. Chiral molecules have the same chemical properties and are subjected to an elaborate system of nomenclature in order to distinguish them. Firstly, the use of (R) derived from *rectus* or right and (S) derived from *sinister* or left is based on a clockwise and counterclockwise priority sequence respectively. The priority is established, when the exact shape of the enantiomer is known, on the basis of the four groups attached to the chiral carbon. These are prioritized in descending order from 1 to 4 based on their descending atomic number. When the two attached atoms are the same, as is usual in natural products thus precluding prioritization in the first instance, then this is decided on the priority of the atoms attached to each such atom. For example, in the pheromone of the dermestid *Trogoderma* spp. (Fig. 25a, b) the chiral carbon (-C-14) is attached to a hydrogen (atomic weight 1), and to three carbons (6 each) which are prioritized as follows. The methyl group contains only hydrogens (3x1) rendering it lower in priority than the other two (C-13 and C-15) as each is bonded to a carbon (C-12 and C-16) respectively. As the latter is bonded to hydrogen and the former to another carbon C-11, the priority sequence becomes: C-13 > C-15 > methyl > H giving a clockwise sequence and an absolute configuration of (R) (Fig. 25c). In the event, only this enantiomer (14R)-14-methyl-*Z*-8 hexadecanal is biologically active as a pheromone, and the (14S)-isomer is inactive. The presence of a double or a triple bond is dealt with by treating each bonded atom as duplicated and triplicated respectively and the above system of prioritization is applied. Where double bonds are polysubstituted the above rules are again applied and molecules are designated *E* when the two highest priority groups are on opposite sides, and *Z* when on the same side.

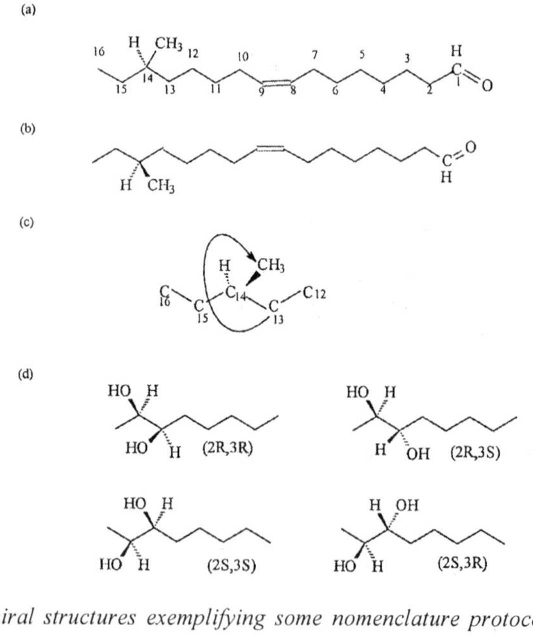

Figure 25. Chiral structures exemplifying some nomenclature protocols; details in text. Compiled from Stevens [3.60] and reprinted with kind permission of Kluwer Academic Publishers.

The presence of more than one chiral centre producing 2^N isomers from N chiral centres allows diastereoisomerison or the presence of stereoisomers that are not enantiomers (Fig. 25d). Accordingly, 2,3-octandiol has (2R,3R) and (2S,3S) as mirror images and thus enantiomers designated stereo-isomers: similarly the (2R,3S) and (2S,3R) forms are also enantiomers and are designated erythro-isomers. However, neither the (2R,3R) nor the (2S,3S) form is the mirror image of either the (2R,3S) or the (2S,3R) forms. Each member of a pair is the diastereomer of the members of the other pair. Unlike enantiomers, diastereomers differ in their physical properties and are separable by distillation, chromatography, and crystallization. Insects may distinguish between diastereomers of which usually only one is biologically active. Chirality is not confined to pheromones of male Lepidoptera or of orders other than Lepidoptera but it is also displayed by pheromones of female Lepidoptera. Epoxides such as the gypsy moth pheromone exhibit chirality around the two ring carbons. Also various hydrocarbon pheromones are branched. [for more details see 3.60, and references therein].

Among insect pheromones at least ten broad categories may be distinguished according to function: sex attraction; aphrodisiacal; epideictic; aggregation and antiaggregation; alarm; trail-marking, territorial marking; regulation of sexual maturity; and in social insects, regulation of queen-rearing.

SEX PHEROMONES

Bombykol or (*E-Z*)-10,12-hexadecadien-1-ol, a sex pheromone of the silk moth *Bombyx mori*, was the first pheromone extracted and identified from an insect; the yield was 12 mg from 400,000 virgin females [3.26, and references therein]. However, this remarkable achievement by Karlson and Butenandt [3.26] set a pattern such that work on corresponding identifications from other species rested after the isolation of a single compound. Furthermore, the following three reported identifications of pheromones were incorrect. Ultimately, the results of field bioassays cast doubt on the adequacy of single compounds as attractants, and Wright was soon advocating the idea, on grounds derived from information theory, that multicomponent pheromones would provide a richer information source [3.65].

The first verification of a lepidopteran multicomponent pheromone was from the southern armyworm moth *Spodoptera eridania*, involving (*Z*)-9-tetradecenyl acetate and (*Z,E*)-9,12-tetradecadienyl acetate [3.24]. The following year synergism was reported between (*Z*)-9- and (*Z*)-11-tetradecenyl acetate of the bouquet of the tortricid moth *Adoxophyes* spp., the male of which responded behaviourally only to the mixture, and not to either constituent presented individually. Subsequently, the need was established for the appropriate ratio of the mixture. The ultimate analysis, showing that single compound pheromone systems are rare, was provided by Silverstein and Young [3.56] (Lepidopteran sex pheromones were reviewed in 1988 [3.60] followed by a vastly amplified check list in 1992 [3.2]).

Female sex pheromones

Of more than 130,000 lepidopteran species, pheromones have been identified in the latest check list [3.2] from some 1,220 species, overwhelmingly from a few families, especially Tortricidae, Noctuidae, and Pyralidae. These utilize mono-or diunsaturated primary aliphatic alcohols, acetates, aldehydes, ketones and other hydrocarbons. This list is too extensive to readily summarize but if the earlier (1988) list is taken as a representative sample, the following profile emerges. The first three categories of chemicals accounted for 92% of pheromone components and 86% of the compounds employed. Although carbon numbers ranged from 10 to 21, compounds with 12, 14 and 16 carbons and their acetates represented 82% of pheromone constituents. Many compounds recurred in the pheromonal bouquet of a range of species. *Z*- and *E*-oriented carbon to carbon double bonds represented 68% and 29% of the total respectively, and terminal double bonds account for the other 3%. Positions 9 and 11 represented 57% of total positions occupied by double bonds (64% in acetates), and positions 3 and 5 accounted for 60% of the sitings of the terminal methyl. Families Tortricidae, Noctuidae and Pyralidae represented 73% of the species recognized as using female sex pheromones. The former employed aldehydes, alcohols or their acetates with 12 or 14 carbons, with the single exception of 13 carbons in a branched configuration of a 12-carbon alcohol acetate.

In 45 species of Noctuidae, 91% of components were 12, 14 or 16 carbons long and frequently occurring constituents were (*Z*)-9-tetradecenyl and (*Z*)-11-hexadecenyl entities. All species of Hadeninae were associated with variants of the latter mostly the acetate, with (*Z*)-11-hexadecenal as the principal constituent in eight out of ten of these moth species. However, five of six *Spodoptera* spp. were associated with (*Z,E*)-9,12-tetradecadienyl acetate. In general, closely related species either share a common constituent or use compounds with similar structures, and several individual compounds are employed repeatedly within families [3.61, and references therein].

Specificity

Specificity of effect is achieved by different methods according to whether the pheromone is single - or multicomponent. If the former, minor structural deviations result in a substantial loss of activity. Thus, the most potent geometrical isomer of bombykol is 100-fold less active than the pheromone. In contrast, when the pheromone is a multicomponent mixture, substantive changes such as altered number of carbons and position of the double bond in, for example, the sex pheromone of *T. ni*, elicited only a ten-fold weaker behavioural response from the male; in the field potency was decreased by only 10%. Such plasticity in multicomponent mixtures suggests greater scope for the evolution of new bouquets, in turn associated with speciation.

Female sex pheromones of the primitive moth *E. cicatricella* comprise (*Z*)-4-hepten-2-one, and the methylcarbinols (2R)-heptan-2-ol and

(2R)-(*Z*)-4-hepten-2-ol; subtraction of only the latter abolished attraction to the male in a field trial. The three structures are chemically very similar and presumably are of a common biosynthetic origin. Such structures are not known from the suborder Ditrysia that represents more than 95% of lepidopteran species, but similar structures are known from the Monotrysia including the Nepticuloidea. This indicates evolution of pheromone structure within the Lepidoptera, entailing synapomorphy with non-ditrysian moths (synapomorphy = a homologous character in two or more taxa that occurs only in the nearest common ancestor and not in other ancestors or taxa) [3.66].

Insight into pheromone evolution within the Ditrysia is provided by reference to the ermine moth (*Yponomeuta* sp.) pheromone, tetradecyl acetate. Essentially, of nine sympatric ermine moth species, all but one emit Δ^{11}-unsaturated acetates specifically, (*Z*)-11-tetradecenyl, (*E*)-11-tetradecenyl and (*Z*)-11-hexadecenyl acetates: species specificity is enabled by *E*/*Z* ratios. The sole exception, *Y. rorellus*, employs a saturated acetate, tetradecyl acetate, as the primary component. This simple pheromone probably evolved through reduction rather than by the alternative route of Δ^{11}-desaturation, evolving independently several times [3.38, and references therein].

Male sex pheromones

Pheromones of male Lepidoptera are much more diverse than those of the females, as their structures range from those with a few carbons such as organic acids, to complex heterocyclic compounds. The evidence for male-produced sex pheromones comes from far fewer species (less than 100) and a specific function has been established for even fewer. Suggested functions include: inhibiting female escape reactions (aphrodisiacal); deterring other males from mating; initiating oogenesis; and perhaps repelling other males and predators. Detailed behavioural data in favour of an aphrodisiacal role are few, perhaps because copulation would follow even in their absence. However, there is little doubt about the significance of a male-produced aphrodisiac in the milkweed butterflies (Danainae). Males visit withered plants containing pyrrolizidine alkaloids (PAs) acting as precursors for the biosynthesis of the alkaloid ketone, 2,3-dihydro-7-methyl-1H-pyrrolizin-1-one or danaidone. Secreted from hair pencils that brush the female antennae during flight, danaidone facilitates mating by promoting landing by the female and inhibition of escape-flight. Indeed the male of phylogenetically unrelated taxa such as danaines, ithomiines, and arctiids utilize dihydropyrrolizidine derivatives. The precursors for those of arctiids are sequestered from the larval stage whereas those of ithomiines are sequestered, like those of the danaines, from the plant exudate rich in PAs. Thus, danaidone is characteristic of many danaine genera and of the *Utetheisa ornatrix* corematal secretion [3.10]. Unlike female sex pheromones, some of those produced by the male may be detected by humans, frequently as a pleasant fragrance. Methyl-2-epijasmonate with methyl jasmonate (jasmine), ethyl *trans*-cinnamate (cinnamon) and vanillin are among such compounds [3.62, and references therein].

The relationship between the presence of disseminating structures for male lepidopteran pheromones and the probability of contact with closely related species has been examined. Two possibilities, adaptive and incidental, have been examined to explain the origin of such pheromones. Essentially, the former proposes that their evolution has adaptive significance by facilitating mating with insects adapted to the same niche, and the latter suggests them as an incidental consequence of isolation. The test employed the sharing of a common host plant by two species of the same genus (congenerics), to compare the frequency of male pheromone incidence with that in species not sharing a common host. The five lepidopteran families and corresponding male pheromone structures were: Pyralidae, abdominal hair pencils; Yponomeutidae, abdominal coremata; Tortricidae, forewing costal fold; Noctuidae, abdominal brush organs; Ethmiidae, hind wing hair pencils. Within each family, there was a significant positive correlation between host overlap and presence of such pheromone-releasing structures. Furthermore, combining all taxa revealed that 53% of congenerics sharing a common host were thus equipped, compared with 28% of those not sharing one. Such data sustained the adaptive model of events and indicated that male pheromones evolved to preclude cross-specific attraction between congeneric species with broad overlap in space, time, and pheromonal chemistry. The underlying mechanism was not envisaged as natural selection but as sexual selection that avoided interpopulational matings producing progeny of lesser fitness, and selected males with preferred characteristics. In support of this was claimed: the considerable morphological diversity in structures expressed by male as compared with female pheromones; their infrequent occurrence; and the lack of association between their complexity and the phylogeny of the Lepidoptera [3.44].

Sex-pheromone biosynthesis

Radiotracer studies established the fatty acid oleic acid, as a precursor to n-nonanal, a component of the male sex pheromone of the greater wax moth *Galleria mellonella.* In vivo labelling studies of the spruce budworm *Choristoneura fumiferana* also showed that the lipid (*E*)-11-tetradecenyl acetate was synthesized specifically in the pheromone-producing gland, and was degraded in-step with pheromone release to serve as a precursor to the pheromone (*E*)-11-tetradecenal. Such an acetate, less reactive than the aldehyde pheromone, would facilitate steady continuous production and accumulation of the precursor to yield, on demand, the pheromone by a one-step reaction. This would preclude inadvertent release on the one hand, and large sudden metabolic expenditure for full-scale biosynthesis on the other. Host terpenes, nerol, geraniol, myrcene and limonene, may act as precursors for a component of the pheromone of the cotton boll weevil *Anthonomus grandis. A. grandis* absorbs myrcene and limonene through the cuticle and oxidizes them to alcohols. Thus, myrcene is utilized as a precursor in the synthesis of two distinct pheromones; a sex attractant in *A. grandis* and an aggregation pheromone in bark beetles (see below) [3.25, and references therein].

In general, ditrysian sex-pheromone constituents are aliphatic, 10-18 even-numbered carbons long, with a terminal functional group (aldehyde, alcohol or acetate), and with one to three double bonds, in the *Z* or *E* configuration. Despite the extraordinary range of compounds produced within this framework, recent evidence indicates that the biochemical steps producing them are relatively few. Comprehensive data are available for the redbanded leaf roller moth *Argyrotaenia velutinana* (Tortricidae), and *T. ni* (Noctuidae). In the former species, maximal attraction of males is elicited by a female pheromone comprising: (*Z*)-11-tetradecenyl acetate (*Z*11-14:Ac), (*E*)-11-tetradecenyl acetate (*E*11-14:Ac), tetradecyl acetate (14:Ac), dodecyl acetate (12:Ac), (*Z*)-9-dodecenyl acetate (*Z* 9-12:Ac), (*E*)-9-dodecenyl acetate (*E* 9-12:Ac), and 11-dodecenyl acetate (11-12:Ac). Acetate represents the starting molecule for biosynthesis of the constituents, through tetradecanoate as intermediary, with significant incorporation into hexadecanoate and octadecanoate but not longer molecules. Tetradecanoate is derived by chain shortening through β-oxidation of hexadecanoate. Then the actions of *E*- and *Z*-11 desaturases form the double bond at the 11-position giving rise to large amounts of both isomers. There is no evidence for the action of an isomerase to explain production of the *Z*11:*E*11-14 ratio of 92/8. So, preferential selection by the gland of the *Z*-isomer, from the available phospholipids was proposed to explain the observed ratio. Unused isomeric phospholipids, seem to be converted to triglycerols to provide a fresh pool of non-isomeric precursors, avoiding depletion of any isomer before a new cycle of selection. Additional chain-shortening, followed by reduction and acetylation, produces the minor substituents of the blend. In *T. ni* a similar enzymatic effect occurs except that the absence of an *E*11-isomer is matched, and accounted for, by the absence of an *E*11-isomerase; the *Z*11-isomer is produced through the action of an *Z*11-isomerase. Furthermore, in this species, desaturation occurs before and not after chain-shortening (Fig. 26) [3.49, and references therein].

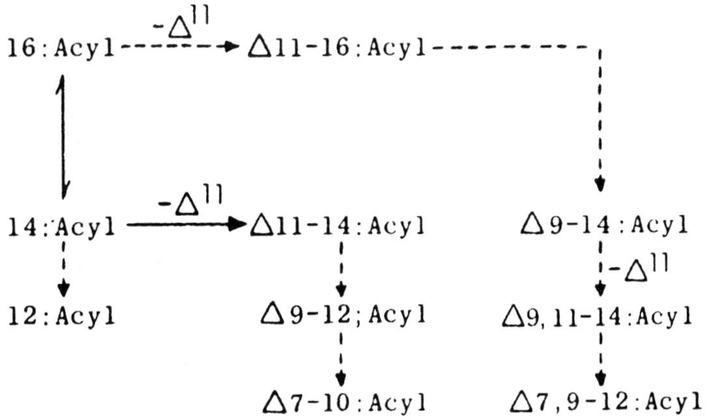

Figure 26. Pathways employed in biosynthesis of sex pheromones by Lepidoptera to produce precursors through action of unique Δ^{11}-desaturases, and chain shortening. Redrawn after Roelofs and Brown [3.50].

Such data on enzymatic derivations, in addition to indicating the presence of significant minor components of pheromones, have facilitated construction of a phylogenetic framework for pheromones of the Tortricidae. The approach taken was that new specific pheromones would evolve by the smallest possible number of changes in their biosynthesis. Mammalian systems have given evidence for only four desaturases: 4-, 5-, 6-, and 9- specific to the carboxyl group. A 9-acyl desaturase occurs in insects accounting for the abundance of oleic acid and palmitoleic acid. Pheromones of many Lepidoptera can be accounted for by invoking an unique Δ^{11}-desaturase. Furthermore, the evolution of chain-shortening enzymes to act on 11-16:Acyl intermediates would explain the presence of *Z*9-14:OAc and *Z*7-12:OAc in primitive species. Subsequent displacement of the Δ^{11}-desaturase to function after chain-shortening from 16- to 14-carbon intermediates would facilitate the formation of very many compounds found in the tortricids (Fig. 26) [3.50].

A novel synthetic mechanism obtains in the biosynthesis of the major sex-pheromone component, (*Z*)-9-tricosene (*Z*9-23:Hy), of the female housefly *Musca domestica*. Incubation of male and female microsomes with radio-labelled Acyl-CoAs afforded equal amounts of labelled pheromone and surprisingly of CO_2; hydrocarbon formation in algae, plants and other animals is associated with release of CO. The intermediate aldehyde was formed only in the presence of O_2 and NADPH, suggesting the involvement of a cytochrome P-450 polysubstrate monooxygenase [3.48].

The blend

A significant advance in the understanding of pheromone effects was the discovery of the importance of the sex-pheromone blend; maximal effect being achieved only when the components are presented in the blend that is naturally emitted [3.36]. Flying males respond to the blend and not to the influence of individual constituents perceived at different times or with differential acuity. Thus, in *T. ni* and other species, upwind flight followed by source location is evinced in a flight tunnel in response to the authentic blend in concentrations 100- to 10-fold less than other blends. In the field, the oriental fruit moth *Grapholitha molesta* exhibits wing fanning at twice the distance from a complete blend as compared to the most efficacious incomplete one. These and other data specifically preclude the idea that the flight response of a downwind male is elicited by the major component of the blend, and that minor components fine-tune subsequent responses such as courtship. Rather, the addition of minor components significantly enhances male responses to the major constituent at lower concentrations by eliciting flight to such concentrations that were inactive alone [3.4, and references therein].

Perception of the blend

Individual components of the blend are detected by neurons specifically tuned to them, and neurons sensitive to different components may exist within the same sensillum. As each neuron has a characteristic amplitude (see Chapter 5) its sensitivity is readily ascertained following exposure to puffs of purified components. Thus, antennal sensilla trichodea of the wild silk moths, *Antheraea polyphemus* and *A. pernyi*, contain, in different combinations, neurons tuned to one of the three pheromone components (*E,Z*)-6,11-hexadecadienyl acetate (Ac_1), (*E,Z*)-6,11-hexadecadienal (Al) and (*E,Z*)-4,9-tetradecadienyl acetate (Ac_2). In both species, most large sensilla (ca 80%) contained two cells, one sensitive to Ac_1, and the other to Al; small sensilla frequently had cells of each of the three types [3.41].

Significance of the blend

Early experimental verification of the importance of blend ratios was provided by the demonstration that the smaller tea tortrix and the summer fruit tortrix, *Adoxophyes* spp. both produced a mixture of (*Z*)-9 and (*Z*)-11-tetradecenyl acetate. A different blend of these ingredients was detected in the female of these species: a 63/3 ratio for the former when coupled with (*E*)-11-tetradecenyl acetate and 10-methyldodecyl acetate (ratio: 40/20), as compared with a 90/10 ratio of (*Z*)-9- and (*Z*)-11-tetradecenyl acetate for the latter [3.60, and references therein]. In species that are already widely separated geographically, and free of pressure to evolve distinct pheromone constituents, the same pheromone ingredients may occur, as in Japanese chafers [3.32].

Genetics of the blend

In addition to differences in pheromone blends between species are differences between different populations of the same species as, for example, in the European corn borer moth, *Ostrinia nubilalis*. In a population in Iowa 11-tetradecenyl acetate is represented by a *Z/E* ratio of 97/3 which was also attractive to most of the U.S.A. and European populations tested in the field. A *Z/E* ratio of 4/96 was represented in a New York State population, and preferential responses have been elicited by ratios of 97/3 and 2/98 from Pennsylvanian populations suggesting the presence of *Z*-sensitive and *E*-sensitive strains; isoenzyme analyses indicated that interbreeding did not readily occur between them. Nevertheless, it did take place and the ensuing blend (*Z/E*, 35/65) seemed to be controlled by Mendelian inheritance involving a single pair of allelles [3.29, and references therein].

Although males of the *E*- and *Z*-strains are not distinguishable morphologically, they are separable on the basis of their electrophysiological responses to pheromone constituents. Specifically, when three populations types existing in New York State, bivoltine *E*(BE), bivoltine *Z*(BZ) and univoltine *Z*(UZ), were investigated, the amplitude of the *Z*-cell action potential in BE males was equivalent to 35% of summed amplitudes from *E*- and *Z*-cells: in *Z*-males it was 65%, and in hybrids 50%. Responses of UZ males were indistinguishable from those of BZ males. The frequencies of male phenotypes in the F_2 generation, and in maternal and paternal backcrosses, were consistent with control by one Mendelian gene with two alleles. The genes controlling pheromone production and perception are independent involving respectively autosomal genes and a gene on the sex chromosome [3.4, and references therein].

Hormonal regulation of sex-pheromone production

The corn earworm *Helicoverpa zea* and other moths produce sex pheromones according to a diel periodicity. This is controlled by a brain factor, a peptide hormone produced in the suboesophageal ganglion (SOG) of both males and females and released through the corpora cardiaca (CC). This pheromone biosynthesis activating neuropeptide (PBAN) was isolated from two sets of some 2,500 suboesophageal complexes dissected from both sexes of *H. zea* (Hez-PBAN). This comprises 33 amino acid residues with a molecular weight of 3,900, with the pentapeptide fragment 29-33 (FXPRL-amide) representing the minimum sequence required for biological activity. The C-terminal amide was also essential for activity being 1000-fold more potent than the analogue with a free carboxylic acid group. The amino acid sequence of PBANs from *H. zea*, *B. mori* (Bom-PBAN), and the gypsy moth *Lymantria dispar* (Lyd-PBAN) express ca 80% homology and exhibit the C-terminal pentapeptide, but are not significantly homologous with any other fully characterized invertebrate or vertebrate peptide hormone. However, there is a pronounced similarity with the partial amino-terminal sequence of the melanization and reddish coloration hormone (MRCH) isolated from *B. mori* (see below). Two to four picomoles of synthetic Hez-PBAN elicited characteristic levels of sex pheromone from ligated *H. zea* females. The fact that Hez-PBAN also elicited sex-pheromone production in six other moth species, with different pheromone chemistry, suggested that PBAN activates an early step in their biosynthesis. This may occur at or before the synthesis of fatty acids or in the reduction by reductases of fatty acids to alcohols. At the level of the pheromone gland, PBAN triggers pheromone biosynthesis by opening calcium channels on the plasma membrane, thus eliciting cAMP production [3.47, and references therein]. As detailed knowledge of this process might suggest methods for interruption and, thus, for a novel form of pest control, it has been the subject of detailed investigations.

Three groups of PBAN cells, identified by use of polyclonal antibodies, occur along the ventral midline of the SOG. Axons from the most posterior (labial) of these project to the CC and extend to the oesophageal nerve. The other two groups send axons to the maxillary nerve, and thence to the CC. Furthermore, the maxillary axons also extend to the ventral nerve cord (VNC) from where they innervate all the segmental ganglia. These associations prompted the suggestion that the SOG continuously secretes PBAN that is released from the CC into the haemolymph. An alternative suggestion implicated the terminal abdominal ganglion (TAG) and claimed that pheromone production required an input from nerves posterior to this, as antibodies did not consistently detect PBAN in the haemolymph. This, however, was discounted by the fact that neither severing the VNC nor excision on the TAG prevents pheromone production [3.47, and references therein]. Nevertheless, it is clear that PBAN takes effect through various routes.

It proved necessary to sever the VNC of *L. dispar* posterior to the thoracic ganglia to obtain a high pheromone titre under various experimental conditions. Accordingly, these ganglia inhibit pheromone build-up, in turn suggesting a neural pathway for PBAN. In contrast, PBAN in *A. velutinana* seems to elicit the release of a peptide from a male accessory structure the bursa copulatrix, that promotes pheromone production. In the true armyworm *Pseudaletia* (=*Leucania*) *unipuncta*, PBAN production and efficacy seems to be regulated by JH but reproductive maturity, known to be stimulated by JH, could account for this [3.47, and references therein].

PBAN also directly affects biosynthetic pathways of pheromones as in *A. velutinana* where PBAN may promote either fatty acid synthesis through enhanced supply of substrate, or activity of acetylCoA carboxylase, the primary enzyme in the pathway. In *H. zea*, PBAN acts directly upon the pheromone gland either to enhance the availability of substrate or to stimulate the activity of a biosynthetic enzyme. In the cabbage armyworm *Mamestra brassicae*, PBAN seemed to affect fatty acid synthesis, rather than pheromone synthesis from the precursor [3.51, and references therein]. In the cotton leafworm *Spodoptera littoralis*, PBAN dose-dependently and directly stimulates pheromone production in the isolated gland within 45 min of incubation, peaking after 3hr at levels 8-10-fold greater than in controls. Detection of PBAN in haemolymph by antibodies during darkness, when pheromone is released, indicated a humoral path to the gland. Use of radiolabelled precursors indicated that PBAN affects pheromone biosynthesis after production of the final intermediate, the alcohol of the pheromone (*Z,E*)-9,11 tetradecadienyl acetate. The suggested step was at the reduction of Acyl moieties, which is consistent with data from *B. mori.* Essentially, the early step catalyzed by PBAN is at the level of fatty acid synthesis in some species, and in others at or before the reduction of Acyl precursors [3.40, and references therein].

Melanization and reddish coloration hormone (MRCH)

In *B. mori* this melanization-inducing peptide (Bom-MRCH) is identical to Bom-PBAN-I (33 amino acids as distinct from Bom-PBAN-II, 34 amino acids). Furthermore, Bom-MRCH induces cuticular melanization in the larvae of *P. separata* and *S. litura*, and also sex-pheromone production by *S. litura* female. Accordingly, PBAN regulates two key phenomena each at a different life stage. Treating the 3.5-day old female of *Heliothis peltigera* with synthesized fragments of Hez-PBAN indicated that omission of the first eight N-terminal amino acids did not diminish pheromonotropic activity, but removal of a further four and eight amino acids did so. Essentially, PBAN activity required the presence of the amino acids between nine and 13 and the C-terminal. In terms of melanotropic activity, omission of the first eight residues from the N-terminal end had no significant effect but removal of the next four significantly decreased it to control levels, corresponding to the effect on pheromone biosynthesis; also correspondingly, the C-terminal amide was required for activity. In contrast, was the requirement of PBAN 26-33 NH_2 and

PBAN 28-33 NH2 for melanotropic but not for pheromonotropic activity. Also the fragments PBAN 26-33 NH2 and 28-33 NH2 derived from the C-terminal were required for melanotropic but not for pheromonotropic activity. The mobility of synthesized C-terminal short peptides to elicit pheromonotropic activity was attributed to their exposure to proteolysis, implying that a substantial region of the molecule protects against this process [3.1, and references therein].

Orientation to the Pheromone Source

An early contribution to our understanding of insect flight to a pheromone source came from R. Wright who visualized the break-up of the odour plume into filaments that would be more dispersed with increasing distance from the source [3.65]. Thus, an insect flying upwind to a source could ultimately locate it by differentiating between plume regions with short and long time intervals between filaments. This laid the foundation for subsequent investigations of plume ultrastructure using ionized air. Experimental evidence for the importance of stimulus intermittence came from the demonstration that the summer fruit tortrix male *Adoxophyes orana* did not fly upwind in a uniform pheromone cloud, but readily did so to a point source in it. Confirmation of this effect came from males of *G. molesta* that failed to evince zigzag upwind flight in a continuous pheromone cloud, but did so when these were pulsed with clean air [3.4, and references therein].

Filaments in plumes may also protect the pheromonal signal from background chemical noise. Detection of the pheromone blend of the female *T. ni* was unaffected when the appropriate inhibitor was released 5 cm crosswind or 10 cm upwind, creating an overlapping plume downwind. Significant inhibition occurred only when the inhibitor was released from the same or an abutting point source, such that filaments from pheromone and inhibitor were detected simultaneously [3.37].

Comparing the walking of flightless and flight forms of the cowpea weevil *Callosobruchus maculatus*, indicated that the former always reached the pheromone source, as the response of flight forms was significantly decreased by continuous stimulation. Thus, intermittent, on-off, or flickering stimulation by pheromone may be necessary for this form, perhaps due to habituation [3.35].

A moth flying upwind controls its speed, direction and flight altitude by reference to the apparent movement of ground structures. Such control must result from interplay of the only two responses available to it, change in course angle relative to the wind line, and change in speed through the air. Resultant mechanisms are optomotor anemotaxis (using visual information to influence direction and speed of upwind progression), and internally-programmed or self-steered counterturning. Steering may result from either yawing or body rolling, with force modifiable either by alterations in wing-beat frequency, or by body angle to the ground (pitch angle). Lift is another influencing factor. Ground images, especially patterns, influence optomotor anemotaxis but whether external stimuli influence self-steered counterturning is unclear. In any event, it seems that pheromone concentrations set the tempo for such responses [3.4].

Although some moth species exhibit a looped flight, most employ a zigzag pattern i.e. side to side flight across the wind line. Zigzagging seems to be a fundamental flight response in the absence of pheromone, the presence of which elicits upwind flight. After losing contact with a plume, moths exhibit a casting flight, involving both counterturning and changes in course angle that may lead to straight upwind flight. An obviously efficacious mechanism involves a continuum from narrow to wide zigzagging that would be affected by the frequency of encounter with filaments governed by pheromone concentration; this would be useful in rapid shifts of wind direction.

Opinions vary as to the significance of counterturning. The fact that moths respond to pockets of clean air suggested that male flight patterns were modulated by reaction latencies to pheromone filaments and clean air pockets, in addition to filament frequency. Thus, filaments would elicit upwind surges and an encounter with clean air generated by turbulence would elicit casting (zigzagging without upwind displacement) and counterturning. This designated casting flight as a pheromone mediated reaction [3.5, and references therein].

In contrast, another model attributes the zigzag element to inability of the male to employ optomotor cues in order to advance upwind in the presence of pheromone. The zigzag, viewed merely as a course correction, was affected by blend composition as synthesized pheromones and incomplete blends elicited convoluted routes, in contrast to a direct route to the calling female. Also the use of two-dimensional recording of tracks was judged as less appropriate than a three-dimensional record that would take account of vertical undulations. When these were recorded the track was indicative of a looping flight without zigzags. High wind speed also elicited straight flights which did not support the concept of an intrinsic counterturning programme [3.64, and references therein].

However, the metronome-like quality (similar timing) of counterturning in wind, in windless conditions, and in casting within pheromone-free air, all argue for the presence of programmed counterturning. Furthermore, manipulating the frequency of pheromone pulses in a plume indicated that counterturning suppression and expression of an upwind flight surge were governed by pulse duration and size. Accordingly, plumes that comprise a rapid sequence of pheromone pulses will elicit a prolonged upwind, straight-flight surge. Infrequent pulses would trigger the casting and counterturning responses [3.13, and references therein].

In forests, insects may search in all directions, apparently because wind direction is very variable there. Furthermore, it becomes more fruitful to search parallel to, rather than across the wind, if its direction swings rapidly over a range of ±30°C from the mean; resultant plumes will generate a smaller cross section along the wind rather than across it. Then it may be more efficient to fly downwind and parallel to, rather than across the wind [3.53].

Although Lepidoptera are by far the most extensively studied group in the context of sex pheromones, those of other insect groups should not be overlooked. Aphids (Hemiptera) are economically significant pests that are difficult to control on account of their mobility, remarkable reproductive ability, and resistance to many synthetic insecticides.

Aphid sex pheromones are synthesized by sexual females in glandular epidermal cells under plaques on the tibiae of the hind legs. In some 15 species of the subfamily Aphidinae studied to date, the principal constituents are (4aS,7S,7aR)-nepetalactone and (1R,4aS,7S,7aR)-nepetalactol (Fig. 27). The ketone, which may be reduced to the alcohol, is long established as a constituent of catmint *Nepeta cataria* (Labiatae). Citronellol is a widespread if minor constituent of aphid sex pheromones but, with no behavioural or electrophysiological activity, it may represent a precursor for their synthesis. As only the former two compounds are known from most species studied, then blends must be significant for recognition of the ca 2,200 species. Nevertheless, as some species employ a sole constituent then several species are likely to utilize the same pheromone [3.18, and references therein].

(a) (b)

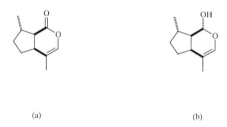

Figure 27. Structures of the principal components of aphid sex pheromones. (a) (4aS,7S,7aR)- nepetalactone and (b) (1R,4aS,7S,7aR)-nepetalactol. After Hardie et al. *[3.18].*

AGGREGATION PHEROMONES

They are best documented from the Coleoptera and especially from bark beetles (Scolytidae), the most destructive pests of trees grown for timber and pulp in the Northern Hemisphere, destroying billions of cubic feet of timber annually. In polygamous species, the host tree is selected principally by the male beetle, and in monogamous species by the female; this step is designated primary attraction. Typically, the colonizing beetle bores through the outer bark and excavates a gallery in the phloem layer. Subsequent release of aggregating pheromone draws in both sexes (secondary attraction), facilitating assembly of the sexes and attack in sufficient numbers to overcome the resistance of the stressed tree. Overall, the host selection and mate detection processes may be viewed as a long behavioural sequence comprising: flight initiation; dispersal; orientation towards the odour source, landing; establishment in the host; and mating. In this catenary process, orientation steps may entail upwind anemotaxis, close-range orientation to the source, followed by landing and tunnelling. In addition to mass recruitment, the aggregation pheromone also shifts attack from one part of the tree to another, and from one tree to another, thus minimizing beetle overcrowding and mortality. Overall, the decisive positive chemical cues, including the presence of monoterpene hydrocarbons (MT), are provided initially by the host and then by 'pioneer' beetles, but the green leaf volatiles (GLVs) emitted by broadleaved plants repel coniferous bark beetles [3.54, and references therein].

Interactions between these cues may extend to synergism. In the simplest situations, ipsdienol alone attracts the double-spined spruce bark beetle *Ips duplicatus*; and frontalin is the sole attractant for the male southern pine beetle *Dendroctonus frontalis* (Fig. 28). In contrast, the California five-spined ips *I. paraconfusus* requires a three-component pheromone mixture of ipsenol, ipsdienol, and *cis*-verbenol. A measure of the intimate association between host and beetle-produced volatiles is provided by the pheromone biosynthetic pathway in *I. paraconfusus*: the male converts the host volatile myrcene to ipsdienol which is in turn converted to ipsenol (Fig. 28); for other examples of detoxifying hydroxylations, see [3.54]. The specificity of this pathway in *I. paraconfusus* is indicated by the fact that the female is unable to similarly convert myrcene. Such conversions are mediated by gut microorganisms, as elegantly confirmed when the male *I. paraconfusus* was fed a medium containing streptomycin; this prevented the conversion of myrcene to ipsdienol and ipsenol. In contrast, as conversion of α-pinene to *cis*-verbenol was unaffected, this step seems independent of microorganisms [3.11, and references therein] (Fig. 28).

Ethanol is also significant. Generated naturally by anaerobic fermentation in the moist phloem of trees, this compound may represent: the major signal of attraction for scolytid beetles such as *Hylecoetus dermestoides* and *Xylosandrus germanus*; a synergist of host odours for *Hylurgops palliatus* and *Hylastes* spp.; a synergist of host odours and/or attractant pheromones for *Dryocoetes autographus*, *Trypodendron* spp. and *Gnathotrichus* spp.; and a dosage-dependent synergist of host odours and/or attractant pheromones for pest species of low aggressiveness in attack such as *Leperisinus* spp. and *Tomicus* spp., and possibly the black turpentine beetle *Dendroctonus terebrans* and the red turpentine beetle *D. valens*, as well as for major pest species such as the Douglas-fir beetle *D. pseudotsugae*. However, pest species of high aggressiveness towards healthy trees, such as the southern pine beetle *D. frontalis* and the spruce bark beetle *I. typographus,* initiate their primary attack independently of the presence of ethanol [3.28].

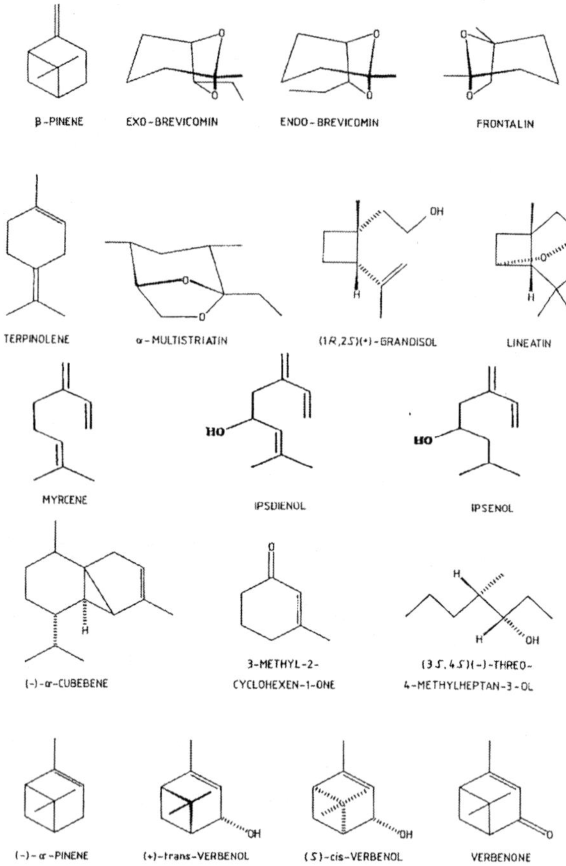

Figure 28. Configurations of some beetle pheromones and relevant plant kairomones. Pheromones (in other than the left column of figure): exo-brevicomin, endo-brevicomin sex attractants, and frontalin, an aggregation pheromone in Dendroctonus *spp.; multistriatin, antiaggregation pheromone in the smaller elm bark beetle* Scolytus scolytus*; grandisol, aggregation pheromone in* Pissodes *spp. and sex attractant in the cotton boll weevil,* Anthonomus grandis*; lineatin, aggregation pheromone in* Trypodendron lineatum*; ipsdienol and ipsenol, aggregation pheromone in* Ips *spp.; 3-methyl-2-cyclohexen-1-one, antiaggregation pheromone in the Douglas-fir beetle,* D. pseudotsugae*;4-methylheptan-3-ol, aggregation pheromone in* S. scolytus*;* trans-*verbenol, aggregation pheromone in* Ips *spp., and* cis-*verbenol, aggregation pheromone in* Ips typographus*; verbenone, antiaggregation pheromone at elevated concentrations in* Dendroctonus *and other genera. Relevant tree kairomones (see left column): β-pinene, a precursor for trans-myrtanol (see Fig. 29); terpinolene, kairomone for the larger pine shoot beetle* Tomicus piniperda*; myrcene, precursor for ipsdienol; α-cubebene, host synergist for 4-methylheptan-3-ol; α-pinene, precursor for trans and cis-verbenol and verbenone. For further details, see Schlyter and Birgersson [3.54].*

The timing of the release of aggregation pheromones provides an insight into their origins. Unlike sex pheromones, aggregation pheromones are not released immediately after eclosion as their production is associated with more mature adults that have fed on the host. Possible exceptions are provided by *D. frontalis* and *D. terebrans* whose pupae apparently store the monoterpene α-pinene in a conjugated form. This is deconjugated and metabolized into the pheromone *trans*-verbenol in the adult. More usually, aggregation pheromones that are terpene alcohols seem to be derived from host precursors sequestered by the adult only after feeding. This is indicated by the absence of such pheromones in newly-emerged adults.

However, de novo synthesis of pheromones by forest beetles of one sex may be more widespread than pheromone production from detoxification. The male double-spined spruce bark beetle *Ips duplicatus* produces de novo ipsdienol (Fig. 28) and *E*-myrcenol, as the two components of its pheromone, when constructing the nuptial chamber. Females and males of the western pine beetle *Dendroctonus brevicomis* exhibit de novo production of their pheromones in the form of exo-brevicomin and frontalin respectively (Fig. 28). In this monogamous species, the female is the pioneer and the release of her pheromone, together with the host monoterpene myrcene, attracts the male. Release of his pheromone elicits mass attack by both males and females. Such aggregation is subsequently inhibited by verbenone (Fig. 28) that inhibits attraction for at least 10 species of bark beetles.

Derived by oxygenation from α-pinene (Fig. 28) either by exposure to air (autoxidation) or by metabolism by yeasts and beetles, or by a combination of these, verbenone is indicative of a stressed and unsuitable host for scolytids requiring a fresh host; in contrast, it may serve as a kairomone for longhorn beetles (Cerambycidae). More potent than verbenone is (+)-ipsdienol (Fig. 28) that, produced by male *D. brevicomis* and by competing *Ips* species, serves to terminate the attack. Such a sequence of events was depicted as a function-shift from sex attraction to signalling the presence of exploitable trees, rare and dispersed, thus eliciting aggregation [3.54, and references therein].

Synergism

In the spruce bark beetle *I. typographus*, the aggregation pheromone comprises three compounds, 2-methyl-3-buten-2-ol (MB), *cis*-verbenol (cV) and ipsdienol (ID). The male emits more MB and cV before than after mating, and when ID production is increased MB and cV interact synergistically, yielding large catches in baited traps but with a low male representation (21-31%). To satisfy synergism, the combined potency of compounds should be significantly larger than their sum, which was fulfilled by the effect of MB and cV (standard ratio 50:1) on trap catches. Each was essential for the effect as the omission of either abolished the response; ID conferred a significant additive effect to a medium dose of MB/cV (0.1mg/day) (Fig. 29).

Catch distribution in northerly traps was consistent with upwind migration as the predominant wind direction was southeast/south. Distribution of trap catches was also consistent with the hypothesis that MB acts as a close-range landing attractant and cV as a long-range orientation constituent; increased concentrations of cV inhibited male landing. Interestingly, this contrasts with the roles visualized for lepidopteran pheromone constituents (see above). The effect in bark beetles could be relevant in the switching of mass attack from one tree to another. Nevertheless, the data need to be interpreted with caution as trap catches were always small in response to single compounds [3.55].

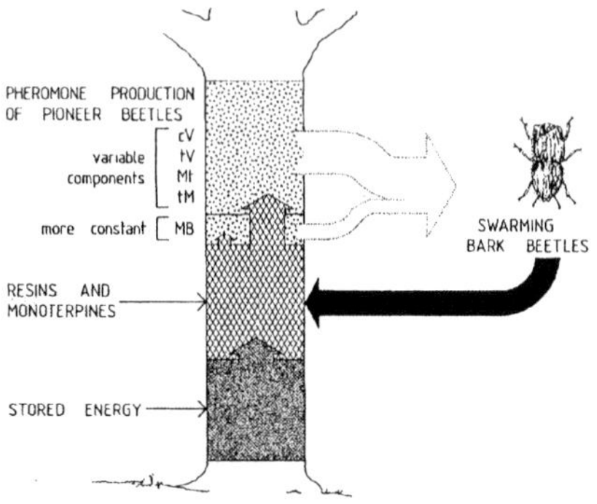

Figure 29. Schematic representation of roles of constituents of the pheromone of the spruce bark beetle Ips typographus *in its interactions with trees. cV,* cis-*verbenol; tV,* trans-*verbenol; Mt, myrtenol; tM,* trans-*myrtanol; and MB, 2-methyl-3 buten-2-ol. Compiled from Schlyter* et al. *[3.55].*

Methods to investigate synergism include combining chromatographic fractionation with bioassays entailing binary splitting, additive combination, and subtractive combination. The relative efficiency of these methods was compared by a computer programme calculating the probabilistic number of tests required to isolate two to five synergists from extracts with any number of major fractions, containing any number of compounds. For detection of three synergists the three methods required 23, 30 and 20 tests respectively. But the binary splitting method required 11 gas chromatography (GC) fractionation passes compared with only two, or at most four, for each of the other two methods. As multiple passes elicit loss of biological activity, associated with thermal decomposition and adsorbtion, the latter two methods were specifically compared. The subtractive approach was always superior, i.e. required fewer tests [3.12].

Also in *I. typographus*, amounts and proportions of aggregation pheromone constituents may vary considerably. The overall amounts secreted ranged over three orders of magnitude for each component. Most beetles had very small amounts (ca 100 ng) of MB and cV which were independent of each other. However, amounts of cV covaried with those of other pinene alcohols and also varied between beetles taken from different spruce trees. Mated males had less MB than unmated ones, and ipsdienol and ipsenol, which modify beetles responses to MB and cV, were found only in mated males. Weight variations were small and did not account for observed differences [3.6].

Chiral specificity

Chiral specificity by scolytids has been established by the demonstration that exposing *I. paraconfusus* to the two optical isomers of α-pinene elicited secretion of different verbenols: (+)-*trans*-verbenol from (R)-(+)-α-pinene and (+)-*cis*-verbenol from (S)-(-)-α-pinene. Furthermore, attraction of the pine engraver *I. pini* to its pheromone (-)-ipsdienol was interrupted (a statistically significant decrease in trap catches) by its antipode [3.56]. Chemospecificity of some bark beetles extends to the chirality of host monoterpenes. Such discrimination between plant kairomones is not otherwise known. Thus, chemoperception by *D. valens* of host trees (*Pinus* spp.) is most influenced by (S)-(-)-β-pinene whereas 92% (S)-(-)-α-pinene was inactive; 75% (R)-(+)-α-pinene from sugar pine *P. lambertiana*, a sympatric host species for *D. valens*, was attractive. The antipode (S)-(-)-α-pinene interrupted the response to (R)-(+)-α-pinene. Were such stereospecificity to occur widely in Scolytidae, then enantio-selectivity could constitute a significant element of host selection by bark beetles [3.21].

Aphid aggregation pheromones

Aphids aggregate in the field and winged forms of at least three species, the turnip aphid *Lypaphis erysimi*, the cabbage aphid *Brevicoryne brassicae*, and the bird cherry-oat aphid *Rhopalosiphum padi*, produce aggregation pheromones, as yet unidentified that attract other winged forms. Sex pheromones from sexual females also serve aggregation of asexually reproducing, conspecific winged females.

OVIPOSITION PHEROMONES

Encephalitis and Bancroftian filariasis are serious diseases of man transmitted by mosquitoes (*Culex* spp.). The female lays eggs in a group or egg raft, and the gravid *C. torsalis*, vector of St. Louis encephalitis in the U.S.A., prefers to oviposit in water already containing such rafts. Ether-washed egg rafts contain a volatile attractant, the oviposition pheromone, released from droplets forming on the egg apex.

Preliminary analyses identified the pheromone as a fatty acid with asymmetric carbons at the 5 and 6 positions; the other three isomers are inactive. The natural pheromone was identified by chiral GC as identical with (-)-(5R,6S)-6-acetoxy-5-hexadecanolide (Fig. 30). Although volatile it was desirable to increase its vapour pressure to increase the range of attractancy. Chain shortening abolished activity but maintaining chain length and replacing the acetoxy with the trifluoracetoxy gave 6-trifluoroacetoxy-5-hexedecanolide, that was active and five-fold more volatile. Maintenance of activity was assigned to the fact that although the electrophilicity of the ester-carbonyl carbon would be increased by the trifluoromethyl group, the interaction between the oxygenated region of the molecule and the pheromone receptor would be unaffected. Nevertheless, this analogue was of no field significance as it readily hydrolyzed to an inactive form. However, the heptadecafluor analogue was active. In this, the fluorine was substituted for all hydrogens except for two methylene groups adjacent to the acetoxy, thus minimizing interference with the oxygenated region [3.16].

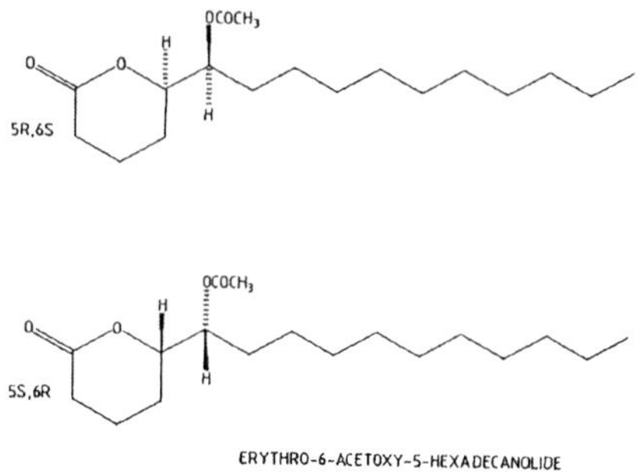

Figure 30. Configurations of the mosquito oviposition pheromone. After Dawson et al. *[3.16].*

SPACING (EPIDEICTIC) PHEROMONES

In some phytophagous species, spacing pheromones affect the extent of colonization, or exploitation of a plant resource. Essentially, they stimulate migration from such a resource when population density is optimal. Much of the evidence for the existence of these pheromones stems from Prokopy's work on the apple magot fly *Rhagoletitis pomonella.* This female oviposits in fruit of different species, and in hawthorn only one larva reaches maturity irrespective of the number laid. After oviposition the female deposits a pheromone on the fruit, deterring subsequent females from laying eggs there. Originating from the hind gut, the pheromone builds up in the lumen prior to its release by the dragging action of the ovipositor. A water-soluble protein, with a molecular weight of less than 10,000, the pheromone is biologically active for several days after deposition. As the first egg laid in hawthorn is usually the one to hatch, it is likely that the oviposition-deterring pheromone gives that larva an early start over competing conspecifics. Such pheromones, identified from related species including the Mediterranean fruit fly *Ceratitis capitata*, are also termed host marking pheromones (HMPs) [3.45].

Among Lepidoptera the cabbage white butterfly *Pieris brassicae* also seemed to apply a HMP to egg batches (ca 30 eggs) and leaf surfaces. This deterred additional oviposition for some two days, an effect assigned directly to three novel avenanthramide alkaloids isolated from egg wash. Subsequent quantification, however, revealed that their concentrations (below 0.15μg/leaf) were too low to explain the effect (1.5μg/leaf required). This led to the proposal that miramides might elicit deterrence chemicals from brassica leaves [3.7].

Aggregation by bark beetles, considered earlier in this Chapter, also involves epideictic pheromones. Their effect is to regulate conspecific competition that otherwise would decrease brood output per female. Several species have evolved olfactory responses leading to spacing as mediated by inhibition of attraction. Population densities and distributions of the western pine beetle *D. brevicomis* were such that population density ranged from 5.9 to 23.2/0.1m^2, a distribution that avoided overcrowding. Corresponding observations on *D. frontalis*, and *D. pseudotsugae* confirmed that the attack pattern was more uniformly spaced than random. A nearest neighbour analysis of the attack pattern of *I. typographus* indicated a minimum allowed distance of 2.5 cm between attack sites, although the angular direction to the four nearest neighbours seemed random. In regard to olfactory regulation of spacing, the arrival of male *D. frontalis* attracted by the female pheromone, a mixture of frontalin and *trans*-verbenol, and by α-pinene from the host tree, stimulates female beetles to release verbenone that deters other males from landing. Further inhibition is provided by *endo*-brevicomin released by the male, inhibiting landing by both males and females. Thus, each female generated a territory sufficiently large around the attack hole to maximize survival of her offspring.

Although this model of events was widely accepted, the experimental data did not extend to a quantification of verbenone released from naturally infested hosts, nor to a comparison of this from ageing hosts with a constant rate of decline. Such a comparison was undertaken with Scots pine *P. sylvestris* damaged by the European pine shoot beetle *Tomicus piniperda*. Beetles in flight-barrier traps near felled trees decreased in number compared with those in perforated traps loaded with three attractive monoterpenes: -(+)-α-pinene and (-)-β-pinene; (+)-Δ^3-carene; and terpinolene. Furthermore, behavioural bioassays demonstrated that verbenone, occurring in the hindgut of *T. piniperda*, significantly decreased the attractiveness to both sexes of a constant release of host monoterpenes. Finally, Porapak-entrapped verbenone increased geometrically in quantity from female-infested logs, such that after six days it was 12-fold larger than from a control; logs infested by males and females released about half as much as female-infested logs. Such data were consistent with the view that flying beetles ascertain host suitability [3.11, and references therein].

PHEROMONES OF SOCIAL INSECTS

Pheromones of social insects may be characterized by three main features: multifunctionality or one pheromone performing several functions; multiple cueing or multiple pheromones eliciting a single behavioural response; and integration of olfactory information with that from other sources of communication.

Multifunctionality

Although not unique to social insects, multifunctionality is most richly documented from them. Defined as pheromonal parsimony, it is very evident in the social life of the honeybee *Apis mellifera* especially as mediated by the queen pheromone, 9-oxo-*trans*-2 decenoic acid (hereafter 9-ODA). This originates from her mandibular gland and has two primer and three releaser activities that may be enhanced by other compounds. As primer and ingested by workers, it precludes their becoming queens by inhibiting both their ovarial development and their queen rearing behaviour. Injecting 9-ODA into a worker's body cavity elicits the former but not the latter effect, indicating the latter is mediated by cuticular chemoreceptors. The fact that the queen must dispense about 0.1 mg 9-ODA/worker/day and that it is ingested by workers suggests they perceive it by taste. As releaser they also perceive it by olfaction both inside the hive leading to retinue formation, and outside the hive leading to location of the swarming queen by workers flying upwind. On arrival at the new site, 9-ODA is not adequate to settle the swarm and *trans*-9-hydroxy-2 decenoic acid (9-HDA), a second mandibular gland pheromone, takes effect. Detected by workers over only a short distance, it causes them to dispense their own Nassanoff gland scent that includes geraniol, nerolic acid, geranic acid and citral, in various isomeric forms. This attracts other workers leading to the formation of a stable colony. For completeness, the mandibular gland contains about another 30 substances and the secretion of queen's Kosehevnikov's gland is also attractive to workers.

In the third releaser effect 9-ODA, emitted by the virgin queen on her nuptial flight, attracts the male and stimulates it to copulate; large numbers of flying drones are attracted to tethered virgin queens, extracts of queens, or synthetic queen substance. Nineteen analogues of 9-ODA prepared by synthesis and rigorously purified, failed to attract drones to a balloon; in contrast 100 μg 9-ODA attracted an average of 456 drones in 3 min. So 9-ODA plays a pervasive role in the social organisation of the hive and may represent the epitome of pheromonal parsimony [3.9, and references therein].

Care by workers of the honeybee brood entails: feeding the larvae; capping their cells; and thermally regulating the vicinity. Ten fatty esters present in drone larvae just before capping, and in smaller concentrations in worker larvae, elicited capping when experimentally presented in natural proportions. Capping, at a rate indistinguishable from that assignable to a combination of all 10 esters combined, was elicited by the following individual esters at concentrations representing three to six larval equivalents: methyl oleate (MO), methyl linoleate (ML), methyl linolenate (MN) and methyl palmitate (MP). However, the parasitic mite *Varroa jacobsoni* perceives MN and MP before workers do so, enabling it to enter a few hours before capping. Thus, a brood pheromone serves also as a kairomone [3.33].

Pheromonal parsimony is widespread in other Hymenoptera especially ants. Their response is greatly influenced by such contextual features as: location of the responder, whether inside or outside the nest; behaviour in which the recipient is engaged when stimulated; time of day; interaction of the pheromonal signal with other cues; caste, age, physiological state, and sex of the recipient. Concentration effects are also significant. Thus, 4-methyl-3-heptanone, the principal alarm pheromone of the American harvester ant, *Pogonomyrmex badius*, elicits attraction at small concentrations, but an alarm frenzy at large concentrations. If these persist, workers dig at the source consistent with the rescue behaviour elicited by workers buried in sand after a nest cave-in [3.63, and references therein].

Kin recognition

Esters are also significant components of the Dufour's gland secretion that, in ground dwelling bees, provides the brood cell with a hydrophobic lining. The nest entrance is also marked by this secretion which, to be effective, should be both species- and individual-specific. Halictine bees display differing levels of sociality apparently mediated by pheromones associated with the Dufour's secretion. The secretion contains, in addition to esters, macrocyclic acid lactones (C_{16}-C_{21}) and paraffins (C_{17}-C_{27}), amounting to 27 compounds from the genus, and 25 in one species *Evylaeus marginatum*.

A computer algorithm was used to assess the number of components, from a total of 27, necessary to distinguish two to 12 bees involving various degrees of relatedness. Essentially, 50 replicates of 50 simulated "bees" were generated in which each "bee" was assigned 27 components in proportions consistent with empirical means, standard errors and a normal distribution. Variances were partitioned between: genera within the family Halictidae; species within each genus; nests of *E. malachurus*; random individuals within a nest of that species; full-sibs within a population of that nest assuming, for this particular comparison, an equal mixture of full-sibs and half-sibs from an infinite number of fathers. A typical simulation compared two bees, I versus II, each with five pheromonal components A, B, C, D, and E, amounting to: 35% *v.* 40%; 25% *v.* 30%; 20% *v.* 8%; 15% *v.* 20%; and 5% *v.* 10% respectively. Using the two most abundant components would not discriminate between the two species, but using the other three components (C,D,E *v.* D,E,C) would do so. Thus, three compounds were enough to permit a distinction. Similarly, 12 sympatric species using five components could be distinguished on the basis of the five. Moreover, two nestmates could be distinguished by six compounds, and 12 by 14, although the latter value may be conservative if paternal lineages are significantly less than infinite (Fig. 31).

A second algorithm was used to ascertain how complex a pheromone bouquet should be by removing one component at a time, in an order determined by random numbers. As before, more components are needed to distinguish more individuals. Thus, two species are distinguishable by five components, and 10 by nine; sympatric species are distinguishable by four components but nine must be secreted; 10 individuals are distinguishable by six components but 19 must be produced. Overall, it is the relative intensity of each component that conferred specificity. Such data suggested two possible encoding mechanisms. In the first, the blend serves as a unit and is compared with templates encoded in the brain, so the degree of specificity is the principal feature transmitted. In the second, the blend is examined in terms of constituent concentrations such that kin recognition is slower; the species is delineated by concentrations of the major components, and variations of minor ones discriminate individuals. The analysis favoured the latter mechanism. Species discrimination was feasible using five to nine compounds, which corresponds to the range present in many Lepidoptera. Significantly larger numbers of components are required to discriminate between individuals, explaining the occurrence of such larger numbers of components in social insects. Within the Halictidae, the solitary *E. leucozonius* and the social *E. marginatum* exhibit 6 and 25 components respectively. It was concluded that kin recognition based on only pheromones was feasible when the number of components was very large, and the number of individual insects was not large [3.20] (Fig. 31).

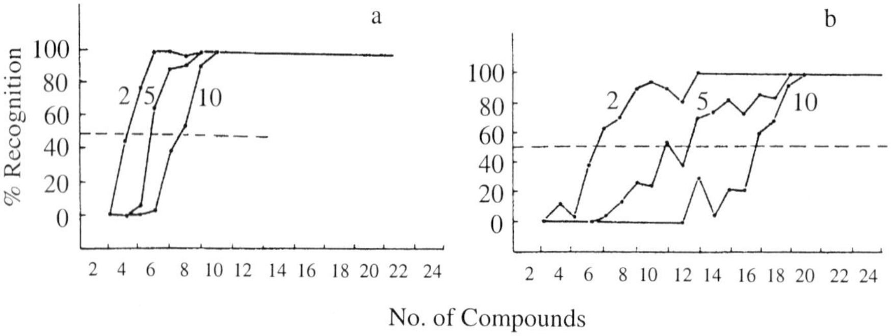

Figure 31. The minimum number of pheromone constituents necessary to distinguish between: a) species; b) conspecific individual insects. Redrawn after Hefetz and Graur [3.20].

Very large numbers of nestmates elicit two further discriminatory devices. Firstly, a general nest odour to which all individuals contribute and that, in turn, approximates to each one. Alternatively, lines and individuals might be distinguished by a chemical cue introduced by the queen [3.31]. The issue of colony-specific bouquets was addressed in an examination of the cuticular hydrocarbon composition of colonies of the social wasp *Polistes metricus*. Of 25 compounds identified, major constituents were: *n*-pentacosane; 9-, 11-, 13-, and 15-methylentriacontane; 13,17-

and 15,19-dimethyltriacontane; and 11,15- and 13,17-dimethylentriacontane. Subjecting the data to discriminant analysis indicated that composition of cuticular hydrocarbons; especially methyl-branched ones may: distinguish queens and workers of a colony from those of others, which is consistent with previous data for the genus; distinguish male wasps as having larger amounts of alkenes and lesser amounts of methyl-branched hydrocarbons than female ones; and distinguish very clearly between males of various colonies on the basis of six constituents, three being alkanes [3.31].

Similarly, cuticular hydrocarbons of ants served to distinguish monogynous (having one queen) from digynous and trigynous colonies. 6,10-Dimethylhexacosane occurred in smaller proportions in monogynous as compared with polygynous colonies; 5-methylnonacosane was present in lower proportions in digynous than in trigynous colonies; and 8-methylhexacosane of varying proportions in monogynous, was of larger and yet larger proportions in digynous and trigynous colonies respectively. Such data strongly suggest that colony bouquet is greatly influenced by the queen, either directly by labelling, or indirectly by eliciting it from workers [3.14, 3.46]. Such labels may be indicative of monogynous carpenter ants (*Camponotus* sp.) that, removed as pupae and reared separately in the absence of queens, are more tolerant of each other than non-relatives. In the presence of a queen, they equally attack unfamiliar kin and non-kin [3.23, and references therein].

Overlapping defence and sex attractant functions

Some compounds established as defensive in function also serve as sex pheromones and the wide distribution of this phenomenon indicates that it has arisen independently several times [3.9, and references therein]. In the Heteroptera, the mullein bug *Campylomma verbasci* produces from the metathoracic scent gland butyl butyrate and (*E*)-crotyl butyrate which are characteristic defence allomones, but they also serve as female-produced sex pheromones. The metathoracic gland of the bean bug *Riptortus clavatus* also produces the characteristic defence compounds (*E*)-2-hexenyl (*E*)-2-hexenoate, (*E*)-2-hexenyl (*Z*)-3-hexenoate, and myristyl isobutyrate that, as a mixture, serve as a male-produced sex attractant.

Among Coleoptera, the defensive tergal gland of the staphylinid *Aleochara curtula* produces a mixture of aliphatic aldehydes, hydrocarbons and benzoquinones. Of these, (*Z*)-4-tridecene, dodecanal, and (*Z*)-5-tetradecanal in low concentrations, and in combination with a female-produced sex pheromone, release copulatory responses of the male; this is inhibited by higher concentrations of gland constituents and a defensive function prevails.

In Hymenoptera, the obviously defensive role of sting-associated glands (poison and Dufour's glands) have evolved, in queen ants for example, to release sexual behaviour of the male. Thus, virgin queens of *Formica lugubris* employ as a sex-pheromone undecane, amounting to some 90% of the Dufour's gland secretion, that also serves as a defensive allomone in this and in many other arthropod species. Given the genesis of this compound by the Dufour's gland, there can be little doubt that undecane's original role was defensive, and that sex attraction evolved secondarily [3.9, and references therein].

Alarm pheromones

Early evidence for the existence of alarm pheromones came from the observation that the freshly-excised sting of the honeybee elicited agonistic behaviour in workers. Characteristic of Isoptera and eusocial Hymenoptera, especially Formicidae, such alarm pheromones are also found in Homoptera, Hemiptera and in the Class Arachnida. The unifying feature seems to be the capacity to elicit aggregation in response to a trauma ameliorated by a collective response. Typically, attacks by predators on such aggregations elicit an alarm signal leading to rapid dispersal of the aggregation, that may reform under the influence of aggregation pheromones. In eusocial species, nest invasion elicits production of alarm pheromones that rapidly attract members of the soldier or worker castes to attack the intruder, stinging it and marking it with pheromone to elicit further attacks. Exceeded in number by only sex pheromones in the arthropods, alarm pheromones may represent the key evolutionary advance necessary for survival of the colonial life style in insects [3.8, and references therein].

In some reports, alarm is used synonymously with recruitment as the effect is frequently the same. In the lower termite *Zootermopsis nevadensis*, odour trails derived from secretions of the sternal gland connect to breaches of the nest wall; recruited workers both repel invaders and repair the breach. In higher termites that forage outside the nest, the trails recruit workers to new food sources. So, termite evolution modified alarm recruitment to foraging recruitment, as nests gradually provided an insufficient source of food [3.22].

A recurring theme in the literature on alarm pheromones is the extent of their overlap with defence pheromones, to such an extent that alarm pheromones are believed to have evolved from chemical defence [3.22]. Abdominal droplets secreted by aphids were originally viewed as defensive in function, but alarm components occur in the secretion. Dispersal in response to such secretions is reported from many species of aphid, 14 of which exhibited dispersal in response to a synthesized aphid alarm pheromone. Overlap is also evident from data on stingless bees *Trigona* spp., in which alarm responses are elicited by their cephalic secretions, that also elicit other behavioural responses. The role of alarm pheromones in the nest of the honeybee is well established, as is the occurrence of such pheromones in virtually all species of ant in which they were sought [3.8, and references therein].

Alarm pheromone glands

Alarm pheromones are frequently produced by glands that originally served a different purpose. Such multifunctional glands in insects include frontal, mandibular, metathoracic, abdominal, Dufour's, sting, poison, shaft, and anal glands. Aphids secrete alarm pheromones from cornicles, tubular structures located at the abdominal tip. Also in the Hemiptera, alarm pheromones are produced by larvae and adults, frequently from dorsal abdominal glands. Larval and adult bedbugs produce alarm pheromones from dorsal abdominal and ventral metathoracic glands respectively. Among Isoptera, the frontal gland of termite soldiers, once thought to produce only defensive secretions, is established as the source of the alarm pheromone. In formicine ants, products of poison, mandibular, and Dufour's glands serve as alarm pheromones. The exploitation of such glands for an additional purpose is consistent with the principle of parsimony raised in the previous section. It indicates that some compounds acting as alarm pheromones originally may have served a different function, entailing a function-shift.

Dispersal and attack

Dispersal is the general response of presocial insects to alarm pheromones; eusocial forms evince a range of responses usually including attack. Dispersal, instantaneous with pheromone release, spares the aggregated insects from predatory attacks. In aphids, this involves their walking to the other side of the leaf in response to low doses (0.02 ng), and falling off in response to significantly higher ones. Immature aphids seem less sensitive than adults to alarm pheromones suggesting that the risk in cessation of feeding is greater than that from predation, but their sensitivity is enhanced by vibratory stimuli. The first stadium larva of the sugarcane woolly aphid *Ceratovacuna lanigera* is equipped with prominent frontal horns that confer significant protection against predators. The presence of tending ants decreased aphid dispersal in response to their alarm pheromones that elicited attack on predators by the ants. A few presocial forms will stay to defend their brood from aggressors [3.18, and references therein].

Although some eusocial forms exhibit dispersal, especially when away from the nest, attack is the usual response to their alarm pheromones. Specific reactions include, snapping and biting, stinging, and producing defensive secretions that may include additional alarm pheromones to label the intruder. Some eusocial forms, as the honeybee, are actively attracted to sources of alarm pheromone that is, of course, the converse of the repellent effect involved in dispersal, or equivalent to a function-shift. Such attraction causes only soldier ants to leave the nest in large numbers and follow a territorial trail to the source, and causes honeybee workers to assemble a mass attack [3.8].

Chemistry of the compounds

Many aphid species utilize terpenoids and (*E*)-β-farnesene is utilized by species in more than a dozen aphid genera. In contrast, the monoterpene, α-pinene, is the principal alarm signal for the vetch aphid *Megoura viciae*. Monoterpenes, especially β-unsaturated aldehydes, also predominate among the alarm pheromones of true bugs, Hemiptera. (*E*)-2-Hexenal is an alarm signal in families such as Pyrrhocoridae, Cimicidae, Pentatomidae, and in Acanthosomidae in which it also serves a defensive role. Monoterpene hydrocarbons, such as limonene and terpinolene, predominate in the identified alarm signals of termite soldiers. Eusocial bees utilize alkyl and alkenyl acetates such as isopentyl acetate by the honeybee; this species also employs 2-heptanone, a product of the mandibular gland, as do stingless bees *Trigona* spp. This characteristic releaser of alarm behaviour also occurs in the defensive secretion of cockroaches *Platyzosteria* spp. and beetles of the genus *Dyscherius*.

Rather more elaborate molecules, including the unique animal product, 4-methyl-3-nonanone, are produced by ants. Many myrmicine genera utilize 3-alkanones and ponerine species employ pyrazines. Formicine species include terpenoids such as citral and citronellal as alarm signals in the products of their mandibular glands. Studies with radiolabelled tracers have shown that citral and citronellal are biosynthesized by conventional acetate condensation associated with the mevalonic acid pathway. Citral is suggested as a key compound in the synthetic pathway of several alarm pheromones as stereospecific reduction would yield citronellal, and oxidative metabolism would yield 6-methyl-5-hepten-2-one, 4-methyl-2-hexanone and 2-methyl-4-heptanone, alarm pheromones of dolichoderine ants. Serine, glycine and histidine are established as jointly contributing to the formation of formic acid that serves both a defence and an alarm function.

Such duality in function is commonplace among social insects: citral and citronellal also serve as defence compounds; the alarm pheromone of the little fire ant *Wasmannia auropunctata* (Myrmicidae), 2,5-dimethyl-3-isopentylpyrazine, also defends against predators; as mentioned above, 2-heptanone, an alarm signal for the honeybee is also defensive in function; *n*-undecane an alarm signal for *Formica* species, is also a principal defence component of the Dufour's gland secretion. Furthermore, it is commonplace to find compounds that are alarm pheromones in social arthropods acting as defensive allomones in less evolved species. Such compounds include 2-hexenal, citral, 3-nonanone and 2-tridecanone. Apparently, their action as defensive allomones was an evolutionary precursor to their subsequent function as alarm pheromones [3.8, and references therein].

Tandem running behaviour provides another insight into such evolution. Reported from a range of species, this behaviour is commonest in the primitive subfamily Ponerinae. In *Pachycondyla obscuricornis* only one nest mate at a time is recruited by close tactile contact with the leader ant. If contact is interrupted the leader immediately stops, moving forward only when contact resumes. Secretions of the leader's pygidial gland, located between the sixth and seventh abdominal tergites, are the chemical bond between leader and follower. In solitary foragers, tandem running occurs during only nest emigration. In *P. laevigata*, an obligate predator on termites, foraging is associated with group raids organized by a pygidial gland secretion from scout ants. Thus, a dietary shift from dispersed to aggregated food (termites) elicited a function-shift in the pheromone, from recruitment to mass communication through orienting signals. In the subfamily Myrmicinae, tandem-calling in the genus *Leptothorax* derives from the poison gland of the sting apparatus. Raising the gaster and releasing a droplet of the poison gland secretion, attracts nestmates and initiates tandem running [3.22, and references therein]. Clearly, function-shift underlies many evolutionary relationships within social insects.

Function-shift in pheromone evolution

This was subjected to a rigorous examination within the superorder Amphiesmenoptera that comprises Trichoptera (caddisflies) and Lepidoptera (butterflies and moths); the former, with only some 10.000 species or 7% of Lepidoptera, served as a reference group for an analysis of pheromone systems of the latter. This detailed phylogenetic analysis employed the following seven characters: long-distance pheromone present; female-produced pheromone; sternum V glands; short-chain pheromones; male wing-fanning; terminal glands; and long-chain pheromones. As caddisflies synthesize in their abdominal glands chemicals that are repellent or toxic to other invertebrates, this raised the question of whether pheromones evolved from defensive secretions, or the converse. It seemed relevant that sternum V glands, that produce a defensive secretion and are present in most lower taxa, are never retained in taxa that produce sex pheromones in terminal glands. This was interpreted as favouring the concept of sex pheromones originally serving a defensive function, and specifically that

> the communicative function of sex pheromones evolved from compounds originally used for defence in ancestral moths and caddisflies [3.39].

The physiological mechanisms underlying such functions-shift are unknown but some pointers are available. A larval secretion plays a key role in the ontogeny of the sex pheromone of the southern masked chafer *Cyclocephala lurida*. The adult male is attracted to the adult female, the male and the female larva, and to hexane extracts of each. Ultimately, the male as adult becomes unattractive to other males but the female retains her attractiveness. This system suggests the evolution of sex pheromones from larval compounds subsequently lost by the male [3.19].

In regard to biochemical mechanisms, monoterpenes are established as reversible, competitive inhibitors of acetylcholinesterase (AChE) from a vertebrate and an insect source (Chapter 2 [2.49, 2.35 respectively]). Extending this approach to pheromones associated with more than 100 insect species established the same effect for pheromones ranging in function from sex attraction to aggregation to alarm [3.52]. For example, 2-heptanone, the alarm pheromone of the dolichoderine ant *Conomyrma pyramica* and also of the honeybee, elicited a K_i value of 0.38mM. This is consistent with a defensive function. Such a neurotoxic effect is consistent with the observation that high concentrations of 2-heptanone elicit rapid and erratic movements in workers of *C. pyramica* and lurching when near the source [3.63]. All seven pheromones bioassayed for toxic effects *in vivo* paralyzed and killed the non-adapted test insect *T. castaneum*.

In terms of sex attractants, it is relevant that one AChE inhibitor, (*Z*)-7-dodecenyl acetate, is a constituent of the bouquet of 147 lepidopteran species and another, (*Z*)-11-hexadecenyl acetate, is synthesized by 189 species. So this defensive effect could have been widespread originally.

A more stringent approach compared the susceptibility to 2-heptanone, *in vivo* and *in vitro*, of two species: the wax moth *G. mellonella*, that as a bee hive dweller would be adapted to 2-heptanone; and *Ephestia kuhniella*, the Mediterranean flour moth, a non-adapted species. They represent the subfamilies Galleriniae and Phycitinae respectively (Family Pyralidae). *G. mellonella* displayed no evidence of paralysis in the presence of 2-heptanone concentrations as high as $80\text{x}10^4$ppm, whereas *E. kuhniella* was paralyzed, and killed with an LD_{50} value of $55\text{x}10^4$ppm: AChE purified to homogeneity from *Galleria* was 47% less inhibited than that from *Ephestia* as judged by K_i values [3.27]. Such data indicate how an insect species may tolerate a defensive secretion; repeated exposure could lead to exploitation of this as a recognition cue.

REFERENCES

3.1 Altstein, M., Ben-Aziz, O., Gabay, T., Gazit, Y. and Dunkelblum, E. (1997) Structure-function relationship of PBAN/MRCH. In Cardé, R.T. and Minks, A.K. (eds.) *Insect Pheromone Research: New Directions*. Chapman and Hall, New York, pp. 56-63.

3.2 Arn, H., Tóth, M. and Priesner, E. (1992) *List of Sex Pheromones of Lepidoptera and Related Attractants*. IDBC, Montfavet.

3.3 Baer, R.G., Berisford, C.W. and Hermann, H.R. (1976) Bioassay, histology, and morphology of the pheromone-producing glands of *Rhyacionia frustrana*, *R. rigidana*, and *R. subtropica*. *Ann. Entomol. Soc. Amer.* **69**, 307-310.

3.4 Baker, T.C. (1989) Sex pheromone communication in the Lepidoptera: new research progress. *Experientia* **45**, 248-262.

3.5 Baker, T.C. and Vickers, N.J. (1997) Pheromone-mediated flight in moths. In Cardé, R.C. and Minks, A.K. (eds.) *Insect Pheromone Research: New Directions*. Chapman and Hall, New York, pp. 248-274.

3.6 Birgersson, G., Schlyter, F., Bergström, G. and Löfqvist, J. (1988) Individual variation in aggregation pheromone content of the bark beetle, *Ips typographus*. *J. Chem. Ecol.* **14**, 1737-1761.

3.7 Blaakmeer, A., Hagenbeek, D., Van Beek, T.A., De Groot, A.E., Schoonhoven, L. M. and Van Loon, J.J.A. (1994) Plant response to eggs vs host marking pheromone as factors inhibiting oviposition by *Pieris brassicae*. *J. Chem. Ecol.* **20**, 1657-1665.

3.8 Blum, M.S. (1985) Alarm pheromones. In Kerkut, G.A. and Gilbert, L.I. (eds.) *Comprehensive Insect Physiology, Biochemistry, and Pharmacology* vol. 9. Pergamon Press, Oxford, pp. 193-224.

3.9 Blum, M.S. (1996) Semiochemical parsimony in the arthropoda. *Ann. Rev. Entomol.* **41**, 353-374.

3.10 Boppré, M.S. (1978) Chemical communication, plant relationships, and mimicry in the evolution of danaid butterflies. *Entomol. Exp. Appl.* **24**, 264-277.

3.11 Byers, J.A. (1989) Chemical ecology of bark beetles. *Experientia* **45**, 271-283.

3.12 Byers, J.A. (1992) Optimal fractionation and bioassay plans for isolation of synergistic chemicals: the subtractive-combination method. *J. Chem. Ecol.* **18**, 1603-1621.

3.13 Cardé, R.T. and Mafra-Neto, A. (1997) Mechanisms of flight of male moths to pheromone. In Cardé, R.T. and Minks, A.K. (eds.) *Insect Pheromone Research: New Directions*. Chapman and Hall, New York, pp. 275-290.

3.14 Carlin, N.F. and Hölldobler, B. (1986) The kin recognition system of carpenter ants (*Campanotus* spp.) I. Hierarchical cues in small colonies. *Behav. Ecol. Sociobiol.* **19**, 123-134.

3.15 Crema, R. (1984) Correlation of aphid sex pheromone gland number with ovarian development. *Experientia* **40**, 569-570.

3.16 Dawson, G.W., Mudd, A., Pickett, J.A., Pile, M.M. and Wadhams, L.J. (1990) Convenient synthesis of mosquito oviposition pheromone and a highly fluorinated analog retaining biological activity. *J. Chem. Ecol.* **16**, 1779-1789.

3.17 Faustini, D.L., Post, D.C. and Burkholder, W.E. (1982) Histology of aggregation pheromone gland in the red flour beetle. *Ann. Entomol. Soc. Am.* **75**, 187-190.

3.18 Hardie, J., Pickett, J.A., Pow, E.M. and Smiley, D.W.M. (1999) Aphids. In Hardie, J. and Minks, A.K. (eds.) *Pheromones of Non-Lepidopteran Insects Associated with Agricultural Plants.* CABI, Wallingford, pp. 227-250.
3.19 Haynes, K.F., Potter, D.A. and Collins, J.T. (1992) Attraction of male beetles to grubs: evidence for evolution of a sex pheromone from larval odor. *J. Chem. Ecol.* **18**, 1117-1124.
3.20 Hefetz, A. and Graur, D. (1988) The significance of multicomponent pheromones in denoting specific compositions. *Biochem. Syst. Ecol.* **16**, 557-566.
3.21 Hobson, K.R., Wood, D.L., Cool, L.G., White, P.R., Ohtsuka, T., Kubo, I. and Zavarin, E. (1993) Chiral specificity in responses by the bark beetle *Dendroctonus valens* to host kairomones, *J. Chem. Ecol.* **19**, 1837-1846.
3.22 Hölldobler, B. (1984) Evolution of insect communication. In Lewis, T. (ed.) *Insect Communication.* Academic Press, New York, pp. 349-377.
3.23 Hölldobler, B. (1995) The chemistry of social regulation: Multicomponent signals in ant societies. *Proc. Natl. Acad. Sci.* **92**, 19-22.
3.24 Jacobson, M., Redfern, R.E., Jones, W.A. and Aldridge, M.H. (1970) Sex pheromones of the southern armyworm moth: isolation, identification, and synthesis. *Science* **170**, 542-544.
3.25 Jurenka, R. and Roelofs, W.L. (1993) Biosynthesis and endocrine regulation of fatty acid derived sex pheromones in moths. In Stanley-Samuelson, D.W. and Nelson, D.R. (eds.) *Insect Lipids: Chemistry, Biochemistry and Biology.* University of Nebraska Press, Lincoln pp. 353-388.
3.26 Karlson, P. and Butenandt, A. (1959) Pheromones (Ectohormones) in insects. *Ann. Rev. Entomol.* **4**, 39-58.
3.27 Keane, S. (2001) Isolation, purification, and inhibition by monoterpenes of acetylcholinesterase from an adapted species *Galleria mellonella*, and a non-adapted species *Ephestia kuhniella.* Ph.D Thesis, National University of Ireland, Dublin.
3.28 Kliemetzek, D., Köhler, J., Vité, J.P. and Kohnle, V. (1986) Dosage response to ethanol mediates host selection by "secondary" bark beetles. *Naturwiss.* **73**, 270-272.
3.29 Klun, J.A. and Maini, S. (1979) Genetic basis of an insect chemical communication system: the European corn borer. *Environ. Ent.* **8**, 423-426.
3.30 Landolt, P.J., Heath, R.R., Millar, J.G., Davis-Hernandez, K.M., Dueben, B.D. and Ward, K.E. (1994) Effects of host plant, *Gossypium hirsutum* L., on sexual attraction of cabbage looper moths, *Trichoplusia ni* (Hübner) (Lepidoptera: Noctuidae). *J. Chem. Ecol.* **20**, 2959-2974.
3.31 Layton, J.M., Camann, M.A. and Espelie, K. E. (1994) Cuticular lipid profiles of queens, workers, and males of social wasp *Polistes metricus* say are colony specific. *J. Chem. Ecol.* **20**, 2307-2321.
3.32 Leal, W.S., Kawamura, F. and Ono, M. (1994) The scarab beetle *Anomala albopilosa sakishimana* utilizes the same sex pheromone blend as a closely related and geographically isolated species, *Anomala cuprea. J. Chem. Ecol.* **20**, 1667-1676.
3.33 Le Conte, Y., Arnold, G., Trouillier, J. and Masson, C. (1990) identification of a brood pheromone in honeybees. *Naturwiss.* **77**, 334-336.
3.34 Levinson, H.Z., Levinson, R., Kahn, G.E. and Schafer, K. (1983) Occurrence of a pheromone-producing gland in female tobacco beetles. *Experientia* **39**, 1095-1097.
3.35 Lextrait, P., Biemont, J.-C. and Pouzal, J. (1994) Comparisons of walking locomotory reactions of two forms *of Callosobruchus maculatus* males subjected to female sex pheromone stimulation (Coleoptera: Bruchidae). *J. Chem. Ecol.* **20**, 2917-2930.

3.36 Linn, C.E. Jr., Campbell, M.G. and Roelofs, W. L. (1986) Male moth sensitivity to multicomponent pheromones: critical role of female-released blend in determining the functional role of components and active space of the pheromone. *J. Chem. Ecol.* **12**, 659-668.

3.37 Liu, Y.-B. and Haynes, K.F. (1992) Filamentous nature of pheromone plumes protects integrity of signal from background chemical noise in cabbage looper moth, *Trichoplusia ni*. *J. Chem. Ecol.* **18**, 299-307.

3.38 Löfstedt, C. (1991) Evolution of moth pheromones. In Hrdy, I. (ed.) *Insect Chemical Ecology*. SPD Academic Publishing, The Hague, pp. 57-73.

3.39 Löfstedt, C. and Kozlov, M. (1997) A phylogenetic analysis of pheromone communication in primitive moths. In Cardé, R. and Minks, A.K. (eds.) *Insect Pheromone Research: New Directions*. Chapman and Hall, New York, pp. 473-489.

3.40 Marco, M.P., Fabriàs, G. and Camps, F. (1997) PBAN regulation of sex pheromone biosynthesis in *Spodoptera littoralis*. In Cardé, R.T. and Minks, A.K. (eds.) *Insect Pheromone Research: New Directions*. Chapman and Hall, New York, pp. 46-53.

3.41 Meng, L.Z., Wu, C.H., Wicklein, M., Kaissling, K.-E. and Bestman, H.J. (1989) Number and sensitivity of three types of pheromone receptor cells in *Antheraea pernyi* and *A. polyphemus*. *J. comp. Physiol.* **165**: 139-146.

3.42 Percy, J. (1979) Haemocytes associated with basement membrane of sex pheromone gland of *Trichoplusia ni* (Lepidoptera: Noctuidae): Ultrastructural observations. *Can. J. Zool.* **56**, 238-245.

3.43 Percy-Cunningham, J.E. and MacDonald, J.A. (1987) Biology and ultrastructure of sex pheromone-producing glands. In Prestwich, G. D. and Blomquist, G. J. (eds.) *Pheromone Biochemistry*. Academic Press, New York, pp. 27-75.

3.44 Phelan, P.L. and Baker, T.C. (1987) Evolution of male pheromones in moths: Reproductive isolation through sexual selection? *Science* **235**, 205-207.

3.45 Prokopy, R.J. (1981) Oviposition-deterring pheromone system of apple maggot flies. In Mitchell, E.R. (ed.) *Management of Insect Pests with Semiochemicals*. Plenum Press, New York, pp. 477-494.

3.46 Provost, E., Riviere, G., Roux, M., Bagneres, A.-G. and Clement, J.L. (1994) Cuticular hydrocarbons whereby *Messor barbarus* ant workers putatively discriminate between monogynous and polygynous colonies. Are workers labeled by queens? *J. Chem. Ecol.* **20**, 2985-3003.

3.47 Raina, A.K. (1997) Control of pheromone production in moths. In Cardé, R. and Minks, A.K. (eds.) *Insect Pheromone Research: New Directions*. Chapman and Hall, New York, pp. 21-30.

3.48 Reed, J.R., Vanderwel, D., Choi, S., Pomonis, J.G., Reitz, R.C. and Blomquist, G.J. (1994) Unusual mechanism of hydrocarbon formation in the housefly: Cytochrome P450 converts aldehyde to the sex pheromone component (*Z*)-9-tricosene and CO_2. *Proc. Natl. Acad. Sci.* **91**, 10000-10004.

3.49 Roelofs, W.L. (1995) Chemistry of sex attraction. *Proc. Natl. Acad. Sci.* **92**, 44-99.

3.50 Roelofs, W.L. and Brown, R.L. (1982) Pheromones and evolutionary relationships of Tortricidae. *Ann. Rev. Ecol. Syst.* **13**, 395-422.

3.51 Roelofs, W.L. and Jurenka, R.A. (1997) Interaction of PBAN with biosynthetic enzymes. In Cardé, R. and Minks, A.K. (eds.) *Insect Pheromone Research: New Directions*. Chapman and Hall, New York, pp. 42-45.

3.52 Ryan, M. F., Awde, J. and Moran, S. (1992) Insect pheromones as reversible competitive inhibitors of acetylcholinesterase. *Invert. Reprod. Develop.* **22**, 31-38.

3.53 Sabelis, M.W. and Schippers, P. (1984) Variable wind directions and anemotactic strategies of searching for an odour plume. *Oecologia* **63**, 225-228.

3.54 Schlyter, F. and Birgersson, G.A. (1999) Forest beetles. In Hardie, J. and Minks, A.K. (eds.) *Pheromones of Non-Lepidopteran Insects Associated with Agricultural Plants.* CABI, Wallingford, pp. 113-148.

3.55 Schlyter, F., Birgersson, G., Byers, J.A., Löfqvist, J. and Bergström, G. (1987) Field response of the spruce bark beetle, *Ips typographus*, to aggregation pheromone candidates. *J. Chem. Ecol.* **13**, 701-716.

3.56 Seybold, S.J. (1993) Role of chirality in olfactory-directed behavior: aggregation of pine engraver beetles in the genus *Ips* (Coleoptera: Scolytidae). *J. Chem. Ecol.* **19**, 1809-1831.

3.57 Silverstein, R.M. and Young, J.C. (1976) Insects generally use multicomponent pheromones. In Gould, R.F. (ed.) *Pest Management with Insect Sex Attractants.* American Chemical Society, Washington, DC, pp. 1-29.

3.58 Sreng, L. (1984) Morphology of the sternal and tergal glands producing sexual pheromones and the aphrodisiacs among the cockroaches of the subfamily Oxyhaloinae. *J. Morphol.* **182**, 279-294.

3.59 Sreng, L. (1985) Ultrastructure of the glands producing sex pheromones of the male *Nauphoeta cinerea* (Insecta, Dictyoptera). *Zoomorphol.* **105**, 133-142.

3.60 Stevens, I.D.R. (1998) Chemical structures and diversity of pheromones. In Howse, P.E., Stevens, I.D.R. and Jones, O.T. (eds.) *Insect Pheromones and their Use in Pest Management.* Chapman and Hall, London, pp. 135-179.

3.61 Tamaki, Y. (1988) Sex pheromones. In Kerkut, G. and Gilbert, L.J. (eds.) *Comprehensive Insect Physiology, Biochemistry, and Pharmacology* vol. 9. Pergamon Press, Oxford, pp. 145-191.

3.62 Weatherston, J. and Percy, J.E. (1977). Pheromones of male Lepidoptera. In Adiyodi, K.G. and Adiyodi, R.G. (eds.) *Advances in Invertebrate Reproduction* vol.1. Peralam-Kenoth, Kerala, 295-307.

3.63 Wilson, E.O. (1971) *The Insect Societies.* Belknap Press, Cambridge, Mass., U.S.A.

3.64 Witzgall, P. (1997) Modulation of pheromone-mediated flight in male moths. In Cardé, R.T. and Minks, A.K. (eds.) *Insect Pheromone Research: New Directions.* Chapman and Hall, New York, pp. 265-274.

3.65 Wright, R.H. (1958) The olfactory guidance of flying insects. *Can. Ent.* **90**, 81-89.

3.66 Zhu, J., Kozlov, M.V., Philipp, P., Francke, W. and Löfstedt, C. (1995) Identification of a novel moth sex pheromone in *Eriocrania cicatricella* (Zett.) (Lepidoptera: Eriocraniidae) and its phylogenetic implications. *J. Chem. Ecol.* **21**, 29-43.

CHAPTER 4

THE CHEMORECEPTIVE ORGANS: STRUCTURAL ASPECTS

The fundamental chemoreceptive organ of the insect is the sensillum commonly called sensory hair, sensory peg, sense organ, or sensory receptor. Sensillum is the least ambiguous term and receptor is the least appropriate one, because in pharmacological and biochemical usage a receptor has a different meaning; it describes a macromolecule with sites having properties to recognize specific molecules or drugs (See Chapter 6). Sensilla are distributed on a range of insect structures including the antenna [4.22], mouthparts, and limbs especially tarsi [For more recent reviews with emphasis on the antenna, see 4.26, 4.27, 4.14].

The early classification of insect sensilla was of necessity based on external morphology as revealed by light microscopy [4.23]. This recognized five categories of sensilla to which a further five were subsequently added giving 10 principal categories ordered in terms of the extent of their cuticular extensions as follows: s. trichodea, long, slender and hair-like; s. chaetica, long, heavy and thick-walled ; s. basiconica, shorter and peg-like and s. auricillia, similar but 'rabbit-eared'; s. styloconica, sharply-tipped pegs on a stylus; s. coeloconica, pegs in shallow pits; s. ampullacea, pegs in deep pits; s. campaniformia, dome- or button-shaped; s. placodea, flattened plates; s. scolopalia and scolopophora, subcuticular; s. squamiformia, scale-like (Fig. 32) [4.21, 4.27, and references in both].

The difficulty with any system of classification based on external morphology is that it does not readily relate to function, as sensilla from different categories may serve the same function e.g. olfaction. Slifer, using crystal violet stain, distinguished between thick- and thin-walled sensilla in which, as a rule, the former were perforated by a single pore and served as contact chemoreceptors, and the latter were penetrated by multiple pores and were olfactory receptors [4.23]. But as multiple pores are not associated with any single category of the 10, this has prompted further attempts at categorization (see below).

The advent of the electron microscope facilitated more detailed information on sensillar ultrastructure and a steady stream of reports on a wide range of insects indicated a basic, common pattern of construction comprising: (1) cuticular components most usually expressed as hair-like extensions; (2) sensory neurons; and (3) associated cells.

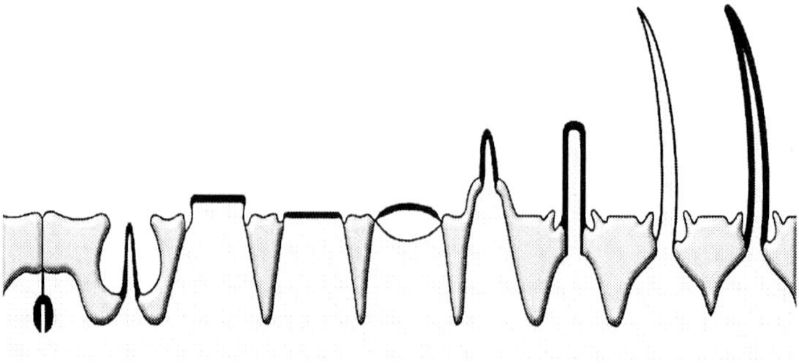

Figure 32. Outline drawings of the principal types of sensilla with increasing cuticular expression from left: sc, scolopophora; c, coeloconic; a ampullaceous; cp, campaniform; p, placoid; s, styloconic; b, basiconic; t, trichoid; ch, chaetica; scale-like or squamiform is not included.

As external cuticular parts are most accessible for study, much detailed data are reported from them especially the pores. Typically, sensilla with no pores, aporous (AP) and with one neuron are sensitive to mechanical stimuli; however, some no-pore sensilla may be hygro- or thermosensitive. Sensilla with one pore (uniporous, UP) or with multiple pores (multiporous, MP) are chemoreceptive with contact and olfactory functions respectively [4.26, 4.27, and references in both]. The single pore, when brought into physical contact with a chemical stimulus, funnels molecules to relatively unbranched dendrites. In addition and more widespread are dual-function gustatory/mechanoreceptive organs, the first detailed account of which was provided by Lewis in 1954 [4.15]. Essentially, this cuticular hair is thick-walled and grooved, tapering to a blunt tip with a single terminal or sub-terminal pore which may have a closure apparatus. A viscous sticky substance exuding from the tip has been reported as α-glucosidase. Distally, the cuticular sheath or scolopale divides the lumen of the peg into channels isolating mechanoreceptive from chemoreceptive dendrites. At the pore, the scolopale appears as an invagination of the cuticle and is shed during ecdysis.

In contrast, the walls of olfactory sensilla are perforated by n x 10^3 pores (10 to 100 μm in external diam) that filter molecules from the airstream. Pore densities vary from three to eight/μm^2, in the pheromone-sensitive sensilla of the silk moths *Bombyx mori* and *Antheraea pernyi*, to 35/μm^2 in the olfactory peg of the face fly *Musca autumnalis*. Such pores lead to a chamber, the pore kettle, where puffs of stimuli may be accommodated and concentrated. Pore tubules, 100 to 200 A° diam extend toward the lumen of the sensillum that contains much-branched dendrites. Three pore systems may be distinguished:

(1) the pore funnel of the surface leads to a narrow pore canal that widens into a pore kettle from which are subtended several pore tubules;

(2) pores give rise to spoke canals leading to the lumen, usually in longitudinally-grooved sensilla. If there is more than one peg lumen, pore tubules connect with the outermost one, and radial canals penetrate to the innermost one;

(3) dense osmiophilic pore strands extend from the pores to the lumen, perhaps as an adaptation to dry climates.

THE SENSORY NEURON

Usually bipolar, this comprises three distinct regions, dendrite, cell body or perikaryon, and the axon that conveys impulses to the central nervous system. The dendrite is the source of much attention on account of its primary position in chemoreception. It comprises: an outer segment or distal dendritic process (DDP); the constricted ciliary region; and a thicker inner segment or proximal dendritic process (PDP) (Fig. 33).

The typical DDP contains filaments, vesicles, and only longitudinal microtubules that increase in number from dendrite termination to about mid-length. There they decrease in number to multiples of about eight in the narrower ciliary region, so-called because its microtubules are frequently arranged in the ciliary pattern as nine peripheral doublets with or without a central pair, i.e. 9 x 2 + 2 or 9 x 2 + 0 but 36, 45 or 54 x 2, and 0 also are recorded. A thin, porous, electron dense membrane or ciliary collar, that in fly larvae may be a fibrillar mass, lines the circumference of peripheral tubules. The outer member of each doublet lacks the arms that are associated with motile cilia. Ciliary microtubules lead to a centriole-like structure the basal body (of which there may be two, one above the other), comprising nine sets of peripheral triplet tubules. This body gives rise to banded rootlets, the ciliary rootlets. The similarity of this ultrastructure to that found in cilia of lower Phyla has promoted the suggestion, unverified and perhaps unverifiable, that insect sensilla are derivable from cilia either intrinsic to arthropods, or originally associated with ectoparasitic Protozoa.

The role of microtubules is unambiguous in mechanoreceptors in which they form a tubular body, possibly derived from folding of the DDP, of 50-100 microtubules but sometimes much fewer, arranged in parallel and embedded in an electron dense matrix (Fig. 33). Physical contact with the sensillum compresses the tubular body, deforms the transducer membrane, and initiates nerve impulses. The cytoplasm of the PDP may have a distal concentration of mitochondria and many small vesicles.

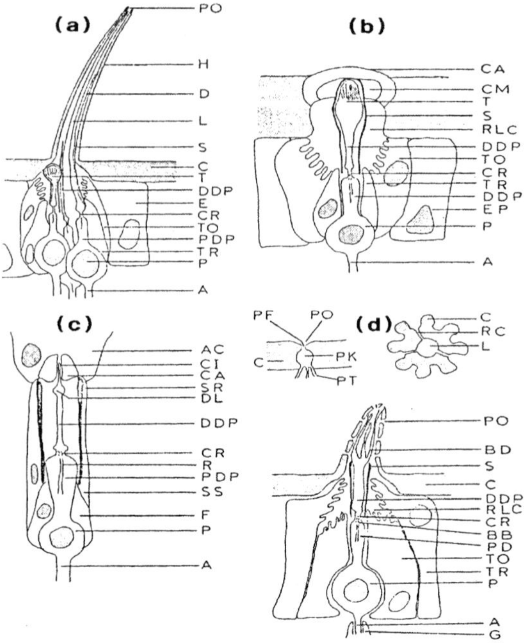

Figure 33. Schematic representation of the ultrastructure of: (a) UP dual-function gustatory/mechanoreceptive trichoid sensillum; (b) AP campaniform mechanosensitive sensillum; (c) UP coeloconic sensillum; and (d) MP trichoid sensillum with inset of pore tubule system (left), and end-on view. A, axon; AC, attachment cell; BB, basal bodies; BD, branched dendrites; C, cuticle; CA, cap; CI, cilium; CM, cap membrane; CR, ciliary region; D, dendrite; DL, dilation; DDP, distal dendritic process; E, epidermal cell; F, fibrous sheath cell; G, glial cell; H, hair; L, lumen; P, perikaryon; PF, pore funnel; PK, pore kettle; PO, pore; PT pore tubule; PDP, proximal dendritic process; R, ciliary rootlets; RC, radial canal; RLC, receptor lymph cavity; S, scolopale; SR, scolopale rods; SS, scolopale sheath cell; T, tubular body; TO, tormogen cell; and TR trichogen cell. Compiled from Behan and Ryan [4.6].

The perikaryon located just below the surface of the cuticle has a large spherical nucleus with finely dispersed chromatin that has been associated with high metabolic activity. Abundant cytoplasmic organelles are: mitochondria; longitudinal microtubules; rough endoplasmic reticulum (RER); free ribosomes; vesicles; Golgi and onion bodies. Mitochondria and microfibrils are also abundant in the axons. The number of sensory neurons is usually in the range of one to 10 but placoid sensilla may be innervated by 35 bundles of two to 35 neurons, each bundle ensheathed by a scolopale [4.6].

ASSOCIATED CELLS

Surrounding the neuron are two or more sheath cells that have been named trichogen and tormogen if two in number. A Schwann cell surrounds the nerve fibre, glial cells surround the perikarya and axons, and a neurilemma cell surrounds the PDP region. Alternatively, cells additional to trichogen and tormogen cells (numbered I and II respectively) are designated accessory cells III, IV and V as appropriate. Usually however, the innermost cell is termed the thecogen cell as it enwraps the bundle of sensory neurons from their perikarya to the base of the scolopale. Individual neurons are also enwrapped by cytoplasmic extensions of this cell as is the proximal part of each individual dendrite, presumably for electrical insulation.

The contents of the thecogen cell of the uniporous, gustatory chemosensilla (UP GCS) of the ventral sensory field on the maxillary palps of the American cockroach *Periplaneta americana* have been studied in detail. This cell is conspicuous for the abundance of granules that are scarce in the other enveloping cells. The structure of these electron-dense granules indicated that they were lysosomes which was confirmed by acid phosphatase cytochemistry. Coated pits occurred on the thecogen cell plasma membrane where it borders the receptor lymph cavity. Use of the fluorescent dye Lucifer yellow indicated very extensive fluid-phase endocytosis from the cavity into the cell, followed by digestion in the endosome-lysosome pathway. Possible functions attributed to lysosomes and endocytosis included: purging the receptor lymph of physical impurities ingested following probing by the maxillary palps; participation in catabolic processes associated with embryonic development and moulting; and direct involvement in the chemoreceptive process by removal and metabolism of stimulus molecules from the receptor lymph to preclude continuous stimulation. Such endocytosis would be highly efficient if receptor-mediated [4.22].

The trichogen cell forms the hair wall, the hair, and the scolopale; the tormogen cell forms the socket and enwraps the trichogen cell from near the perikaryon to the cuticle. After forming the hair, the trichogen cell withdraws from the hair lumen down to the hair base or below, filling the spaces with the liquid or 'sensillum liquor' that it secretes. Processes of this cell surround the cilium and extend microvilli into the receptor lymph cavity. This vacuole lies within the tormogen cell and is formed by that cell's withdrawal after it has formed the socket. The receptor lymph cavity extends up to the hair base and is continuous with the sensillum liquor in the hair peg. Labyrinthine connecting canals and lamellae are characteristic of both enveloping cells. Septate desmosomes connect them to each other and to epidermal cells. In the moth *A. polyphemus* a very tight contact, sufficient to preclude entry of ionic lanthanum, is formed between the apical membrane of the tormogen cell and the socket cuticle of olfactory s. trichodea. The contact is characterized by numerous hemidesmosome-like structures that could electrically isolate the receptor lymph cavity and might also attach the sensillum to the cuticle [4.13].

There are variations of this pattern. For example, in the carrion beetle *Necrophorus* there are four enveloping cells of which: the first secretes the scolopale; the trichogen cell secretes the porous cuticle; the tormogen cell makes the hair base; and the fourth cell secretes the sensillum base and ring wall [4.10]. Consequently, there is some dissatisfaction with the use of the terms trichogen and tormogen as the cells are not truly homologous sheath cells. An alternative notation may be preferable employing the terms inner, intermediate, outer and basal sheath cell. Thus, the thecogen is the inner or number I cell; the number II cell is the trichogen or intermediate cell; the tormogen cell is usually number III and becomes the outer sheath cell; and cell IV enwraps the basal sheath cell (Fig. 34).

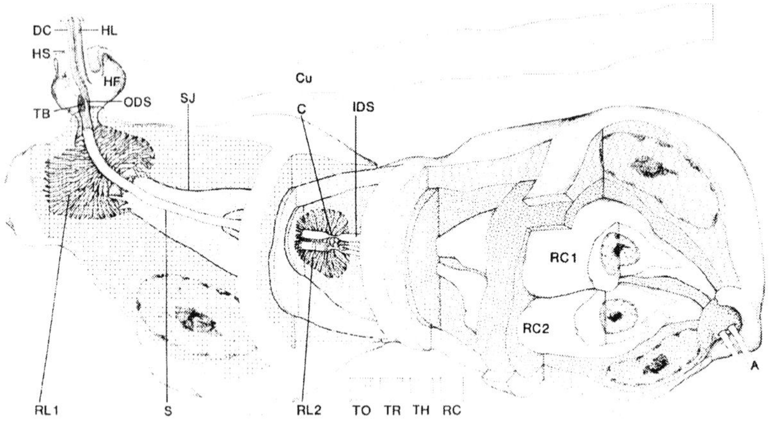

Figure 34. Schematic cutaway representation of the ultrastructure of a dual-function multiporous (MP) olfactory/mechanoreceptive trichoid sensillum indicating the relative positions two receptor cells (RC1 and RC2) and cells (TH, thecogen; TO, tormogen; and TR, trichogen). A, axon; C, ciliary region within inner receptor lymph space; CU, cuticle; DC, dendritic canal; HF, hair follicle; HL, hair lumen; HS, hair shaft; IDS, inner dendritic segment; ODS, outer dendritic segment; RLI, receptor lymph space in the tormogen cell; RL2, receptor lymph space in the trichogen cell extending forward; S, scolopale; SJ, septate junctions; TB, tubular body indicative of a mechanosensitive function for that receptor cell (RC2). Compiled from various sources.

The foregoing account is necessarily condensed, but the absence of an analysis of ultrastructural detail in the original sensillar categories (trichoid etc.) reflects the lack of systematic relationships, as each original category embraces several functional types. For example, trichoid sensilla may be exclusively mechanoreceptive, gustatory and mechanoreceptive, or olfactory and mechanoreceptive in function. As the original categories do not enable a sensillar classification based on function, this necessitates the use of ultrastructural features.

SYSTEMATICS OF SENSILLA

Some associations between structure and function are long established and beyond argument:

(1) AP sensilla are never chemosensitive but may be mechano-, hygro- or thermosensitive;

(2) UP sensilla are contact chemoreceptors, and with a tubular body are dual-function, i.e. gustatory and mechanoreceptive;

(3) MP sensilla function as olfactory receptors. The three pore systems comprising this category are described (Fig. 35).

A new typology proposed by Altner in 1977 [4.2] is not a new typology where the first two categories are concerned, as it simply reaffirms their validity. Noteworthy, however, was the care taken to study individual sensilla by inserting a cactus needle near each sensillum from which single-unit electrophysiological recordings were made (details in Chapter 5). The lesion facilitated identification of the sensillum during subsequent scanning and transmission electron microscopy (SEM and TEM respectively). Thus, ultrastructural detail and sensillar function could be reconciled without ambiguity. The real thrust of the proposal was for a subdivision of category 3 into: (1) single-walled (SW) sensilla having pore tubules, dendrites branched and unbranched, and exclusively olfactory in function; (2) double-walled (DW) sensilla having secretion-filled spoke canals, unbranched dendrites, and functioning as olfactory and/or thermo-hygrosensitive sensilla, the latter having a grooved wall (Fig. 35). However, these two categories had been explicitly recognized by Slifer in 1970 [4.24].

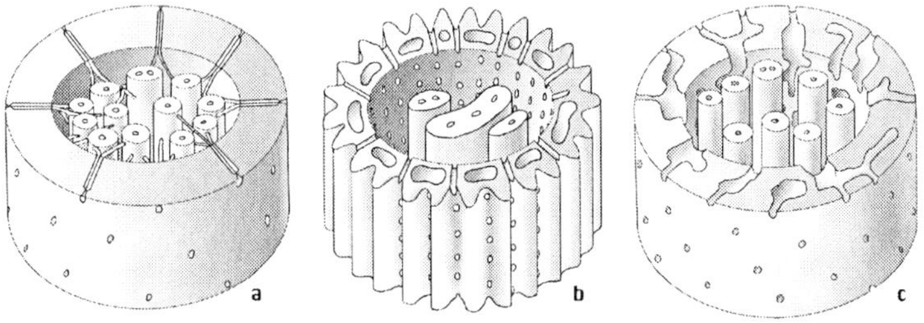

Figure 35. Schematic representation of the structures of multiporous (MP) sensilla: a) single-walled with spoke canals leading to closely-packed pore tubules; b) grooved, double-walled with secretion-filled, spoke canals, separated by lumina extending from the receptor lymph cavity above the tormogen cell, leading to dendrites, and characteristic of thermo-hygrosensitive sensilla with chemoreceptive neurons; c) irregular, double-walled as a variation of b). Redrawn after Altner [4.2].

In 1980, a very detailed review of the ultrastructure and function of the chemosensilla by Zacharuk reaffirmed the widely-accepted classification of the three categories and subdivided the third category of MP sensilla into those with a pitted surface (MPP), and those with a grooved one (MPG); the latter, either had pore kettles and two sets of wall channels, or did not have kettles and had a simple ring of such channels [4.26]. Essentially, this is a return to the essence of Slifer's original subdivisions.

No scheme readily accommodates sensilla found in the cave-dwelling beetle *Speophyes lucidulus*. The "vesicule olfactive" comprises two or three chambers atop one another with the uppermost opening to the exterior. Contents of each chamber include various sensilla such as, pine cone sensilla, claviform, black and star-shaped pegs some of which might perceive the infra-red spectrum [4.7].

The issue of whether sightless, cave-dwelling beetles sustain modification of other sensory organs was considered in *Speonomus hydrophilus* (Catopidae) [4.8]. Twelve sensory organ types were represented by seven types of trichoid, two types of basiconic, a bottle-like, and a campaniform sensillum; in addition four peg sensilla possessed tubular bodies. Of trichoid sensilla, two types (A and B) contained only one sensory cell, and as this had a tubular body they were viewed as exclusively mechanoreceptive. As trichoid type A are distributed all over the antennal surface, and as trichoid type B are short and parallel to the antenna such that deformation of the antenna would bend them, their functions were designated external mechanoreceptors and proprioreceptors respectively. Trichoid type C and the bottle-like sensilla contain respectively four and one additional sensory cells, without a tubular body. Although sections were not obtained at the distal tip, the presence of a terminal pore was presumed and hence these sensilla were designated gustatory in function. The two basiconic types are distinguished by only the hair shafts. Similarities in the absence of terminal pores, in the two sensory cells, and in the presence of the three sheath cells in each, indicated they are typical olfactory sensilla.

All grooved sensilla are aporous and all contain three bipolar sensory cells with the trichogen cells developing extensive lamellae containing numerous mitochondria. Between the lamellae occur secretory droplets that extend into a ring- shaped channel in the shaft of the sensillum, and are extruded externally through transverse channels. Electron-lucid near the trichogen cell, this secretion is electron-dense in the shaft. Another channel accommodates the dendrites of the sensory neurons. The grooved sensilla were assigned a chemoreceptive function which, given the absence of multiple pores, implies gustation. The sensilla of this study resemble those of other epigean Coleoptera and are compatible with existing categories of sensilla. This study also indicated that although cryofixation was more difficult than chemofixation, its successful use offered better cell preservation, better definition of the cytoplasm and microtubules [4.8].

Coleoptera provide another example of a unique sensillum from the male of the western corn rootworm *Diabrotica virgifera virgifera* [4.20]. The five distal segments of the male antenna possess ventrally aligned pits containing cup-shaped sensilla, each with a rich supply of pore tubules characteristic of olfactory sensilla. Unusually, however, the thin epicuticle lacks pores. Although associated with all pits, such sensilla are not confined to them but occur singly or in pairs elsewhere on the antenna. Also associated with but not restricted to each pit is a Class III dermal gland comprising three cell types: an inner secretory cell producing, concentrating, and transporting the secretion; cells forming a ductule overlying the secretory cell; and an intermediate or intercalary cell type having a microvillar surface and fibrillar matrix around the cuticle lining. Staining indicated that the secretion was consistent with an esterase that might function in metabolizing excess pheromone. Despite the paucity of sensory dendrites and absence of pores, the abundance of pore tubules so close to the very thin sensillar cuticle (ca 10 nm) was taken as indicative of a function in olfaction and probably in pheromone perception [4.20].

Perhaps a fruitful direction for future investigations on sensilla rather than repeated descriptions of ultrastructure, would be experiments with radiolabelled ligands to investigate their specific binding to the integral parts.

EVOLUTION OF SENSILLA

Collembola

Sensillar evolutionary relationships are not easily discerned but some insight may be derived from the studies of Altner and colleagues on sensilla of the Collembola (Apterygota), wingless and thus primitive insects. The collembolan postantennal organ (PAO), originally and erroneously considered a homologue of the insect eye, frequently takes the form of a fold or gutter-like depression. In *Onychiurus* species of the *armatus* group, a sensory function is indicated by: the presence of cuticular pores; one sensory and several enveloping cells; and the perikaryon of the sensory cell lying within the brain. Enveloping cells are represented by one inner, several outer, and one basal, in contrast to the usual arrangements in insect sensilla of two or three enveloping cells [4.3] (Fig. 36).

The inner cell of *Onychiurus* is homologous with that of more evolved insects as it surrounds the ciliary region but the homologies of the other cells are problematical. Although the outer enveloping cells could have arisen from the division of an original trichogen cell, this remains a speculation without data on the ontogeny of the organ. In addition, the *Onychiurus* sensory cell possesses two ciliary structures, rather then the usual one, emerging from the proximal dendritic segment. Accordingly, the dendrite comprises two outer segments arising from a single inner one. The presence of pores in the cuticular wall and of dendritic branches beneath an electron dense substance are indicative of a chemoreceptive or hygroreceptive function. Also unusual is the location of the sensory cell perikaryon within the brain.

Apparently unique is the absence of a fluid-filled lumen, due to the extensions of the enveloping cells into the cuticular space; also absent is a cuticular sheath [4.3] (Fig. 36).

A more extensive survey of this PAO within the Collembola suggested that evolution of this organ was associated with transition of the Onychiuridae to an edaphic mode; in tandem with surface enlargement the organ sank into a groove, and the sensory cell perikaryon, losing contact with the epidermis, was integrated into the protocerebrum. Truly primitive features were likely to be the presence of two cilia, and the intrusion of enveloping cell processes into the outer receptor lymph cavity. The latter condition is consistent with minimal and electrogenic ion transport [4.12].

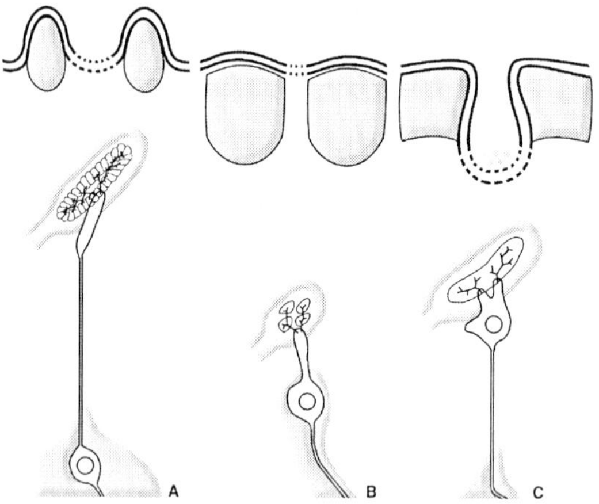

Figure 36. Schematic representation of the collembolan postantennal organ (PAO) in A) Onychiuris *sp.; B)* Hypogastrura socialis*; and C)* Isotoma olivacea *in terms of: (above) configuration of cuticular pore systems; and (below) ramifications of the sensory cell, and its location in relationship to the protocerebrum (shaded). Redrawn after Altner and Thies [4.3].*

Odonata

Additional data are available from a consideration of UP sensilla of damselflies (Odonata) as representatives of the Palaeoptera, the phylogenetically oldest group of the winged insects (Pterygota) [4.5]. The four selected sensilla, located on the tip of the larval maxillary palp of *Agrion puella* and *Ischnura elegans*, were of two types A and B with pore diameters of 0.22 μm and 0.12 μm respectively, and with five and two sensory cells respectively. Each UP sensillum has three sheath cells of which the innermost, sheath cell 1, surrounds the receptor cell over its total length. Cell 1 is enclosed distally by sheath cells 2 and 3 but only the latter reaches the cuticle with its distal parts. Unusually, each sheath cell contained two centrioles. The usual terms of dendrite sheath cell, trichogen cell and tormogen cell were not applied to these three cells as: each term implies a specific function e.g. hair shaft and socket formation which did not occur; and there is contradictory evidence regarding homologies of such cells in other sensilla, such as their degeneration during ontogeny and the occurrence of additional sheath cells.

The principal difference observed in these UP sensilla, as compared with the conventional structure of taste hairs, is the lack of any outer cuticular structure. Other sensilla with pores opening into flat, undifferentiated cuticle occur on leg appendages of the horseshoe crab, *Limulus* and in the ventral organ of the housefly larva. The Palaeoptera sensilla are assumed to be contact chemoreceptors on account of: the single cuticular pores connecting the dendritic tips to the external medium; and the lack of differentiation in dendritic outer segments (DOS=DDP). A unique feature was the presence, between apical cuticle and sheath cells, of a separate, extracellular cavity connected to the exterior by small cisternae and canals. As a similar system is reported from Collembola (Apterygota), its presence in these damselfly species is consistent with their primordial position in the Pterygota.

Odonata and Ephemerida represent the phylogenetically oldest living pterygote insects and their sensilla have both primordial and highly-specialized features. Specifically, the lack of outer cuticular differentiation and the existence of three separate lymph cavities seem ancient, if typical, features, of which the most primitive seems to be the absence of the hair shaft in typical taste hairs. These sensilla are viewed as an early phylogenetic stage in sensillar development [4.5].

Lepidoptera

Although lepidopteran sensilla conform in general to the categories checklisted earlier, the data are derived mainly from the Ditrysia, the most highly-evolved subdivision, containing 99 % of all lepidopteran species. Very few studies relate to families of non-Ditrysia although it is obviously desirable to have descriptions from the base of the lepidopteran cladogram. A relatively recent investigation on the Nepticulidae has revealed an entirely new type of sensillum [4.25].

These small insects (3-10 mm wingspan) have antennae 1.0-2.5 mm long comprising an enlarged scape, short pedicel, and a flagellum bearing 15-70 flagellomeres, almost invariably more on the male than on the female antenna. Conventional sensillar types represented include: s. trichodea; s. chaetica type I, 10-18 μm long, with a blunt tip having a single pore and thus consistent with an UP, thick-walled sensillum with combined chemo- and mechanoreceptive functions; s. chaetica type II, 12-17 μm long, arranged in sets of 12 in a ring corresponding in position to the ring of scales and covered by them, without a terminal pore or dendrites in the lumen and thus, a mechanoreceptor; s. coeloconica with a peg 2-4 μm long and with longitudinal grooves.

The novel sensillum is the s. vesiculocladum (or small branched blister, SV) occurring only in the Nepticulidae, and typically comprising varying numbers of tubular, thin-walled, cuticular extensions from a shallow depression, with up to five branches extending downwards and parallel to the principal axis of the antennae. Between adjacent branches lie one s. chaeticum, type II, and scales (Fig. 37). In some species the number of branches decreases to four, three, and very commonly to one; unbranched forms also occur. As the longitudinally-grooved wall is penetrated by many pores, the SV was assigned to the SW category of Altner. Sexual dimorphism is pronounced as there are two SVs in all male flagellomeres except the apical, and this sensillum is absent from several female antennal segments yielding a functional surface area of at least 3:1 for male and female respectively (Fig. 37).

SV was claimed as quite different from most other sensillar types by comprising several individual sensilla although this feature is shared with the "vesicule olfactive" as reported from the *S. lucidulus* [4.7]. What seems clear is that the SV is a SW, porous sensillum that, given the scarcity of s. trichodea (25-46 in Nepticulidae compared with 710-29,150 in Ditrysia), may be seen as a functional replacement of the latter by the former. This, taken in conjunction with the pronounced sexual dimorphism, has been interpreted as consistent with pheromone perception. Phylogenetically, the five-branched condition is viewed as the primordial state within Nepticulidae with decreased branching occurring secondarily. The evolution of such branched sensilla in place of s. trichodea might reflect a constraint for space for sensory neurons in the small antenna [4.25].

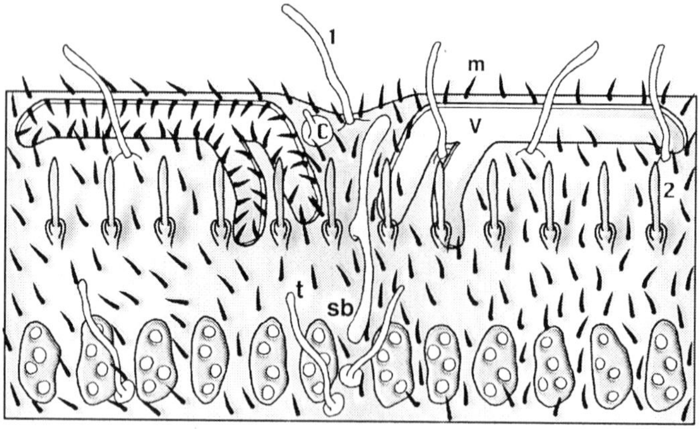

Figure 37. Outline drawing of the sensillum vesiculocladum V and a single branch of it sb, unique to the non-ditrysian Nepticulidae (Lepidoptera). m, microtrichia; 1, s. chaeticum type I; 2, s. chaeticum type II; c, s. coeloconicum; t, s. trichodeum. Redrawn after van Nieukerken and Dop [4.25].

SENSILLA OF IMMATURE INSECTS

As many deleterious effects of insects are attributable to the larval stage, a knowledge of larval sensilla clearly desirable. As insect eggs are usually laid near a host plant, larval olfaction may be confined to short-range attraction to food or to aggregation and trail pheromones. In the event, data on larval sensilla are rather scanty [4.28, and references therein, serving as a primary source for the following account].

Of the 10 sensillar types associated with adult insects all, except the squamiformia, are recognized in larvae in which they comprise three functional types: AP and mainly mechanosensilla (MS) but also including thermo-hygrosensilla (T-HS); UP acting either as gustatory chemosensilla (GCS) or as chemomechanosensilla (GC-MS); and MP serving as olfactory chemosensilla (OCS). With the exception of scolopalial sensilla, the basic design of sensilla of immatures is consistent with that for adults in comprising cuticular specialization, innervation by one or more sensory cells, and the presence of three or four accessory cells.

Aporous Mechanosensilla

Touch, physical pressure, movement, stretch, vibrations and contraction all serve to alter the position of the cuticle of these sensilla. This, in turn, transmits the stimulus through the dendritic sheath to the membrane of the tubular body (defined above), deformation of which initiates electrical activity.

Sensilla scolopalia and the related scolopophora, the latter abundant on the larval mandibles of a range of orders, are exceptional for their subcuticular location. The scolopophoran cuticle is directly innervated by blind-ending sheaths enclosing the dendritic endings with the point of attachment frequently marked externally by a depression. The scolopalian dendrites end in a homologue of the dendritic sheath of exterosensilla, the scolopale cap. This lies just short of the cuticle and is frequently connected to it by an attachment cell rich in microtubules. However, this connection is lost in some sensilla scolopalia, wherein the capped dendrite extends through the attachment cell into a matrix associated with bundles of sensillar cells.

Thermo-hygrosensilla

These peg-like sensilla enclose the dendritic terminations of hygro-sensitive neurons having dendritic tightly-packed microtubules in an array resembling that of the tubular body. This suggests a similar mode of stimulation i.e. deformation of the cuticle elicited by humidity changes. Thermosensitivity is assigned to the neuron having the dendrite equipped with a lamellated termination.

Chemosensilla

The basic pattern is innervation by a single ensheathed neuron or by a bundle of neurons. However, examples occur of several such bundles in large MP sensilla on the larval antenna of many endopterygote species, indicating the fusion of several sensilla. Plecopteran larvae display an intermediate stage in this process in the form of closely contiguous but not fused pegs. Following entry through a pore or pores, chemical stimuli are conducted through a pore-canal system (UP sensilla) or through a pore-tubule system (most but not all MP sensilla). A sharp distinction is usually drawn in adult insects between OCS and GCS on the basis of the presence of either multiple or single pores respectively. This distinction is weakened in regard to aquatic insects where both kinds of stimuli may be delivered through water. In larval Lepidoptera the distinction is also qualified as GCS may react to olfactory stimuli.

Gustatory Chemosensilla

Although the size of the terminal pore and properties of the adjacent cuticle in such UP sensilla may through the extent of closure affect stimulus access, the inclusions of the pore canal may be of greater significance. In aquatic larvae the pore contents may be a plug of mucous, but plugs of fenestrated fibrils have also been detected that could provide enhanced selectivity of putative stimuli based on interfibrillar spaces, binding sites, and ionic charges in the plug. Consistent with this is the demonstration of variable permeability to Co, Ag, and Pb ions between median and lateral styloconic sensilla of the tobacco hornworm *Manduca sexta* larva, and in such sensilla between larvae of various lepidopteran species. Also variable is the permeability to divalent ions of the pore-canal system near the terminal pore, and of the distal dendritic sheath into the sinus of the peg. Such permeability could enable a two-way traffic to replenish dendritic ions and other constituents.

Olfactory Chemosensilla

Staining responses of a cone sensillum of the larva of the prairie grain wireworm *Ctenicera destructor* indicated the pore tubules as hydrophobic but with a hydrophilic core, properties meriting investigation in corresponding sensilla of other insects. In cone- and flower-shaped sensilla of other larvae, the stimulus-conducting medium is hydrophilic liquor. As pore tubule systems of lepidopteran larvae are not permeable to Co ions, these may serve stimulus filtering.

The Sensory Neuron

This is essentially similar in ultrastructure to that of sensilla of adult insects. However, the usual microtubular arrangement of 9x2+0 in only the basal region, extends in scolopalial sensilla and in some T-HS to the end of the cilium. Multivesicular bodies containing synaptic-like vesicles, apparently for transport from proximal cytoplasm to dendrites, have been identified near ciliary boxes in UP styloconic sensilla of the *M. sexta* larva. The proximal segment may be thicker than the ciliary one and less electron-dense in mechanosensory than in chemosensory neurons.

The number of neurons per larval sensillum ranges from one to 17, usually unbranched, depending on sensillar type. Larval MS are with a single exception usually innervated by a single neuron terminating in a tubular body: scolopalia sensilla by one or two; scolopophorous by two that are devoid of tubular bodies; T-HS by three or four; digitiform by one; UP taste sensilla by two to eight of which one may terminate in a tubular body; MP OCS by from one to nine or 17 in a single unit, in an assemblage of 10-11 units, amounting to 130 neurons all branched.

Associated Cells

The fundamental pattern of three sheath cells and a glial cell usually obtains in larval sensilla. The innermost cell (thecogen or trichogen depending on source) and the glial cell individually enwrap and separate proximal and not distal segments of the dendrite, cell bodies, and axons, except in mosquito larvae and other aquatic insects. Their cell bodies and axons are unsheathed and thus, in direct contact with the haemocoel. The pore canals of the sidewalls of AP and UP sensilla, in addition to the pore tubules of MP sensilla, extend to the liquor-rich sinus surrounding the lamellated regions of the sheath cells, which may be studded by coated vesicles. The sinus may have elevated levels of K ions compared to the haemolymph. In sensilla with only two associated cells, the inner one encloses the ciliary sinus and the outer arcs around the sensillar sinus. Where four such cells occur, the fourth serves as a further intermediate cell.

Morphogenesis and Moulting

Both sensory cuticle and dendritic sheath are detached, shed, and replaced at moulting. The sheath becomes much extended and folded proximally and is evacuated with retraction of the hypodermis. The intermediate sheath cell also elongates to form a process that serves as the sensory cuticle of AP MS and MP OCS, the latter being supplemented by the shed pore tubule system. The outer cuticle or socket that surrounds the sensory cuticle receives contributions from the outer sheath cell, and the entire moulting process is marked by a pore or scar in the fresh cuticle.

Exopterygote insects may add sensilla at moulting usually through increasing the number and length of subsegments of the flagellum. The new sensilla arise from the subdivision of a single epidermal cell into four and eight in MS and CS respectively.

Dipteran larvae are devoid of anterior appendages but they present various organs including dorsal, terminal, and ventral organs that are claimed to develop from embryonic antennal, maxillary, and labial placodes, respectively. However, this is not rigidly adhered to as a mixed segmental origin has been ascribed to the terminal sensilla of the vinegar fly *Drosophila melanogaster*, and the paired terminal sensilla of *Haematobia irritans* have a labial origin. This led to the interpretation that the ventral organ is derived from a maxillary complex and not from a labial source.

Sensillar Distribution

In general, sensillar numbers change at moulting by exopterygotes with the greatest change at the final moult, but remain relatively constant during growth and development of larval endopterygotes.

Antenna

In endopterygote larvae this comprises scape, pedicel and a decreased, one-segmented flagellum, with most sensilla mainly comprising MP and UP types concentrated on the latter two structures. The flagellum of exopterygotes is longer, more segmented, and exhibits rapid increases in sensillar numbers primarily OCS, especially in later instars.

Maxillae

In larval endopterygotes each maxillary palp is equipped apically with five-10 UP pegs, one-three MP pegs, supplemented by one to two UP pegs and one T-HS on the galea-lacinia. As representatives of exopterygotes, gryllids exhibit an abundance of sensilla mainly tactile on the palp, galea and lacinia. Numbers of MS and CS increase and decrease respectively at succeeding moults. The palp tip exhibits the usual concentration of sensilla, that are mainly UP with some MP, while the galeal surface is covered by trichoid sensilla.

Labium

Among endopterygotes, each labial palp of lepidopteran larvae contains only two MS compared with coleopterans exhibiting 12-23 pegs, that are mainly UP except for three-four MP. The labial palps of acridid nymphs contain fewer sensilla than maxillary palps but follow the pattern of increased numbers of CS with each moult. Furthermore, this pattern also applies to the paraglossae of the migratory locust *Locusta migratoria*, the first and fifth instar nymphs of which exhibit 74 and 184 multineuronal sensilla respectively.

Clypeo-labrum

The anterior surface is relatively devoid of sensilla as endopterygotes exhibit only MS, and exopterygotes have a few CS in addition. The epipharynx or posterior surface is richer as: lepidopteran larvae have one pair of GCS and five pairs of MS; elaterid larvae and catopid larvae have five and 10 pairs respectively of multineuronal UP sensilla, together with campaniform and trichoid MS; larvae of

D. melanogaster have two sensillar groups, comprising a total of nine UP (including one unique composite) and AP on the dorsal pharyngeal wall and one composite UP sensillum on the ventral one. The epipharynx of gryllids, acridids and blattids contain many paired UP and AP sensilla.

Mandibles

These are equipped with only MS represented by varying numbers of trichoid and by many campaniform sensilla, supplemented by scolopophorous sensilla, responding to pressures on the mandibular teeth [comprehensively reviewed in 4.27].

Embryological Development of Sensilla

This has been studied in coeloconic sensilla, one AP and one MP, on the antenna of *L. migratoria* [4.4]. 132 hr after oviposition, each type exhibits three sensory and three enveloping cells. Only the outer two enveloping cells are equipped with microvilli, and those of the outermost cell extend beyond the epidermal cells to form a tuft surrounding the DOS of the sensory cells. The inner enveloping or sheath cell secretes the dendritic sheath that sends an irregular extension into the exuvial space containing floccular extensions, and almost to the first embryonic cuticle. Each dendritic inner segment (DIS=PDP) exhibits a cilium and many microvilli-like processes. Enveloping cells are richly endowed with RER, Golgi bodies, and mitochondria. No cells are pigmented.

180 hr after oviposition, it is possible to distinguish four developing cells of which the outermost two form a pit 4 μm deep, 2.7 μm diam, and similar for AP and the MP sensilla. At the bottom of the pit the first and the second enveloping cells surround the dendritic bundle (Fig. 38). Through the apical opening of the pit the dendritic sheath, containing the DOS, projects to and forms a tight connection with the second embryonic cuticle. The epidermal cells are pigmented at this stage.

216 hr after oviposition, the second enveloping cell has initiated formation of a trichogen process that distally splits into projections covered by a floccular deposit. A distinction between the two cell types is facilitated by variation in the number of DOS. AP sensilla have only two in the region of the exuvial space with a third extending to only the tip of the trichogen process; the MP sensillum contains three-four DOS in both the exuvial space and in the trichogen process. In addition to epidermal cells, the outer enveloping cells are now pigmented.

252 hr after oviposition, the receptor lymph cavity is formed by retraction of the second and third enveloping cell from the peg, leaving only the outermost enveloping cell connected to the cuticular socket. Both sensilla are now fully formed and distinguishable, as the lumen of the AP sensillum now contains the dendritic bundles with two DOS surrounded by electron-dense fluid; the third DOS with

finger-like projections comes to lie 1.7 μm beneath the socket. In the MP sensillum, hollow spokes connecting the inner and outer walls are loaded with fluid continuous with the floccular material that encloses the peg.

In the moulting process, differences are detectable in the structure of the trichogen process. In the AP sensillum the process outline is smooth, with microvilli confined to the tip; in contrast, the process of the MP sensillum forms very deep microvilli that enclose microtubules, and vesicles fused to the plasma membrane. The spaces between these projections are filled by the flocculent material that is also continuous with the electron dense core of the spoke channels and, as described above, forms a cap above the trichogen process. Also known as the trichogen sprout, this halved, apical process serves as a mould for the formation of the hair cuticle in the trichoid sensillum of the silk moth. After this is initiated, the halves fuse and the enclosed, extensive array of microtubules provide mechanical support and convey vesicles, secreted in the cell body, of precursor cuticle. Following cuticle formation, the sprout is dismantled and conveyed to the cell body to contribute to apical fold formation (Fig. 38) [4.14, and references therein].

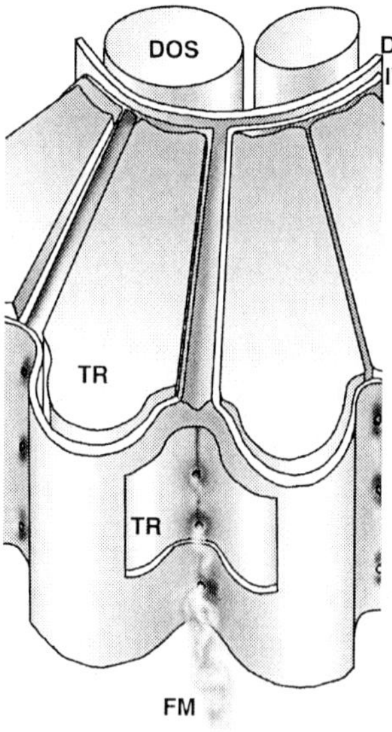

Figure 38. Schematic representation of moulting of a multiporous (MP) double-walled, dual-function (olfactory/mechanoreceptive) sensillum of Locusta migratoria*. DOS=DDP dendritic outer segment; DS, dendritic sheath; IW, inner wall of sensillum; TR, trichogen cell folding to allow deposition of cuticular lining of the spoke canals; FM, floccular material apparently extruded from spoke canals. Redrawn after Ameismeier [4.4].*

SENSILLAR ACCESS

Dimensional Movement

A consideration of internal sensillar relationships may commence with the pore, the point of entry of chemical stimuli. Some such molecules could obviously strike a pore with a direct hit especially in sensilla with relatively high frequency of pores/μm^2 as, for example, the sensilla placodea of *Apis mellifera* with an estimated pore density of 125/μm^2. So given a mean pore diameter of 15 nm, 8.8 % of the surface is porous. In contrast, sensilla trichodea of *B. mori* have only 2-7 pores/μm^2. The comparison does not end there, however, as antennal configuration would obviously affect the target area presented. The *Apis* antenna is slim and unbranched whereas that of *Bombyx* has many branches, each containing thousands of sensilla. Nevertheless, it is obvious that most molecules will not make a direct hit on a pore. Their fate was elegantly considered by the late Max Delbruck who, with his co-worker, established that stimulus molecules not making a direct hit will remain on the cuticle sufficiently tightly not to be detached, but sufficiently loosely to diffuse along the surface and into a pore. The hydrophobic nature of most insect stimuli especially pheromones suggested this was at least physically feasible. For two dimensions, the best estimate of the diffusion coefficient of the sex pheromone bombykol was $D_{min} = 10^{-8}$ (cm^2/s). Available experimental data were consistent with this as they gave a value of $D > 5.10^{-7}$ (cm^2/s), equivalent to a diffusion time of less than 5 ms in sensilla trichodea of *B. mori*. This was in good agreement with reported values of not less than 10 ms for latency i.e. the time interval from discharging the chemical stimulus to the first appearance of an electrical response. In other words, molecules such as bombykol could diffuse in decreased dimensionality, i.e. in two dimensions to a pore after striking the cuticle [4.1].

On entering a pore, a molecule first encounters the sensillum liquor, the composition of which remains to be established fully. It seems to be secreted by the tormogen cell and may be lipoidal. As lanthanum, a water-soluble heavy metal tracer penetrated pores only when all traces of liquor were removed by acetone or by chloroform, a coarse filtering effect was indicated, as well as the more obvious function of preventing desiccation [4.13].

Nearest the pore the liquor is contained in canals and tubules and for a time numbers of tubules were emphasized in reports on sensillar ultrastructure. Apparently, they were viewed as possible extensions of the dendrites and thus, as capable of sensory discrimination. This is no longer a serious question as it is accepted that the tubules are not dendritic tissue; cuticular in origin they lack neuronal properties necessary for transduction.

The DDP, which may be branched or unbranched, narrows at the ciliary region the neurotubular composition of which has already been commented upon. This region is surrounded by the ciliary sinus through which dendrites extend unwrapped and for which no explicit function has yet been suggested.

SIGNIFICANT ORGANELLES

Cytoplasmic organelles reported from the perikaryon include RER, Golgi bodies, mitochondria, vesicles, and microtubules the functions of which are reasonably well understood. Occasionally, there are onion bodies defined as several sets of double-membrane cisternae bounded by a single membrane, and with ribosomes sometimes on the outer but not on the inner membranes; they may form complete or partial circles, ovals, or be flat. Detailed observations suggest that some perikarya may have two onion bodies. The first, consisting of eight double-membrane cisternae bounded by a single membrane with ribosomes, RER strands towards two Golgi bodies (Fig. 39). The second onion body, consisting of three double-membrane cisternae, is bounded on the outer side by a single membrane and on the inner by the nuclear membrane; this body also continues to RER cisternae that extend toward a Golgi body. Cisternae of Golgi bodies have many associated vesicles 80 nm in diameter. Similarly-sized vesicles, apparently aligned along microtubules, are present in the PDP up to the level of the basal bodies and adjacent to the dendritic membrane [4.9]. Similar structures have been reported from locust sensilla where their function was unknown.

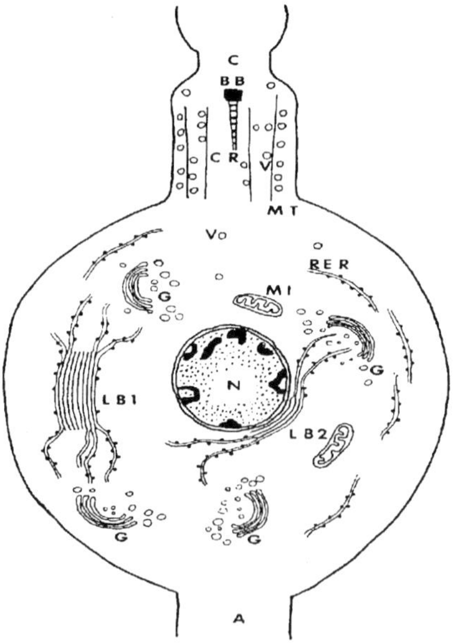

Figure 39. Schematic representation of a lamellar body of a basiconic sensillum of Nebria brevicollis *(Coleoptera). A, axon; BB, basal body; C, ciliary region; CR, ciliary rootlets; G, Golgi bodies; LB1, lamellar body 1; LB2, lamellar body 2; M, mitochondrion; MT, microtubule; N, nucleus; RER, rough endoplasmic reticulum. Redrawn after Daly and Ryan [4.9].*

A universally accepted model of cellular events involving ribosomes is that free ribosomes synthesize soluble proteins; ribosomes bound to endoplasmic reticulum (ER) synthesize proteins for secretion by the cell, and also integral membrane proteins. Bound ribosomes extrude secreted proteins fully into, and membrane ones partly into, the ER lumen. Here, core carbohydrate is attached and proteins are conveyed in vesicles to the Golgi bodies. The core carbohydrate may then be modified, after which the proteins are transported in vesicles to the cell surface, either for secretion or incorporation into the membrane. Applying this model to the present data indicates that the onion body is part of a protein-synthesizing system, and specifically that the ribosomes, bound to its outer membranes and those of the continuous RER, are synthesizing protein into their continuous cisternae. This protein after traversing the closely associated Golgi bodies, travels in vesicles along microtubules to the dendritic membrane for export or incorporation [4.9].

This singular form of RER has now been described from a variety of cells including, for example, the pigmented epithelium of teleosts, amphibians, reptiles, and birds in which they are termed myeloid bodies of unknown function. In neurons, this structure has been called the lamellar body and its distribution in brain cells of guinea pig is as follows: most (51 %) were adjacent to glial processes; less frequently, they were relatable to synaptic boutons (10 %); nerve cell processes (6 %); and granule cells (14 %). Furthermore only 37 % of the brain cells studied were covered by glia but 69 % of cisternae were apposed to glia. Such cisternae were viewed as involved in increased and/or specific protein synthesis, not least because they occur predominantly in cells with enhanced metabolism such as embryonic, immature, and tumour cells. Also lamellar body formation is associated with receptor membrane breakdown in retinular cells of *Limulus*. Accordingly, it would be appropriate to investigate the precise product of this organelle in insect CS and a single unifying term could focus attention on it. In view of its widespread use in relationship to neurons, the term lamellar body seems the most appropriate [4.9].

The receptor lymph cavity is something of an enigma. Its fluid has been assigned various functions: preventing desiccation of dendrites; providing nutrients for them; transmitting dissolved compounds; providing hydrostatic pressure to open the terminal pore of contact CS; providing a route for the return flow of current when a receptor potential (defined in Chapter 5) travels down the dendrite; and coupling with an electrogenic ion pump to provide an additional energy source for the receptor potential. In assessing these possibilities, it seems relevant that most of the MS studied so far have large receptor lymph cavities, indicating that they are not specifically associated with chemoreception. Electron-dense fluid in the receptor lymph and associated cavities is regarded as an acid mucopolysaccharide in the glial lacunar system of *Locusta*. In many sensilla, the RLC and receptacle cavity are continuous. The septate desmosomes between adjacent accessory cells allow passage of ions and molecules along intercellular spaces.

EXTERNAL MORPHOMETRY

If it is accepted that molecular access to the sensillar neuron is an important factor affecting generation of the physiological response, then it is appropriate to consider in detail the physical parameters governing such access. Three types of sensilla trichodea have been quantified from the antenna of two groups of the cabbage looper *Trichoplusia ni* by light microscopy and SEM [4.17]. The first group were from laboratory-reared pupae weighing 180-200 mg, selecting one antenna from each of seven male and seven female moths. The second group were selected from laboratory-reared pupae weighing 120-300 mg to investigate a possible correlation of pupal weight with antennal surface area. Measurements made were of subsegment length and diameter, as well as numbers, dimensions, and proportions of the different sensilla.

Subsegment surface areas and volumes were calculated in both weight groups from the equations:

$$S = \pi\, dl \tag{1}$$

and

$$V = \pi \frac{d^2 l}{4} \tag{2}$$

where $S(\mu m^2)$ is the subsegment surface area, $d(\mu m)$ is the mean subsegment diameter, $l(\mu m)$ is the mean subsegment length and $V(\mu m^3)$ is the subsegment volume. Total values for the entire flagellum were derived by summing over 75 subsegments. Polynomial regressions of subsegment length and diameter, and the number and proportion of sensilla on each subsegment, were derived as functions of subsegment numbers according to:

$$Y = b_o + b_1 x + b_2 x^2 + b_3 x^3 + \ldots + b_m x^n \tag{3}$$

where Y represents the parameter to be estimated, b_0, b_1, b_2 ... b_m are regression coefficients, and x represents the number of subsegments from the base.

Essentially, numbers of sensilla were related to the weight of the pupa or adult, as an antennal area increased by 0.007 mm^2 for each mg of pupal weight. Also there was a significant correlation between the total surface area of the flagellum and adult weight, equivalent to 0.008 mm^2/mg adult weight. This, in turn, led to 19 and 22 additional sensilla/mg pupal and adult weight respectively. In addition, pore densities might act as indicators of types of chemospecificity as at least some sensilla trichodea with three-eight pores/μm^2 were pheromone sensitive, and others with 30-40/μm^2 were responsive to a broader range of olfactory stimuli [4.17].

Morphological Correlates

It has been claimed that sensillar function is determined by: intrinsic surface proteins accounting for specificity and sensitivity; and extrinsic properties that link the neuron to external stimuli. Both qualities arise originally from a single mother cell in the larval imaginal disc. In a single, defined class of sensilla trichodea, on the antenna of the male *T. ni*, precise cell lineage patterns were assessed for distinctive morphological markers for distinct physiological properties. These sensilla have long (15 to 60 μm) cylindrical shafts each innervated by two neurons sensitive to the female pheromone, and differentiated from each other by the amplitude of their action potentials. The larger spike was produced by the cell designated A and the smaller by B [4.16].

These pheromone-sensitive sensilla include at least two classes separable by the spontaneous activity of their receptor neurons and by their relative sensitivity to pheromone concentrations. Specifically, the first class exhibits relatively high spontaneous activity (HSA) ($\overline{x}$= 1.39 impulses/s) in the A neuron, and 1.20 in the B. Further, the A neuron is stimulated by low doses (0.01 μg) of (*Z*)-7-dodecen-1-ol acetate, 93% of the bouquet, but the B neuron is excited by low doses of dodecan-1-ol acetate, 7% of the bouquet. The second class has relatively low spontaneous activity (LSA) or 0.12 and 0.17 impulses/s for the A and B neuron respectively. The A cell is unresponsive to the two pheromone constituents even when doses are increased 10,000-fold. The B cell is excited by doses increased 1,000-fold.

The most readily apparent morphological correlate was total length, as the HSA sensillum was shorter (28.7 ± 2.6 μm; $\overline{x}$ ± SD) than the LSA sensillum (35.3 ± 4.2 μm). In addition, HSA sensilla were straighter, had more cuticular pores, and were preferentially located on the distal margins of each antennal segment. In contrast, LSA sensilla were larger, had fewer pores, and were more uniformly distributed across the antennal segments. All, except the last, of these differences would seem to affect access of chemical stimuli to the neurons, and the latter might be assigned to pattern formation in the imaginal disc. Such different sensitivities and accessibilities undoubtedly affect the heterogeneous response patterns often observed in electrophysiological recordings from sensilla [4.16, 4.18].

Internal Morphometry

The use of freeze-fracture techniques has enabled the comparison of insect gustatory and olfactory sensilla with vertebrate olfactory receptor cells [4.19].

Unlike olfactory receptor cells of vertebrates where 4-40 cilia sprout from one sensory dendrite, insect olfactory and gustatory neurons give rise to only one cilium. The most conspicuous membrane-bound structure at the base of the cilia is the necklace of particles. Insect olfactory cilia have 11 strands of these with an interstitial spacing of 15 nm giving 36 strands/μm. Particle densities were similar in necklaces of vertebrate olfactory cilia. Chemosensory cilia seem to have more strands in their necklaces than mechano- and non-sensory cilia. Regional differences were apparent in insects such as *B. mori*; particle density increased to 1,500/μm^2 in the distal region, nearest to pores and incoming molecules. In contrast, vertebrate particles are evenly distributed perhaps because the whole olfactory sheet is equally exposed to chemical stimuli. Finally, P-face particle densities of insect gustatory cilia are lower than those of olfactory ones, and densities are highest at the extreme tip of the sensillum under the single pore [4.19].

Undoubtedly, the most extensive morphometric analysis to date has been made on sensilla trichodea of *B. mori* and *A. pernyi* by Steinbrecht and colleagues [4.11]. Essentially, reconstructions were made of sensilla from serial sections after freeze substitution. Antennae, frozen by immersion in propane at -180 °C, are transferred into acetone of -75 to -80 °C, and then into acetone containing 2 % (w/v) osmium tetroxide at the same temperature. Thus, artefacts derived from cryoprotectants were precluded, by achieving a freezing rate so high that the ice crystals were too small to interfere with the electron microscopical images.

Bombyx always has two bipolar neurons per sensillum and *Antheraea* usually has two to three. The internal construction of the sensilla trichodea of both species was similar, especially in terms of ancillary cells. Each species contains a thecogen, trichogen and tormogen cell, the innermost thecogen enwrapping the receptor cell soma and PDPs. It also provides a septum separating both cells, except below the ciliary segment where it forms an inner receptor lymph space. The trichogen cell surrounds the thecogen only in the region of the PDPs and, in both species, its most conspicuous organelles are stacks of RER. The tormogen cell encircles the thecogen and trichogen cells in the ciliary region. Processes of the glia cells surround receptor-cell axons and one glia cell may enwrap both axons from different sensilla (s. basiconica). In the adult insect, thecogen cells exhibit little evidence of metabolic activity, unlike the tormogen and trichogen cells that are equipped with extensive RER, lamellae with associated mitochondria, Golgi bodies, and vesicles [4.11].

In terms of morphometrics, the principal differences between the species lay in the size of the sensory hair, in the larger DDPs, and in the volume of receptor lymph within each hair, mainly associated with the larger antenna of *A. pernyi*. In contrast, little difference was detected between PDPs, receptor-cell somata, trichogen and tormogen cells, and the volume of receptor lymph below the hair base.

Within each sensillum, significant differences in volume and surface area existed between the receptor cell somata, dendrites, and initial axonal segments. The apical cell membranes of the trichogen and tormogen cells, bordering the receptor lymph cavity, are deeply invaginated so that their area is at least twice that of the remaining membrane. If, however, the microlamellae are regarded as folds standing up from a smooth base or as invaginations from a smooth outer surface, then the ratios become 14:1 and 47:1 for trichogen and tormogen cells, respectively. Alternatively, if the reference surface is taken as a flat plane between the apical septate junctions, then the enlargement brought about by the lamellae and the invaginated pouch becomes 380:1. Several authors have suggested that these auxiliary cells secrete the receptor lymph which contains ingredients such as hyaluronic acid and/or chondroitin sulphate and esterases: potassium concentrations are four times higher there than in haemolymph. It has also been suggested but not verified, that these cells facilitate the functioning of an electrogenic cation pump in insect sensilla.

The significance of the direct measurements derived by this nonartefactual method goes beyond precise documentation of cell surfaces and volumes. The sensilla trichodea of *A. polyphemus* have a capacitance of ~ 30 pf. Assuming the usual capacitance of biological membranes of 1μ F/cm^2 gives a membrane area of ca 3,000 μm^2. The summated areas (rounded) of the apical cell membranes of the trichogen and tormogen cells of *A. pernyi* are 1365 and 835 μm^2 respectively, or a total of 2100 μm^2. In addition, the electrical resistance at the apical and basolateral cell membranes were previously estimated as 300 and 10-40 MΩ respectively. The morphometric data allow the calculation of specific membrane resistance as 1 x $10^4 \Omega/cm^2$ and 1 x $10^2 \Omega/cm^2$ for apical and basolateral membranes respectively [4.11].

REFERENCES

4.1. Adam, G. and Delbruck, M. (1968) Reduction of dimensionality in biological diffusion processes. In Rich, A. and Davidson, N. (eds.) *Structural Chemistry and Molecular Biology*. Freeman, San Francisco, pp. 198-215.

4.2. Altner, H. (1977) Insect sensillum specificity and structure: an approach to a new typology. In Le Magnen, J. and MacLeod, P. (eds.) *Olfaction and Taste* VI. IRL, London, pp. 295-303.

4.3. Altner, H. and Thies, G. (1976) The postantennal organ: a specialized unicellular sensory input to the protocerebrum in apterygotan insects (Collembola) *Cell Tissue Res.* **167**, 97-110.

4.4. Ameismeier, F. (1985) Embryonic development and molting of the antennal coeloconic no pore- and double-walled wall pore sensilla in *Locusta migratoria* (Insecta, Orthopteroidea). *Zoomorph.* **105**, 356-366.

4.5. Bassemir, U. and Hansen, K. (1980) Single-pore sensilla of damselfly-larvae: representatives of phylogenetically old contact chemoreceptors? *Cell Tissue Res.* **207**, 307-320.

4.6. Behan, M. and Ryan, M. F. (1978) Ultrastructure of antennal sensory receptors of *Tribolium* larvae (Coleoptera: Tenebrionidae). *Int. J. Insect Morphol. Embryol.* **7**, 221-236.

4.7. Corbière-Tichane, G. (1971) Fine structure of an antennal sensory organ ("vesicule olfactive") of *Speophyes lucidulus* Delar. (cave Coleoptera of the Bathysciinae subfamily). *Tissue Cell* **6**, 535-550.

4.8. Crouau, Y. and Crouau-Roy, B. (1991) Antennal sensory organs of a troglobitic coleoptera, *Speonomus hydrophilus* Jeannel (Catopidae): An ultrastructural study by chemical fixation and cryofixation. *Int. J. Insect Morphol. Embryol.* **20**, 169-184.

4.9. Daly, P. J. and Ryan, M. F. (1981) Ultrastructural relationships of the lamellar or onion body in an insect sensillum. *Int. J. Insect Morphol. Embryol.* **10**, 83-87.

4.10. Ernst, K-D. (1969) Die feinstruktur von Riechsensillen auf der Antenne des Aaskäfers *Necrophorus* (Coleoptera). *Z. Zellforsch Mik. Anat.* **94**, 72-102.

4.11. Gnatzy, W., Mohren, W. and Steinbrecht, R. A. (1984) Pheromone receptors in *Bombyx mori* and *Antheraea pernyi*. *Cell Tissue Res.* **235**, 35-42.

4.12. Karuhize, G. R. (1971) The structure of the postantennal organ in *Onychiuris* sp. (Insecta: Collembola) and its connection to the central nervous system. *Z. Zellforsch.* **118**, 263-282.

4.13. Keil, T. A. (1984) Very tight contact of tormogen cell membrane and sensillum cuticle: ultrastructural bases for high electrical resistance between receptor-lymph and subcuticular spaces in silkmoth olfactory hairs. *Cell Tissue Res.* **16**, 131-135.

4.14. Keil, T.A. (1999) Morphology and development of the peripheral olfactory organs. In Hansson, B.S. (ed.) *Insect Olfaction.* Springer, Berlin, pp. 5-47.

4.15. Lewis, C.T. (1954) Contact chemoreceptors of the blowfly tarsi. *Nature* **173**, 130-131.

4.16. Mayer, M.S. and Mankin, R.W. (1985) Neurobiology of pheromone perception In Kerkut, G.A. and Gilbert, L.I. (eds.) *Comprehensive Insect Physiology. Biochemistry, and Pharmacology*, vol 9. Pergamon Press, Oxford, pp. 95-144.

4.17. Mayer, M. S., Mankin, R. W. and Carlysle, T. C. (1981) External antennal morphometry of *Trichoplusia ni* (Hübner) (Lepidoptera: Noctuidae). *Int. J. Insect Morphol. Embryol.* **10**, 185-201.
4.18. Mayer, M.S., Mankin, R.W. and Lemire, G.F. (1984) Quantitation of the insect electroantennogram: measurement of sensillar contributions, elimination of background potentials, and relationship to olfactory sensation. *J. Insect Physiol.* **30**, 757-763.
4.19. Menco, B. Ph. M. and Wolk, F. van der (1982) Freeze-fracture characteristics of insect gustatory and olfactory sensilla. *Cell Tissue Res.* **223**, 1-27.
4.20. Newman, S. M., McDonald, I. C. and Triebold, B. (1993) Antennal sexual dimorphism in *Diabrotica virgifera virgifera* (Le Conte) (Coleoptera: Chrysomelidae): male specific structures, ultrastructure of a unique sensillum, and sites of esterase activity. *Int. J. Insect Morphol. Embryol.* **22**, 535-547.
4.21. Schneider, D. (1964) Insect antennae. *Ann. Rev. Entomol.* **9**, 103-122.
4.22. Seidl, S. (1992) Structure and function of the thecogen cell in contact chemosensitive sensilla of *Periplaneta americana* L. (Blattodea: Blattidae). *Int. J. Insect Morphol. Embryol.* **21**, 235-250.
4.23. Slifer, E. H. (1962) The fine structure of insect sense organs. *Int. Rev. Cytol.* **11**, 125-159.
4.24. Slifer, E. H. (1970) The structure of arthropod chemoreceptors. *Ann. Rev. Entomol.* **15**, 121-142.
4.25. Van Nieukerken, E. J. and Dop, H. (1987) Antennal sensory structures in Nepticulidae (Lepidoptera) and their phylogenetic implications. *Z. zool. Syst. Evolut.-forsch.* **25**, 104-126.
4.26. Zacharuk, R. Y. (1980) Ultrastructure and function of insect chemosensilla. *Ann. Rev. Entomol.* **25**, 27-47.
4.27. Zacharuk, R.Y. (1985) Antennae and sensilla. In Kerkut, G.A. and Gilbert, L.I. (eds.) *Comprehensive Insect Physiology. Biochemistry, and Pharmacology*, vol 9. Pergamon Press, Oxford, pp. 1-69.
4.28. Zacharuk, R. Y. and Shields, V. D. (1991) Sensilla of immature insects. *Ann. Rev. Entomol.* **36**, 331-354.

CHAPTER 5

ELECTROPHYSIOLOGY OF CHEMORECEPTION

Although the morphometric approach concluding the previous Chapter sheds useful light on the possible functional significance of sensillar parts, verification requires direct recordings by electrophysiological methods. A constraint is sensillar size (dendrite 0.3 μm diam) as compared, for example, with squid neurons (up to 1 mm diam). Electrophysiological investigations of insect chemoreception enable the ranking of the relative potencies of various chemical stimuli, rankings that may indicate the behavioural significance of the compounds. Some background information on the properties of the neuron [reviewed in 5.20 and 5.35] will provide a context for the available data.

THE NEURON

The Action Potential

The nerve impulse, spike, or action potential lasting for about 1ms is the signal by which the neuron transmits information. The action potential is rapid, travels the length of the axon, and does so without decrement. An all or none event, so that its amplitude is fixed for a given neuron in the region of approx. +25 mV, it signals through variations in frequency.

Action potentials are generated whenever the membrane is depolarized from the resting potential (frequently -80 mV in insect sensilla) to the threshold, usually about -50 mV. Then membrane conductance for Na ions (q_{Na}) increases and their inward flow through the Na ion channel further promotes the depolarization process, allowing more Na ions into the neuron. The inrush of sodium at one point alters the electrical properties of an adjacent one, eliciting there an inrush of sodium, thus propagating a wave of electrochemical activity. Ultimately q_{Na} increases to more than 100-fold the resting value, and thus exceeds membrane conductance for K ions (q_K). However, depolarization of the membrane also enhances q_K with a lag of less than 1 ms. Then K ions begin to flow out of the neuron, compensating for the inward Na flow, such that q_K again exceeds q_{Na}. Accordingly, the interior becomes increasingly more negatively charged, representing the repolarization phase. Given that the concentration of the bound stimulus governs the extent of the graded receptor potential, which triggers the action potential that varies only in frequency, then, in general, the sensory neuron's concentration/response relationships will be as in Fig. 40.

Channel Characteristics

Essentially, the channel for each ion is visualized as having two functions, a filter and a gate. Situated at the entrance to the channel, the size of the filter precludes entry of inappropriate ions and its negative charge repels anions. Further down the channel is located the gate that closes during the resting potential and opens during depolarization. Visualized as a large, charged protein molecule, this undergoes a conformational change in response to a stimulus-induced membrane potential depolarization. The change makes the gate permeable to the appropriate ion presumably by opening spaces in the molecule through which the ion can fit; inactivation occurs within a few ms. The density of sodium channels is about 50 μm^2 membrane, each channel 0.5 nm in diameter, with a mean distance between channels of 140 nm.

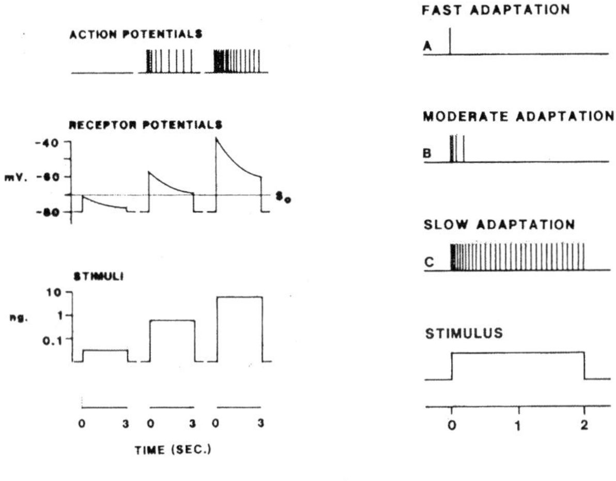

Figure 40. Left, Relationships between stimulus intensities, receptor potentials, and action potentials, characteristic of a chemosensory neuron. Only stimulus intensities eliciting receptor potentials above threshold (S_0) elicit action potentials, the frequency of which reflects the excess. Receptor potentials last as long as the stimulus but amplitudes decline over time (adaptation): action potential amplitudes are fixed and characteristic of each cell type, only their frequency varies. Right, Action potentials may adapt at various rates.

Stimulus-Response Relationships

By governing the magnitude of the receptor potential, stimulus strength is relatable to the frequency of the action potential according to the intensity function (Fig. 40). As action potentials will be generated only when the receptor potential exceeds the threshold, the relationship may be written simplistically as:

$$F = k\ (S - S_0) \qquad (4)$$

or to take account of the Weber-Fechner law (response is proportional to the logarithm of the stimulus amplitude) as:

$$F = k\ .\ log\ (S - S_0)^n \qquad (5)$$

where F is the frequency of action potentials, k is a constant for the family of neurons, S is the actual stimulus intensity, and S_0 is the lowest stimulus that first produces a response, the threshold stimulus. As there must be an upper limit to the frequency of action potentials, the relationship is written more realistically as a power function:

$$F = k\ .\ log\ (S - S_0)^n \qquad (6)$$

where n is constant within each family of receptor cells. Given the physical upper limit constraining the generation of action potentials, n values are usually less than unity thus producing an upper asymptote. Conventionally, this Stevens relationship is converted to a double-logarithmic form:

$$log\ F = log\ k + n\ (log\ S - S_0) \qquad (7)$$

which gives a straight line plot of slope n that may quantify stimulus/response relationships: the more the slope approaches unity the more sensitive the response is to changes in stimulus concentration [5.35] (Fig. 41).

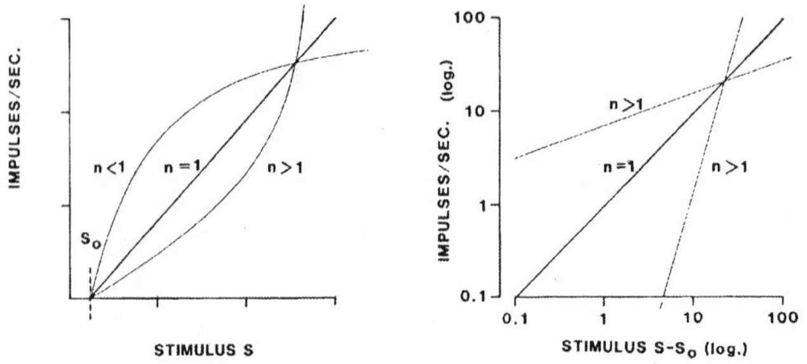

Figure 41. Intensity function characteristic of a chemoreceptive neuron. Left, untransformed values; right, log. transformed values, S_o , stimulus threshold. Redrawn after Schmidt [5.35].

Central Processing

Two principal mechanisms, labelled lines and cross-fibre patterning, have been proposed to explain how a burst of action potentials is interpreted in the central nervous system (CNS). The former suggests that as specific stimuli are detected by only specific neurons, then the arrival of a spike train at a central destination is, sufficient to notify the brain of the identity of the stimulus; the frequency of the action potentials would indicate stimulus intensity. However, stimuli frequently comprise a mixture of often disparate ingredients, some of which would be an adequate stimulus for entirely disparate neurons. In which event, the brain would need to collate simultaneous transmission from different axons into a pattern appropriate to the mixture, which would require cross fibre patterning. It is relevant that pheromones are usually mixtures of compounds. A third mechanism is proposed in the form of temporal patterns in the spike train.

THE INSECT ANTENNA

Countless experiments extirpating the insect antenna have shown that this appendage is crucial for the receipt of many chemical, especially olfactory stimuli. The variety of antennal shapes and sizes elicited the formulation of relationships between antennal shape, outline area, and efficiency through indices such as the adsorption quotient, effective fraction of adsorption, quantum efficiency and filter coefficient. Essentially the sensitivity of olfactory organs was related to their ability to filter out molecules from the carrier medium [5.17].

A controlled investigation by Mankin and Mayer [5.21] used males of *Apis mellifera*, *Bombyx mori*, *Lymantria dispar*, *Plodia interpunctella* and *Trichoplusia ni,* as species having antennae of different shapes. They were exposed in a Plexiglas tunnel to two tritiated sex pheromones. The results were consistent with engineering theory that depicts transport of olfactory molecules as a two-step operation. The first entails transport of the stimulus by convective diffusion to a boundary layer near the surface of the receptive organ; this is velocity-dependent as odourant mixing is improved by turbulence. In the second stage, diffusion across the boundary layer deposits stimulus molecules on the surface according to, as a first approximation:

$$R = CSK \tag{8}$$

where R (moles/s) is the net rate of deposition, C (moles/cm^3) is stimulus concentration, S (cm^2) is surface area, and K (cm/s) is a constant, reflecting deposition velocity.

Values of K varied over a simple order of magnitude, having a mean of 1 cm/s, as compared with the four orders of magnitude in variability of odourant concentration in the field. So stimulus concentration rather than deposition velocity is the key factor affecting stimulus detection. The k value of 1 cm/s is consistent with that for many other chemicals and, if applicable to pheromones, could facilitate the estimation of behavioural thresholds. More powerfully, the data detected no difference between deposition velocities on antennae and inert Parafilm disks, perhaps because their surface composition is not dissimilar [5.21].

THE ELECTROANTENNOGRAM

The first electrophysiological investigation of the insect antenna was that of Chester Roys [5.31] who used tungsten electrodes to record both slow potentials and fast impulses from the cockroach antenna. In a pioneering study that deserves to be better known and acknowledged, he severed the antenna and introduced one electrode in the cut end and another in a joint. Then he measured concentration response relationships elicited by clove oil and 19 other odours. He noted that increasing concentrations increased the amplitude of the slow potentials, and the frequency of spikes (Fig. 42). Thus, he established the pattern and parameters for all subsequent electrophysiological investigations of insect chemoreception, of which the most widely reported is Schneider's [5.36]. Changes in the slow potential of the antenna which, in time, came to be known as the electroantennogram (EAG), frequently amount to 1-2 mV, rarely exceed 5 mV, and were viewed as the sum of many receptor potentials [5.17, 5.19].

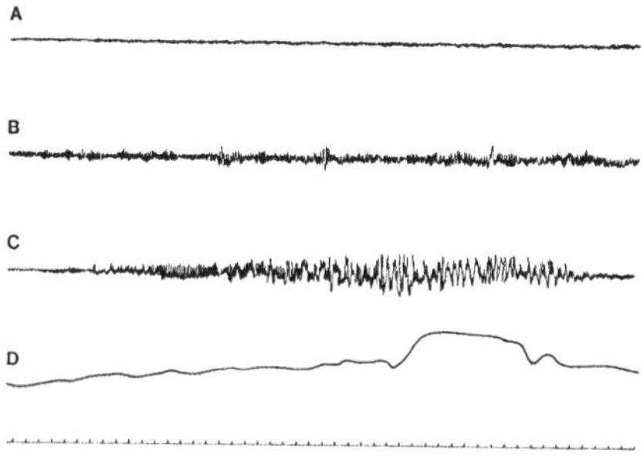

Figure 42. Representation of the first recordings of fast and slow nerve potentials from the insect antenna. Traces on chart recorder are in response to (from top) A, control; B and C, increasing doses of the organic stimulus; D, the slow potential. Redrawn after Roys [5.31].

There is evidence for a gradient in the EAG response along the axis of the antenna of the European corn borer *Ostrinia nubilalis*, when the entire length is stimulated [5.26]. The amplitude increases distally indicating that the potential is preferentially enhanced in one direction. An excised antenna was connected by strips of filter paper to discrete pools of saline at various points along its length, with the distal tip uncut. An Ag-AgCl electrode was introduced into each pool such that potential differences between any two could be recorded and the neutral or reference electrode was connected to the proximal cut end, grounded in a saline pool.

When stimulated by one second, capillary-delivered puffs of the pheromone (*Z*)-11-tetradecenyl acetate (20 μg source), EAG amplitudes decreased along the antennal length towards the reference or indifferent electrode. Furthermore, amplitudes decreased steadily to zero as the reference electrode was brought closer to the point of stimulation. Reversing the position of the reference electrode to the uncut distal end confirmed this response. Finally, when the stimulation jet and the recording electrode were located at the same central point on the antenna, and the reference electrode was moved about, the amplitude difference was largest in the proximal direction. Accordingly, it seems that hyperpolarization of the stimulated region spreads in both directions but is weaker at the uncut distal end. Apparently, the ideal positions for the electrodes are the reference one at the proximal end and the recording one at the stimulation site, which is contrary to established practice by which most investigators insert the electrodes into the cut ends [5.26].

Typically, the EAG comprises a hyperpolarization within 100 ms of stimulus application followed by a much slower recovery (Fig. 41). The amplitude of this depolarization is the most commonly used criterion to assess the potency of the stimulus; a second criterion is half-time of decay of the EAG (τ_δ). Usually, amplitudes elicited by increasing concentrations of biologically relevant test stimuli exceed, by orders of magnitude, the hyperpolarization assignable to a standard chemical. In contrast, corresponding τ_δ values are indistinguishable from those of the standard except at very large stimulus concentrations [5.17].

Relationship of EAG to Olfactory Stimulation

This issue was comprehensively examined by Mayer *et al.* [5.23], using males of *T. ni* and the pheromone (*Z*)-7-dodecen-1-ol acetate (7:12Ac), 99.95 % pure. To restrict the number of responding sensilla, pheromone stimulation was confined to specified regions of the antenna by coupling the dispenser tip to a Pasteur pipette (1 mm diam) that yielded a smoke plume of about the same diameter. The systematic advancement of the tip along the length of the antenna exposed specific regions of it to stimulation; in addition, the antenna was resected yielding segmental stumps of predetermined lengths. Statistical analysis was based on the relationship:

$$R = b_4 + b_5\, n \tag{9}$$

where b_4 and b_5 are regression coefficients, n represents responding sensilla, and R is EAG amplitude (-mV). A morphometric study had previously indicated the types and numbers of sensilla present on each subsegment. Essentially, there was a statistically significant relationship between EAG amplitudes and numbers of sensilla trichodea stimulated, and the dose response relationship fitted a power function. Confining the stimulus to specific regions gave the regression relationship:

$$R = 0.425 + 5.32 \times 10^{-4}\, n \tag{10}$$

for which $r^2 = 0.81$. Resecting the antenna gave the relationship:

$$R = -\,0.458 + 4.02 \times 10^{-4}\, n \tag{11}$$

for which $r^2 = 0.86$. Overall, these data indicate a mean contribution of 0.5 μV from each s. trichodeum in response to a pheromone source of 10 μg.

EAG amplitudes increased linearly with increasing pheromone doses at source, reaching an upper asymptote of about -5 mV in response to 100 μg. However, when such dosages were corrected using calibration equations, to estimate actual concentration, the response was still increasing at 100 μg (equivalent to 1×10^{-7} μ mole/cm^3). In the event, the observed relationships could be described either by: a log-linear function:

$$R = A + B\,log(C) \qquad (12)$$

where A = 7.11 ± 0.86, B = 0.59 ± 0.087, C is the odourant concentration (μ mole/cm^3), and R is the EAG amplitude; or by a power function

$$R = b_3\,(C)^b \qquad (13)$$

where b_3 and b are regression coefficients. The power function was preferred as behavioural responses are relatable to the summed electrophysiological responses of pheromone-sensitive neurons as transmitted to the deutocerebrum. A practical recommendation was that blank responses should not simply be assigned to air as they may also be elicited by water-vapour, chemicals in the room, and plasticizers in the Teflon delivery tube. Such spurious responses were eliminated by purging the lines immediately before the stimulus [5.23].

Uses of the EAG

Roelofs has demonstrated, using several tortricid species and known pheromone structures, that the double bond position and the configuration to which the antenna was most sensitive could be inferred from EAG values. Combined with information on pheromone retention time, using various gas chromatography (GC) columns, this approach rapidly identified sex attractants for many species of Lepidoptera. Further improvement was obtained when use of monounsaturated standards identified the position and configuration of double bonds in doubly unsaturated constituents of pheromones. Traces from just a few adults of the codling moth *Cydia pomonella* were sufficient to show that the retention time of the pheromone matched that of a conjugated 12-carbon alcohol and that the most potent standards had the *trans*-8 and *trans*-10 conformation. From this it was concluded that the pheromone was *trans*-8, *trans*-10-dodecadien-1-ol. This was confirmed later by classical methods. Six steps were emphasized as necessary for a complete study: (1) ascertaining GC retention times of natural pheromone constituents; (2) deriving EAGs of the monounsaturated series including *cis* and *trans* isomers; (3) chemical analysis of GC-collected, EAG-active fractions; (4) synthesis of candidate compounds; (5) comparing their retention times and EAG-potencies with natural constituents; (6) authenticating pheromone activity in laboratory and field behavioural bioassays [5.29, and references therein].

A logical extension of this approach, linking GC retention time of compounds with their EAG potencies, was the development of the electroantennographic detector (EAD). This uses an excised antenna as the sensing element at the outlet of a gas-liquid chromatograph (GLC). The antenna connects with drops of saline at the tip of chloridized silver electrodes leading to an amplifier and chart recorder. The antenna and electrodes are enclosed in a glass tube provided with a humidified air stream to prevent the droplets from drying out. The effluent from the column leaves the oven through an opening 15 cm upstream from the antenna. The result is a simultaneous display of both the capillary column GC trace and the amplitude of the EAG. This has proved invaluable in: clarifying if a pheromone secretion contains constituents with specific numbers of carbons; placing isomers in ranked order of potency; and identifying from the antennal response the presence of compounds in the effluent, too low in concentration to register on the GC trace [5.2].

A further advance was development of an inexpensive, portable device that measured EAG maximum amplitudes using a voltmeter instead of an oscilloscope. Built at a cost of US$25 (1980 values), this device had the obvious advantage of providing supplemental EAG recording systems in a laboratory, but especially in the field. Difficulties to be overcome included the fact that voltmeters usually read voltage changes lasting for several seconds, whereas EAG's peak in a few ms. Secondly, a perpetual feature of EAG traces is baseline drift, usually in the region of 1 mV min^{-1}. Baseline drift was eliminated using a high-pass filter that was followed with a peak detector circuit that recorded maximum input voltage and maintained output voltage at that level [5.5]. Essentially, the base of an excised antenna is pressed into soft wax in a glass dish loaded with sufficient saline solution to cover the base. A chloridized silver wire is inserted into a Pasteur pipette enwrapped with aluminium foil. When the pipette tip touches the saline, capillary action fills the tip, thus contacting the silver wire. Removing the distal antennal segments is following by insertion of the pipette into the antenna. The reference electrode comprises a grounded hook-up wire attached to a chloridized silver wire that is dipped into the saline solution. The aluminium foil surrounding the recording electrode is also grounded. In the event, the first field use of the EAG method occurred within 9 years [5.4]. Given that EAG amplitude is affected by pheromone concentration, and that the insect antenna may be exploited as a pheromone detector, then it was credible that a portable EAG system could quantify pheromone concentrations in the field. Such data are very relevant in assessing the efficacy of mating disruption systems.

In this system developed for use with the pink bollworm moth *Pectinophora gossypiella*, the key features are: (1) the antenna is mounted in a perspex holder in a measuring chamber; (2) three syringes containing pheromone, at dilutions of 1×10^{-6}, 1×10^{-5}, and 1×10^{-4} ml/ml silicon oil, are linked to an electronic drive system; (3) a charcoal filter the position of which precludes entry of ambient pheromone into the system but enables this when removed; (4) an airpump that produces a continuous flow of air over the antenna; (5) an EAG amplifier; and (6) a thermistor. The entire configuration is powered by a car battery linked to a voltage converter. Essentially, the motor drive depresses the piston of each syringe in order of ascending concentration generating an EAG dose-response function. Removing the filter allows the ingress of ambient air the pheromone content of which elicits an EAG, that is quantifiable by reference to the dose-response function. A confounding EAG response may occur if plant odours are taken in but this is overcome by repeating the standard stimuli; these will elicit responses superimposed on that attributable to ambient air, only if that contains no pheromone. In contrast, if ambient pheromone levels are high relative to standards, adding air enriched by the latter will have a negligible effect (Fig. 43) [5.11, and references therein].

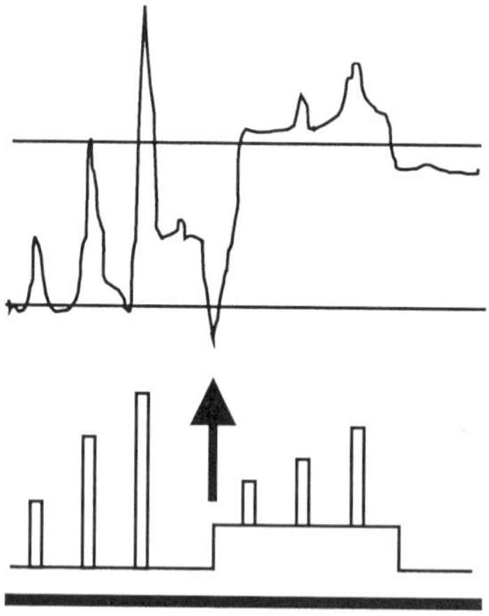

Figure 43. Field recordings of electroantennogram (EAG) traces in response to ambient pheromone with reference to syringe-delivered standard stimuli. At left, vertical bars indicate syringe delivery of standard pheromone concentrations, with EAG responses superimposed. Arrow designates the opening of the filter to access ambient air. At right, corresponding bars and traces indicate no response to the first concentration of standard applied, and increased EAG amplitudes in response to only concentrations two and three. Therefore, the ambient pheromone concentration is higher than that of standard one. Redrawn after Färbert et al. *[5.11].*

CIRCUITRY OF THE SENSILLUM

Of two principal current paths from sensillum hair to haemolymph, the first runs from the circa-dendritic space filled with sensillum liquor, across the extensive microvilli of the tormogen and the trichogen cells and into those cells, and from their basolateral membranes into the haemolymph. The second path crosses the dendritic membrane into the neuron departing it in the soma region to cross the thecogen cell into the haemolymph. The presence of septate desmosomes should preclude current paths between the membranes of adjacent cells. The resistance of the dendritic membrane of sensilla trichodea of the wild silk moth *Antheraea polyphemus* has been estimated from a model as 10^4 Ω.cm^2 approximately, or sufficiently high to enable electronic spread of the receptor potential. This is consistent with the value of 10^4Ω.cm^2 estimated from the resistance associated with the extensive microvilli of all the sensillar cells. A perceived difficulty with this model was that it predicted smaller nerve impulses from shorter hairs which was not supported experimentally. This cast doubt on the applicability of the conventional model: high resistance of the dendritic membrane facilitating electronic spread of the receptor potential, leading to impulse generation at the soma of the neuron. Instead, the dendrite was proposed but not verified as the site of impulse generation [5.8].

In a more comprehensive subsequent analysis, the responses to various stimuli of the sensillum and of a circuit, electrically equivalent to the sensillum and comprising two series of resistances R_1 and R_2 with a capacitor C_2 in parallel to R_2, were assessed. Applying a square current pulse elicited a two-phased response: fast (F) and slow (S), with time constants slower than 50 µs and of the order of 10 ms respectively. R_1 includes the resistance of the sensillar lumen as related to hair length, plus that of the recording electrodes (about 17 MΩ), plus a contribution from the length of the antenna between electrodes (10-40 MΩ). Resistance and capacitance of the sensillar epithelium both proved independent of hair length, indicating they derive from proximally located structures. Factors explored included temperature, lesions, electrical coupling between sensilla, and voltage dependence.

R_2C_2 influenced spike shape as: the time constant of the spike tail response to temperature and polarization matched exactly that of R_2C_2; in current clamp experiments, spikes from short hairs lasted longer than those generated in voltage clamp experiments: epithelial perforation by suction destroyed both generation of action potentials and the membrane giving rise to R_2C_2. This led to the claim that R_2C_2 must originate from epithelial membranes, of which the much-folded apical regions of the neuron and thecogen cell were the most obvious candidates. In regard to spike generation, passive properties of the dendrite were not significant, as amputation of the sensillum not merely failed to decrease spike current but actually increased it. Also spike amplitude did not increase when sensilla were chopped to mere stumps, indicating that dendritic surface area or impedance did not limit spike current. Accordingly, the conventional view of a long-range electronic propagation from the site of initiation of the receptor potential to the corresponding one for the action potential was not sustainable. Taken in conjunction with the fact that dendrite impedance does not limit spike amplitude, this suggested active involvement of the dendrite in spike generation. Such a mechanism would obviate the need for a current loop, linking the initiation sites of the receptor and action potentials, by having to traverse at least five membranes of the following structures: dendrite; apical trichogen cell; outer and inner thecogen cells; and cell body. Those of the apical trichogen and thecogen would be significant barriers [5.7].

Otherwise the properties of the insect chemosensitive sensilla are consistent with the general model for sensory neurons presented earlier. For example, the receptor potential increases to saturation in proportion to the logarithm of the stimulus intensity. However, less efficacious stimuli may elicit curves displaced towards higher stimulus intensities and peaking at lower receptor potential amplitudes. This could be explained either by a smaller proportion of bound receptors or by a smaller increase in conductance per bound receptor. The attenuated fall time of the receptor potential following exposure to high concentrations of stimulus is attributed to the continual presence of stimulus molecules, unaffected by inactivation [5.18, 5.19].

In regard to latency, the time interval between olfactory stimulation and the onset of the electrical response, the values are variable and may be dose-dependent. Average latencies recorded from *B. mori* after stimulation by bombykal at low stimulus intensities were several hundred ms. However, recordings from *A. polyphemus* responding to (*E*)-6,(*Z*)-11-hexadecadienal at high stimulus intensities (3/s) gave minimal latencies of 15 ms for the receptor potential, and only 25 ms for the first nerve impulse [5.18]. Independent verification of such values came from a separate study with the same species [5.32]. Such values are significant for an understanding of the biochemistry of chemical perception (Chapter 6).

Ionic Currents of the Sensillum

The advent of insect neuronal cultures overcome the difficulties, posed by the tightly enwrapped accessory cells, in identifying significant ions. In a seminal study, olfactory receptor neurons (ORNs) from the antenna of male *Manduca sexta* were investigated by whole cell and single-channel voltage clamp techniques [5.46]. Using the former gave mean values of 0.7 GΩ for resistance, a time constant of 25 ms, membrane capacitance of 35 pF and a specific membrane resistance of 25 $K\Omega.cm^2$, with a resting potential of -62±12 mV (mean, SD). Depolarization to -30 mV elicited a fast, transient inward current that, inhibited by tetrodotoxin, was identified as a Na^+ current. This was followed by a voltage-activated, slowly activating, non-inactivating, outward rectifier K^+ current i.e. a pattern consistent with the conventional one for sensory neurons. In addition to the major K^+ current, there was a transient, inactivating K^+ current expressed at the initiation of a voltage change and resembling the "A current" (I_A) reported from molluscan neurons. A third K^+ current was activated by elevating Ca^{++} levels from 0.1 to 5 μM. The fact that no single channel Na^+ currents could be elicited at the same membrane was consistent with electrical inexcitability in somata of *Musca* neurons, *Helix* neurons, and cockroach *Periplaneta americana* motoneurons, and with the electrophysiological data reported in the previous section [5.46].

Sensillar Action Potentials

Following the demonstration of fast potentials by Roys [5.31], the first quantified recordings of fast potentials were from s. basiconica on the antenna of the carrion beetle *Necrophorus*, using tungsten electrodes [5.6]. The recording electrode seems to have been in the interstitial fluid, close to but not inside the s. basiconicum. Thus, the recorded impulses could qualitatively reflect the reverse of the ionic features of the spike within the neuron. Subsequently, a clarification was provided through use of carrion and propionic acid as attractant and repellent stimuli respectively; at the recording electrode the former was associated with a hyperpolarization of the slow potential, and the latter with a depolarization. Furthermore, the polarity of the slow potential was driven from negative to positive by superimposing the repellent stimulus on the attractive one.

Recordings from long sensilla are facilitated by cutting the hair tip and sliding the glass capillary with enclosed silver wire over the stump such that the receptor lymph is directly contacted [5.17]. Most recordings are from single sensilla or single units (single sensillum recording, SSR) that, as outlined in Chapter 4, are likely to contain several neurons. Accordingly, traces reflect their simultaneous activities distinguishable on the basis of amplitude. Rarely, as in some bark beetles, potentials of two different cells exhibit the same amplitude, but the responses may still be distinguished through selective adaptation. This involves following the appropriate stimulus, that has served to adapt responsive cells, with a second the response to which may be assigned to neighbouring cells. Impulse frequencies of 50/s have been recorded from the resting sensillum, increasing 10-fold in response to large stimulus intensities. Tonic patterns are known from species such as *B. mori*, in response to the sex pheromone bombykal, and phasic responses are given by *A. mellifera* in response to the queen substance, 9-oxo-*trans*-2-decenoic acid.

In a development paralleling that of the EAD [5.2], Wadhams succeeded in linking a GC with single cell recordings [5.45]. And corresponding to the development of EAG portable systems, was the derivation of a system for SSR in the field. This measures responses from only one or a few pheromone-sensitive neurons. The key features of this system are: (1) a base with signal meter, battery compartment, control switches, separate circuitry for DC and AC amplification, and a voltage to frequency converter enabling the separate amplification and audio tape recording of the DC and AC signals; (2) arising from the base is a central column with two joystick micromanipulators, the left one bearing the reference electrode in a pipette holder, and the right one carrying a miniature probe and the recording electrode connected to a high-impedance amplifier; (3) the sensillum is cut by the actions of a vertical microknife on the column and is connected to the tungsten recording electrode; (4) a flow tube delivering filtered air to the preparation is mounted on the control column; and a microthermistor mounted above the left micromanipulator and within 2 mm of the preparation records air speed; (5) a computer interface, with associated software designed to analyze long duration electrophysiological signals, samples the AC signal at a rate of 50,000 samples/s creating up to 127 amplitude classes; (6) in addition to DC and AC signals, air-speed readings and operator comments are monitored on a dual channel oscilloscope and stored on a cassette recorder; power for the entire configuration is from a car battery, or conventional power outlets may be employed [5.44].

Experiments were made with sex pheromone constituents from five moth species but especially from the apple sesiid glasswing moth *Aegeria* (= *Synanthedon*) *myopaeformis*, because of the ready accessibility of the sensilla trichodea of the male. In an orchard with suspended pheromone traps and low wind speeds, two receptor cells of this species with large and small amplitudes respectively, exhibited increased firing frequencies associated with relative lengthy exposures (90 s) to ambient air; this was before and following purging the preparation with charcoal-filtered air (80 s). Specifically, the large-amplitude cell evinced 3.6 action potentials/s in ambient air, 2.3/s when purged, and 3.2/s when again in ambient conditions; corresponding values, for the small-amplitude cell were 2.9, 1.1 and 2.8/s respectively. Although the differences from such SSR are relatively small, they might be quickly perceived by a moth equipped with tens of thousands of sensilla. Initiation of responses by the tortricid *Pandemis heparana* was much slower (approximately 1 min) when pheromone was presented 10 m upwind in an open landscape, due to significant fluctuations in wind direction. Such turbulence and the presence of vegetation also dispersed the pheromone plume as judged by co-release of smoke, although the firing rate of the tortricid *Archips podana* gradually increased probably due to a build-up of pheromone. Large variations in wind speed, during quite short time intervals and independent of prevailing wind direction, compromised the reproducibility of field data from this ingenious device [5.44].

Relationships between EAG and Single Unit Recordings

The EAG is a rather poor predictor of single unit responses, and hypotheses based on only the EAG may be erroneous. For example, the EAG dose-response pattern of male of the African armyworm moth *Spodoptera exempta* was identical to two pheromone components, (*Z*)-9-tetradecenyl acetate (*Z*-9-TDA) and (*Z*)-9,(*E*)-12-tetradecadienyl acetate (*Z*-9,*E*-12-TDDA). Furthermore, large concentrations (1 or 10 μg) of each compound adapted the response to the other. This suggested the two constituents as being of similar potency. However, in single unit recordings *Z*-9,*E*-12 TDDA was predominantly the crucial stimulus or 100-fold more potent than *Z*-9-TDA; about 20% of cells were equally sensitive to both. Clearly, the ranked order of equal potency, as suggested by the EAGs, was not sustainable in that study [5.42]. This provides a context for claims that sensilla are or are not tuned to various stimuli, on the basis of only EAGs and without single unit recordings.

PERIPHERAL CODING

Cross-Fibre Patterning

Advantage was taken of the fact that *Periplaneta* sensilla have relatively few sense cells, to examine responses to food odours in the context of peripheral coding. No sensillum responded specifically to only one kind of food. For example, one cell of the two-neuron, pentanol-sensitive type, responded to the odours of banana, apple, orange, lemon, lettuce, grass, bread, and meat up to 2-3 days old. The second cell responded maximally to very rotten meat. Clearly, sensory information to the CNS would be conveyed not by the activity of one cell but by the response pattern from many kinds of neurons. So, *Periplaneta* seems likely to recognize food odours not from cue scents, but from the pattern of stimulated cells i.e. cross-fibre patterning [5.34].

This analysis was taken a stage further when lemon oil, as an example of a complex food odour, was fractionated by liquid chromatography and analyzed by GC combined with mass spectroscopy (GC-MS). Fractions were assayed as odour stimuli at near natural concentrations. (+)- and (-)-Limonene was the main constituent (74%), followed by β-pinene (10%), γ -terpinene (6%), α-pinene (2%), myrcene (1.5%) and sabinene (1.5%). Whilst confirming the reaction types previously described, an investigation of single compounds detected cell types sensitive to α- and β-ionone, eugenol, c-hexylamine, methylheptanone and methylhexanone, and cineole; each compound represents the individually most potent stimulus from a related group. There were significant correlations between the reaction spectra of these various cell types. Accordingly, the alcohol terpene groups did not form a uniform reaction type. Some cells always occurred together as, for example, the ionone with the cyclohexyl amine cell in a smooth-walled sensillum. It was of interest that limonene, the principal constituent, was inactive as a stimulus. Responses to whole oil were less than anticipated from summation, indicating either the absence of summation, or alternatively the action of an inhibitor [5.40]. Taken overall, these studies of *P. americana* receptivity establish conclusively the phenomenon of cells with overlapping spectra and with considerable individual variability between cells of the same group. They also dispense with the need to view such cells as generalists.

An explicit assessment of the signalling system associated with pheromone-elicited trains of impulses to the CNS was made with the pheromone of the redbanded leafroller moth *Argyrotaenia velutinana*, (*Z*)- and (*E*)-11-tetradecenyl acetate (*Z*- and *E*-11-TDA). Sensilla trichodea contain three neurons of which two (A and B) are sensitive to olfactory stimuli with significantly different resting activities of 26 and 19 impulses/10 s respectively. Responses were mainly tonic with an initial phasic element at larger stimulus concentrations. These elicited decreased response frequencies, an effect also noted with EAG amplitudes in other species; latencies also decreased. Increased frequency response was relatable to the log of stimulus intensity with correlation coefficients larger than 0.8, indicating considerable sensitivity to concentration changes.

Z-11-TDA was generally a more effective stimulus for A cells, and *E*-11-TDA for B cells, but there was considerable overlap and no exclusivity i.e. no cell was exclusively stimulated by one ingredient. Yet, the two components (1 μg) elicit mutually exclusive behavioural patterns in the whole insect. Specifically, the former produced rapid walking with wing vibration, brief flights, and efforts to copulate; the latter produced brief wing-buzzing. Plainly, no labelled line system of quality coding could be inferred from such data [5.27].

The effect of various ratios of pheromone constituents, and their isometrical isomers, on single unit responses has been studied in the male moth *Adoxophyes orana*. The s. trichodeum contains two cells: the A cell sensitive to (*Z*)-11-tetradecen-1-ol acetate (*Z*-11-TDA) with a much weaker response to *Z*-9-TDA; the B cell responds to only the main pheromone component *Z*-9-TDA. This situation resembles that in the *A. velutinana* and *Antheraea polyphemus*. In all three species, s. trichodea are pheromone-sensitive, contain two cells, of which the one with the larger spike amplitude is most sensitive to the principal pheromone constituent. In *A. orana*, the pheromone blend ratio is *Z*-9 to *Z*-11, 9/1 but, significantly, neurons were not maximally sensitive to this blend. Apparently, data on pheromone composition is conveyed by the spike trains from A and B cells in the population of responding sensilla. This, in turn, by indicating that such trains are averaged, read, compared, and interpreted in the CNS, contraindicates the labelled line option in these species. Furthermore, the discharge pattern was too irregular to permit a role for a temporal pattern in the response [5.9].

Labelled lines

Other data are consistent with the labelled line hypothesis. S. trichodea of *B. mori* contain a neuron sensitive to bombykol, the principal component of the female sex pheromone, but insensitive to bombykal, a secondary constituent. This is recognized by a different cell of smaller amplitude, a pattern that holds for many other Lepidoptera. In the lepidopteran subfamily Heliothinae, that includes pests of such significant crops as cotton and maize, all species investigated to date employ two compounds as necessary and sufficient to elicit long-range attraction by the male to the female. Furthermore, (*Z*)-11-hexadecenal (Z11-16: Ald, designated A) is the principal component for *Heliothis virescens*, *H. subflexa*, *H. phloxiphaga*, *H. peltigera*, and *H. maritima*, and *Helicoverpa* (=*Heliothis*) *zea*, *H. armigera*, and *H. punctigera*; the minor component conferring species-specificity is either (*Z*)-9-tetradecenal (Z9-14:Ald, designated B), or (*Z*)-9-hexadecenal (Z9-16:Ald, designated C), or (*Z*)-11-hexadecenyl acetate designated D, or (*Z*)-11-hexadecenol (Z11-16OH, designated E). In contrast, *Helicoverpa assulta* employs C and A as the principal and minor components respectively. Each species uses one or two of these four compounds, other than the major and minor components, as inhibitors of attraction, thus serving species isolation [5.25, and references therein].

Receptor neurons detect low concentrations and dose-responsively respond to increasing concentrations of the principal component, but also respond to much larger (by 2-3 log units) concentrations of the minor constituent. However, as the addition of the minor component to the major one elicited no increase in firing rate, it was concluded that the components did not interact at the level of the receptor, i.e. their presence is independently communicated to the brain. There, the signals are read as the relative activities of each type of neuron, or consistent with a labelled line mechanism.

Surprisingly, there was an exception to this pattern in that neurons for the minor component C could not be detected in *H. zea*. This led to the proposal that signals from C are mediated by the B neurons, as these respond secondly to C after B. This implies, but does not prove, the existence of a dual-function neuron as it is possible that C neurons exist, but in numbers too small to be readily detected. A systematic investigation of dose-response relationships exhibited by each type of receptor neuron (A-D) indicated that the labelled line mechanism generally applied. Furthermore, neuronal sensitivities varied according to sensillar location, decreasing from a lateral to a medial position, and from proximal to distal annuli [5.25]. Undue emphasis should not be placed on demarcation between the two coding mechanisms, as cross-fibre patterning may be viewed simply as requiring the mean activity of cells of different types to differ across stimuli. Thus, cross-fibre patterning could resemble the labelled line code, except that several types of neurons are involved.

Temporal Patterns

Whole insect experiments showed that moths would orient to a pheromone source both in a discontinuous odour plume, and when the pheromone was discharged as pulses at various repetition intervals. Therefore, the moths were able to respond efficiently to variations in pheromone concentration. Single cell recordings were made from the pheromone-sensitive s. trichodea of *A. polyphemus*, possessing three cell types each responsive to a different constituent of the pheromone blend. A single brief pulse of pheromone elicited a receptor potential of maximum amplitude followed by the expected decline. A series of pulses faster then the recovery time elicited a potential of smaller amplitude followed by a decline to an intermediate value. The frequency of nerve impulses was affected by stimulus concentration, pulse rate, and position in the sequence of stimuli. Essentially, the strongest response was to the first stimulus with subsequent ones eliciting fewer toward a constant number, equivalent to short-term adaptation. As the repetition rate of pheromone pulses increased, the number of action potentials/stimulus decreased. This would mimic the situation of a moth entering a plume. Accordingly, averaging stimulus intensities has little bearing on reality for a moth. Furthermore, the slowest cell type responds to the major constituent that is detectable from larger distances than the other two components. In such a situation an unduly rapid response may be disadvantageous for the moth. When the two minor constituents are detected the moth is nearer the source and quicker responses of the two corresponding cells enable faster responses, and adjustments to rapid changes in odour concentration in the narrower region of the plume. Although not applicable to all moths, this study shows sensillar sensitivity to the temporal changes in concentration in a pheromone plume [5.32].

Variance

Gustatory sensilla on the labellum of the fleshfly *Sarcophaga bullata* contain three chemosensory neurons giving variable responses when stimulated by sucrose, liver and fish powders. Cell 1 exhibited similar responses (mean firing rate) to all three stimuli; cell 2 was significantly more responsive to fish than to sucrose or liver; and cell 3 responded with a two-fold increase to fish than to either of the other two stimuli. The only factor in the electrophysiological response correlating with behavioural discrimination was the variance of cell 1, that increased more in response to sugar than to fish or liver. It was proposed that, in addition to mean firing rates, variability or variance in responses of related neurons could be a coding mechanism. Accordingly, the most pronounced behavioural responses would be elicited when mean firing rate is optimal, and variance is low [5.24].

Chemosensory variability in response to physiological and environmental factors is emphasized by Schoonhoven and associates. Chemoreceptor sensitivity was influenced by previous diet, time of day, and degrees of hunger. Such plasticity directly affected feeding preference and intensity. For example, the fifth instar larva of *M. sexta* would feed on an otherwise unpalatable plant, following conditioning of the deterrent receptor by incorporating salicin in the larval diet. In addition to emphasizing the need for standardized conditions, such results also indicate the intrinsic variability of the signal to the CNS, as influenced by physiological and environmental factors. This variability is distinguishable from sensory adaptation, as conventional adaptation and disadaptation occur within seconds to minutes. Ageing has contrasting effects, as EAGs increase with increasing age of the spruce budworm *Choristoneura fumiferana* and other Lepidoptera, but decrease by 50% in male blowflies 25 days old. Also in the blowfly, receptor sensitivity is affected by the ovarian cycle entailing increased sensitivity to salt at the onset of vitello-genesis as triggered by juvenile hormone (JH). Light is also significant, as houseflies confined to darkness for 3-6 days exhibit decreased chemosensory responses.

Such changes may be regulated either peripherally or centrally. The former probably involves self-regulation by the receptor cell, with either altered thresholds or decreased numbers of receptor molecules as likely mechanisms. The fact that diet-induced decreased sensitivity to salicins is accompanied by decreasing sensitivity to caffeine, likely to bind to a different receptor, favours the decreased threshold hypothesis, unless a multiplicity of different receptor sites decrease in number. In further support of the threshold hypothesis is the fact that decreased sensitivity persists after moulting, that removes the dendritic ends of chemoreceptors. Such peripheral control is implicated in the modification of behavioural responses, including regulation of feeding in blowflies and host-seeking in the female mosquito.

Central regulation could involve either endocrine or neural control with available evidence favouring the former. Specifically, the distended foregut of a fed locust elicits release of a hormone from the corpora cardiaca that is associated with closing the sensilla, and thus with terminating stimulation. Such data led to the hypothesis that hormones, by governing ion transport in tormogen cells, alter ionic balance in the dendritic liquor and affect receptor sensitivity. There is explicit evidence for an effect of JH on chemoreceptor sensitivity although the direction of change is not consistent. In regard to neural control, the general assumption is that arthropod chemoreceptors are not subject to efferent regulation. The various pathways by which chemically-elicited spike trains are transmitted to the brain are of particular interest [5.38, and references therein].

PROCESSING BY THE CNS

In general, antennal receptor axons terminate ipsilaterally in the deutocerebrum of the brain (for brain outline see Chapter 1, Fig. 1) that contains two neuropiles: the antennal lobe (AL) receiving chemically mediated signals; and the mechanosensory and motor centre (AMMC) or dorsal lobe, receiving mechanically-elicited signals [5.15, and references therein].

The Antennal Lobe: Morphology

Morphologically, the AL resembles the vertebrate olfactory bulb and serves as the first relay complex for antennal afferent fibres. It has been studied in relatively few species, and in depth in mainly the honeybee, locust, cockroach, flies, in various moths including *Antheraea pernyi* and *A. polyphemus*, and perhaps most comprehensively and elegantly in the male *M. sexta*. In that species the brain as a whole and also the AL is surrounded by a perineurial sheath, beneath which lies a 10-40 μm layer of interdigitating glial processes. The AL comprises a central zone of coarse neuropil mainly of deutocerebral neurons surrounded by 64±1 dense glomeruli, to which are confined the synapses between incoming antennal axons and second-order neurons. The glomeruli are, in turn, bounded by three sets of neuronal somata: a large lateral group with two constituent clusters; a smaller dorso-medial group; and a much smaller anterior group. Axons of the antennal nerve connect near to the dorsal pole of the AL in fascicles, insert between the glial processes, and descend into the neuropil where they arborize around the glomeruli. So called because they resemble kidney glomeruli, these are surrounded by glial processes and perhaps are derived from glial cells. Each glomerulus comprises a condensed region of neuropil 50-100 μm diam, with tightly-packed AL axon terminals enclosed by a cortical cup of neurites from receptor axons, and enriched by synaptic vesicles. Usually a single presynaptic element is apposed by multiple (up to seven), postsynaptic ones, the former in a bar-shaped configuration flanked by vesicles, and the latter involved in serial synapses. The neurites of AL neurons may interact with one or all glomeruli of one AL [5.22]. Computer-aided neuroanatomical techniques have mapped the individual positions of these glomeruli in a shell surrounding the coarse neuropil. They comprise three classes of which the larger contains 44 identified through their relative positions; 10 were identified through their proximity to easily identified structures and their size and shape; and nine from their location relative to the previous 10. As the map and organization of these glomeruli is invariant between individual insects, this indicated that the position and function of AL neurons ultimately would be mapped [5.30] (Fig. 44).

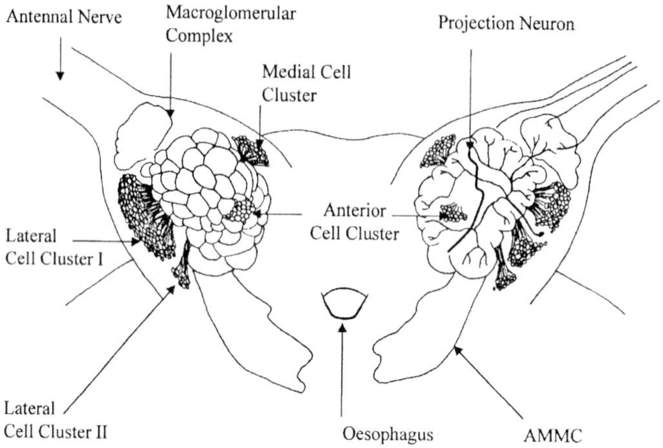

Figure 44. Schematic representation of a frontal section through the brain of Manduca sexta *emphasizing the two neuropils, the antennal mechanosensory and motor centre (AMMC) and the antennal lobe (AL); the latter comprising (left): a male-specific macroglomerular complex; grape-like glomeruli; bounded by three sets of neuronal somata in the form of a lateral cluster of two constituent groups, a dorso-medial cluster, and an anterior cluster. Redrawn after Anton and Homberg [5.1]. Right, a projection neuron (indicated) af the inner antenno-cerebral tract innervating a single glomerulus, and a projection neuron, originating from lateral cluster I, with multi-glomerular arborizations. Redrawn after Homberg* et al. *[5.15].*

Sexual dimorphism is apparent as, in Lepidoptera and Dictyoptera, only the male contains a specific AL structure, the macroglomerulus or macroglomerular complex (MGC) associated with the processing of pheromone-elicited impulses. In the honeybee, the AL of the worker comprises 165±2 egg-shaped glomeruli 18-60 μm diam. In the drone, there are two kinds of glomeruli of which the first, called the glomerular unit (103 glomeruli), corresponds to that represented in the worker. The second kind comprises four glomerular complexes three dorsal and one central, shaped like mushroom caps; 90-105 μm diam, their volumes are 19-37-fold larger than those of the glomerular units and they constitute 21% of the neuropil. Innervation of the AL is similar for drone and for worker except that the diameter of the worker bundle is thinner than the drone's, which contains more axons. The structure of the queen's antennal lobe resembles that of the workers in that there are no glomerular complexes [5.3].

Of the neurons indigenous to the AL, most are amacrine, local interneurons (LNs) with extensive dendrites but no axon exiting the lobe; their somata are located in the lateral and medial cell sets that flank the glomeruli. Also present are output or projection neurons (PNs) with axons projecting into the deutocerebrum, dendrites serving single glomeruli, and somata in all the flanking cell sets. In a third category are centrifugal neurons (CFs) that send axons to the AL and dendritic arborizations beyond it to the protocerebrum (PC) (Fig. 44).

In *M. sexta*, LNs occur in two main categories: Type I arborizes in all or most glomeruli, but not in the MGC, and comprises two subtypes differing in terms of their branching pattern; Type II LNs arborize in most or all glomeruli including the MGC and thus include male-associated cells. In respect of output neurons (ONs), three categories are distinguished: Types I and II each arborize in only single glomeruli but differ from each other in the ramification pattern within them, and in that the somata of Type II are located in the anterior cell set; Type III arborize only within the MGC and accordingly are male-specific. ONs of all three types project into the PC including the calyces of the mushroom bodies (mbs), the lateral lobe of the PC, or other regions within it. Confocal microscopy has visualized of the complex relationships of PNs. Anatomically, the ALs may represent the first order neuropil of the olfactory system, with the second order neuropil represented by the mbs. These are composed of very thin fibres that branch in the pedunculus giving rise to α and β lobes as outputs [5.15].

In a highly original investigation of differentiation, Hildebrand and colleagues trans-sexually grafted antennal disks from the larva to produce gynandromorph moths of *M. sexta*. All transplanted antennae established ipsilateral connections and were characteristic of the donor's and not the recipient's sex. In 60%, the AL was successfully innervated by the nerve from the transplanted antenna without the benefit of guidance from the pupal antennal nerve. Further, the MGC characteristic of only the male was evinced by grafting from the male larva, and was always absent in male recipients of the female graft. Host neurons of female recipients of male grafts gave evidence of male-like interneurons and of sensitivity to the female-produced pheromone, bombykal, a response characteristic of only the male. Furthermore, these cells sent dendritic processes into the MGC-type area of the neuropil. Conversely, in male recipients of female grafts, male-like interneurons were absent and there was no electrophysiological response to bombykal. It is unclear if the dominance of afferent neurons in the differentiation of AL neurons and macroaggregations results mainly from suppression and supplanting, or from merely remodelling the recipients developmental process [5.37].

Formation of Glomeruli

The presence of both antennal sensory axons and glial cells is essential for normal formation of glomeruli. Specifically, the insertion of antennal axons elicits process formation in glial cells followed by migration and glomerular development. In contrast, when antennal extirpation precludes the arrival of sensory axons, the neuropil differentiates only to the extent of a central coarse region ringed by finer material, with the glial cells confined to an outer layer one cell thick. Correspondingly, when sensory input is allowed but the number of glial cells is depleted, glomerular formation was also inhibited.

Further studies with deantennation showed that the entry of fascicles of sensory axons is followed by their deposition around the neuropil to form protoglomeruli. These are knots comprising terminal neurites of sensory axons and PNs, encircled by glial cells, but devoid of synapses. Such protoglomeruli provide a template for formation of fully developed glomeruli. This entails an interaction between the antennal axons and glial cells that elicits glial investment of the protoglomeruli. About one day after their formation, protoglomeruli are invaded by neurites of antennal axons with concomitant formation of synapses. Accordingly, antennal axons put in place the template for glomeruli before seeking the postsynaptic target (Fig. 45) [5.28].

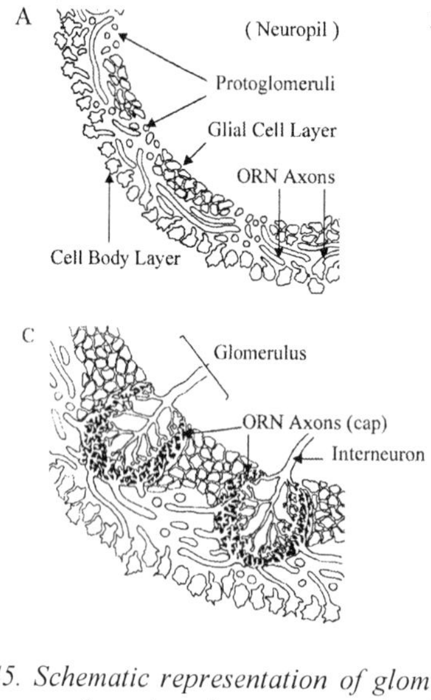

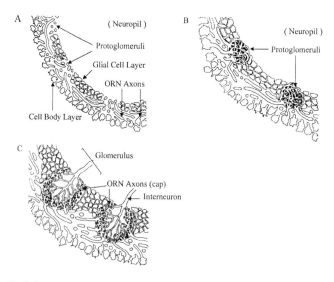

Figure 45. Schematic representation of glomerular development in Manduca sexta. *A) axons grow through and terminate under the glial border to take shape as a circle of protoglomeruli (arrowed) around a core of neuropil; B) protoglomeruli are more defined; and C) ORN axons aggregate as a cap. Redrawn after Salecker and Malun [5.33].*

The Antennal Lobe: Functional Aspects

The AL is the location for very great convergence as exemplified by *M. sexta* where axons from approximately 250,000 ORNs converge into 360 LNs, about 900 PNs, and less than 100 CFs. Such convergence could amplify neuronal signals and enhance signal-to-noise ratios. The AL subdivides into zones particular to classes of olfactory stimuli as for example, pheromones (ipsilateral MGC) and plant volatiles. From the termination rates of various types of ORNs, it seems that each glomerulus may specialize in processing a particular class of stimuli (Fig. 46). This would enable representation of odour patterns by response patterns across a specific cohort of glomeruli. Signal-processing AL neurons may partially distinguish stimulus concentrations as some respond only above a given threshold, without differentiating between concentrations higher than this. Others evince unambiguous concentration-sensitive responses in terms of: total number of spikes; early and mean firing rates; and latency.

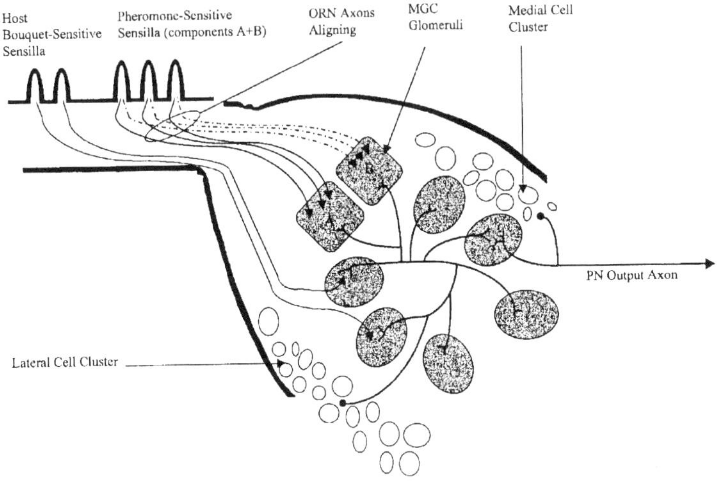

Figure 46. Schematic representation of neuronal connections illustrating convergence in the antennal lobe. This entails ORN (olfactory receptor neuron) axons each serving a specific glomerulus, within which connections are made to local interneurons, and to PNs (projection neurons); pheromone-specific ORNs innervate only the MGC glomeruli sensitive to compounds A and B; and host plant-specific ORNs innervate non-MGC glomeruli. Redrawn after Hansson and Christensen [5.13].

Included among pheromone-sensitive interneurons of *M. sexta* are PNs that link the MGC and protocerebral higher order neurons. Some PNs from the inner antenno-cerebral tract (IACT) or (PIas) receive axons from only neurons specifically tuned to bombykal, the principal constituent of the sex pheromone. Correspondingly, other PIas respond to stimulation by only (*E,Z*)-11,13-pentadecadienal that mimics the effect of another pheromone constituent. Accordingly, the MGC may harbour independent populations of PIas each receiving information from only one category of axon. As both pheromone constituents also elicit responses from other individual PIas, these may be linked to both receptor types.

Other PIas integrate processing of both stimuli through: being inhibited by bombykal; stimulated by (*E,Z*)-11,13-pentadecadienal; and evincing a mixed triphasic response to the natural blend of both. Yet others are inhibited by both pheromone constituents. Such a range of sensitivities among PIas would serve to distinguish blends of pheromonal constituents, separately from the individual components that govern flight responses. Also, stimulus intermittence in pheromone plumes is perceived by PIas that are linked to both types of sensory cell [5.15, and references therein].

A comparison of input and output signals associated with the AL of heliothine moths gave variable results, with particular reference to *Heliothis virescens* and *Helicoverpa zea*. In the former species, the projection neurons (PNs) comprised four categories: the first responded primarily but variably to receptor stimulation by compound A, the principal constituent, such that the less responsive neurons were stimulated by only A but the most sensitive ones responded to both A, and B the minor constituent; the second comprised PNs exhibiting equal responses to receptor stimulation by A and B; and the third receiving signals from both A-sensitive and B-sensitive neurons, but none received input exclusively related to compound B. Therefore, data regarding compound B are always delivered in the context of compound A and at least one neuron serves as a blend specialist, exhibiting a synergistic response to inputs from A- and B-sensitive receptors. PNs of the fourth category were sensitive to compound D, the acetate that disrupts pheromone responsiveness (compounds A-E are identified above, under labelled lines). Staining experiments revealed that receptor neurons sensitive to compounds A and B project into the largest division of the MGC, and those responsive to D project dorso-medially into a different compartment. Accordingly, the inhibitory signal designating a different species is presented through a distinct pathway [5.25].

In contrast, *H. zea*, for which A and C are the principal and minor constituents respectively of the blend and B is the interspecific signal, exhibits only two major categories of PNs. The first, responsive primarily to receptor stimulation by A but with the most sensitive ones PNs responsive also to B, or corresponding broadly to the first category represented in *H. virescens*, indicating that signals elicited by the principal constituent employ a specific pathway or labelled line. The second sensitive primarily to the inhibitor B and also responsive to high concentrations of C, but completely insensitive to A, in turn suggesting a specific line for the interspecific signal B. As at least three neurons exhibit synergistic responses to equal mixtures of A and B, they may serve to identify females of the other species and discontinue attraction.

The PN disposition in *H. zea* resembles in principle that of *H. virescens*, in terms of use of different pathways and MGC compartments for processing signals elicited by the principal and interruption constituents respectively. However, *H. zea* differs by employing the same pathway for signals elicited by both the second attractive constituent and the inhibitor. So, conspecific females respectively are distinguished from sympatric ones on the basis of the ratio of signals along the A- and B-related PIs: a high A:B ratio indicating the conspecific female and a low one the sympatric one [5.25, and references therein] (Fig. 47).

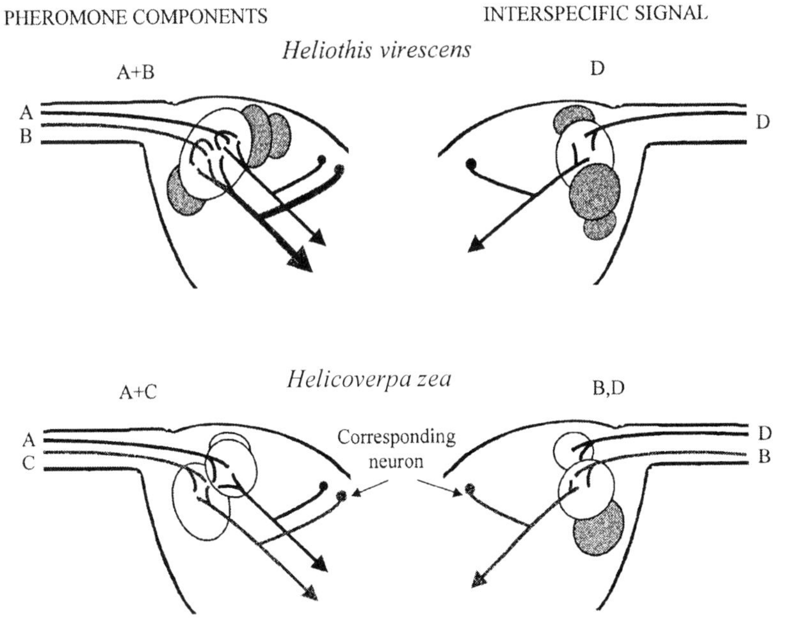

Figure 47. A comparison of Heliothis virescens *and* Helicoverpa (= Heliothis) zea *in terms of recognition of pheromonal constituents indicative of conspecific and interspecific females.* H. virescens *and* H. zea *have peripheral neurons sensitive to their pheromone constituents A and B, A and C respectively, the information from which is processed in their antennal lobes separately from that derived from the interspecific signals D, and B and D respectively. However, in* H. virescens *the input in respect of the second pheromone constituent compound B, is integrated with that from the major component A; in* H. zea, *input from the second pheromone constituent C and that from the interspecific signal B is processed by the same neuron. Redrawn after Mustaparta [5.25].*

Food Odours

As the MGC is linked only to pheromone-sensitive receptor cells, multiglomerular LNs interacting laterally are likely to process impulses elicited by food odours. A highly significant correlation exists between receptor activity during 1 s intervals of stimulation and food ingestion during a 24 hr period. Apparently, spikes are summed algebraically in the ALs, with impulses from quite different receptor types for positive stimuli having about the same force. In contrast, each deterrent stimulus neutralizes 2.5 impulses from feeding stimuli, assuming that firing rates from 1 s intervals are representative of those for long periods. However, impulse frequencies become completely adapted after 60 s reaching levels only 13, 19 and 22% of initial rates in the sucrose, amino acid, and glucosinolate receptor respectively; the adapted rate for the deterrent receptor was 48% of the initial one. Correcting for this abolishes the difference between weightings for positive and negative food stimuli. The resultant wiring model visualized a feeding centre in the suboesophageal ganglion (SOG) where inputs from phagostimulatory stimuli are compared with those from inhibitors (Fig. 48). This centre might comprise just one processing interneuron, based on decision-making affecting behavioural responses in other insects. Overall, the data were viewed as consistent with the hypothesis that food selection in phytophagous insects is governed by avoidance of deterrents [5.39].

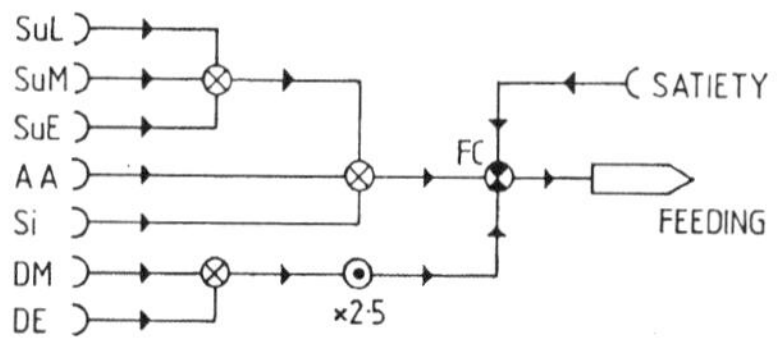

Figure 48. Schematic representation of signal processing by a proposed feeding centre (FC) in the central nervous system. Su, sucrose receptors (L lateral, M medial, and E epipharyngeal); AA, amino acid receptor; Si, glucosinolate receptor; D, deterrent receptors. Redrawn after Schoonhoven and Blom [5.39].

Oscillations in the Antennal Lobe

Recent evidence, principally but not exclusively from the locust *Schistocerca americana*, strongly indicates a role for temporal patterns (oscillations) exhibited by AL activity, in conveying information on the composition of complex odours. Morphologically, the composition of the AL is consistent with this. Specifically, ORNs input both excitatory PNs and inhibitory LNs, that synapse with each other. This design could enable the oscillations, in the form of alternating cycles of excitation and inhibition, and the appropriate connectivity and synchrony that have been observed [5.43].

Odourants presented to the insect antenna elicit differential patterns from responsive PNs in a slow temporal sequence derived from excitation, inactivity, and inhibition, with these elements repeatable. Furthermore, the same stimulus may elicit the above sequence in one PN, and a very different one in another. This variable responsiveness provides a rich source of information expressed as excitatory post-synaptic potentials (EPSPs) and inhibitory post-synaptic potentials (IPSPs), that collectively form a local field potential (LFP) constituting a spatial map of the odour. Furthermore, each individual cycle in the oscillation may vary from previous and succeeding ones by the involvement of different PNs. Indeed, the point at which a particular PN fires is a significant information 'bit' and constitutes the temporal component of the code. In general, up to 10 % of all locust PNs contribute to a particular response. The richness of this response would, of course, generate many oscillations by chance, but this is dealt with by statistical analyses appropriate for signal processing.

In a test of the specificity of information encoded in oscillations, the effects on two PNs of odour from apple and from apple + cherry (1:1) were each compared in terms of raster and cycle-by-cycle plots respectively. In raster plots representing times of firing by dot diagrams and average firing rates by histograms, the patterns elicited by the stimuli were essentially similar. When, however, the cycles were examined the responses could be distinguished: in response to only apple, both PNs simultaneously fired during the first three cycles; in response to apple + cherry, PN1 followed that pattern, but PN2 fired during cycles 2, 3, and 4, enabling discrimination (Fig. 49). This could be manipulated by picrotoxin (PCT), an antagonist of the γ-aminobutyric acid (GABA)-gated chloride channel as injecting PCT into ALs specifically blocked the first IPSPs. This desynchronized the PNs and precluded oscillations in the LFP, but without affecting the odour-specific, slow response of the treated PNs. Therefore, slow inhibition may be facilitated by a different transmitter/receptor system.

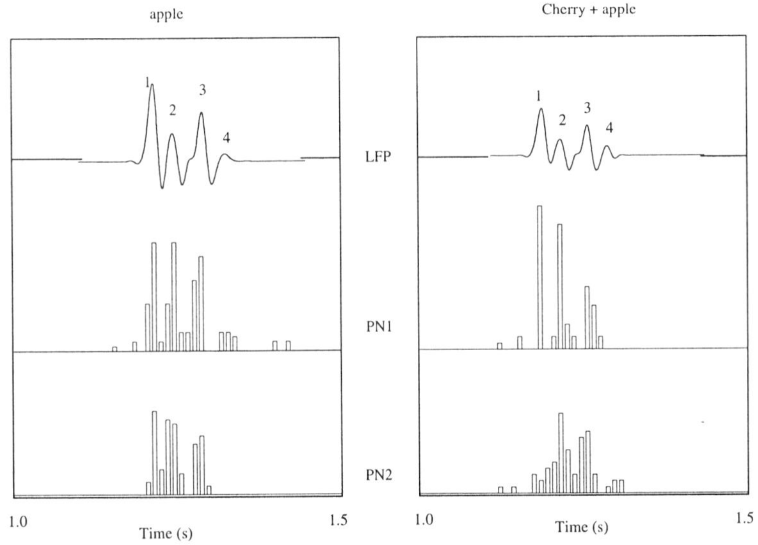

Figure 49. Temporal sequences of projection neuron (PN) activity in terms of the LFP, local field potential, and activity from two designated PNs (PN1 and PN2) in response to stimulation by odour of apple, and of cherry plus apple (1:1). Apple elicits simultaneously firing by both PNs during cycles 1, 2, and 3. Adding cherry, again elicits PN1 firing during cycles 1, 2, and 3, but PN2 fires during cycles 2, 3, and 4, providing a basis for discrimination. Redrawn after Stopfer et al. *[5.43].*

The significance of oscillatory information was explored by recourse to a well-established learning paradigm, the proboscis extension reflex (PER) of the honeybee *Apis mellifera*, which exhibits synchronous oscillations in an LFP. Repeatedly pairing an unattractive odourant with sucrose enables the bee to associate the two, and elicits PER in response to the odourant and conditioning stimulus (see below). Control bees (treated with saline) and trained with the aliphatic alcohols 1-hexanol or 1-octanol subsequently distinguished these compounds from alcohols with different chain lengths, and from geraniol, a structurally different molecule (monoterpene). In contrast, bees PCT-treated, immediately before conditioning, remembered the training stimulus and distinguished it from geraniol, but could not distinguish it from a structurally similar stimulus. Therefore, oscillatory data are a prerequisite for fine-tuned sensory perception. Furthermore, PCT treatment, immediately before conditioning and before testing, elicited the same failure. This indicates that abolition of oscillatory synchronization elicits olfactory ambiguity rather than an altered olfactory representation; the latter, would elicit a result corresponding to that from control bees. That bees distinguish a significantly different molecule in the absence of oscillations is attributed to the identities of the responding PNs, effectively promoting the concept of patterned responses from individual glomeruli as emblematic. When these closely resemble each other, synchronized oscillatory cycles might provide the ultimate discrimination [5.43, and references therein].

Neurotransmitters

In *M. sexta* some PIas, PNs, including some 25 of the dorsal antenno-cerebral tract, and LNs employ GABA. Throughout development of *M. sexta*, GABA levels increase gradually and continuously up to pupation, vanishing during early metamorphoses when the AL is being reconfigured. From pupal stage P6, GABA levels gradually increase such that 18 hr after eclosion the concentration is up to 50% higher. In hemimetabolous insects GABA levels increase uninterruptedly from the larval stage onwards. As a major inhibitory neurotransmitter of insects and other arthropods, GABA seems likely to be associated with inhibitory interneurons and motoneurons of *Manduca*. Thus, injecting GABA into the AL elicited hyperpolarization and inhibition of neuronal firing. This was mediated by increased conductance of chloride, and reversed by bicuculline and PCT, that serve as antagonists of $GABA_A$ subtype receptors in vertebrates [5.16].

However, antennal sensory fibres terminating in the neuropil are cholinergic. Levels of acetylcholine (ACh), choline acetyltransferase, and acetylcholinesterase (AChE) increase dramatically as sensory axons grow from the antenna through the antennal nerves, to the lobes that enlarge substantially. Early deantennation stunts this development and increases the concentration of acetylcholine and choline acetyltransferase. Binding assays with radiolabelled bungarotoxin, a specific ligand for acetylcholine receptors, confirmed their presence. Their development increased gradually throughout metamorphosis and was relatively independent of deantennation [5.14, and references therein]. AChE is detectable histochemically where ORNs arborize with the outer cap of glomeruli, but weaker staining obtains within the MGC. However, one glomerulus in the ventro-median region of the AL stains very strongly, indicating variability in ACh-mediated signalling within the AL [5.14, and references therein].

The biogenic amines serotonin, dopamine, octopamine and histamine all occur in the AL but in relatively few neurons confined to LNs and CFs. Immunoreactivity studies have confirmed the presence of serotonin in the ALs of all insect species studied to date, including *M. sexta* and *A. mellifera*. In the former species, the few serotonin-secreting cells exhibit large cell bodies and varicose terminals. The primary neurite extends through the AL without ramifying, after which it enters the IACT from where an axon extends to the contralateral AL, ramifying in all glomeruli. This neuronal type is established in the larva and continues without modifications through metamorphosis, during which serotonin may enhance neuronal growth. Serotonin increases the excitability of AL neurons in the adult stage with associated broadening of spikes and larger input resistance in neurons. Considerable variability exists between species, as in the locust neurons arborize in the AL and project to the ipsilateral AL [5.16, and references therein].

In regard to dopamine, CFs of moths and honeybee extend dopaminergic fibres to the AL to invest and penetrate the glomeruli. In the cockroach *Periplaneta americana*, all dopamine-immunostained, glomeruli-associated neurons immunoreact with GABA. Dopamine receptors have been identified in AL neurons of the pupal honeybee, suggesting a contribution by this transmitter to AL development. Applying dopamine to the adult PC diminishes access to olfactory memory, and applying it to the AL elevated the threshold for AL sensitivity to olfactory stimuli.

Octopaminergic neurons have also been identified by immunostaining in the ALs of, among other species, *M. sexta* and *A. mellifera*. From the circumoesophageal connective, pairs of fibres climb into each AL and terminate as arborizations with fine blebs. In both species, branches from these fibres innervate the calyces of the mbs, and in the bee, the lateral PC. Branch configuration is consistent with the octopaminergic neurons being ventral unpaired median (VUM) neurons.

Histamine immunostaining is associated in *M. sexta* with a pair of CFs arborizing in and ascending from the ventral midline of the mesothoracic ganglion (Meso), ramifying in the SOG, with each projecting into both ALs (Fig. 50). In contrast, in the honeybee some 35 local AL neurons expressed in small groups are histamine immunoreactive.

More than 60 neuropeptides occur in insects. Allatostatins, allatotropins, FMRFamide-related peptides (FaRPs), kinins and tachykinins have been associated with the AL, especially the LNs and to a lesser extent CFs. Members of the allatostatin family have been identified by immunostaining of LNs having cell bodies in the lateral cell cortex of *M. sexta*. Also in this species, many LNs immunostain for allatotropin; and neurons immunostainning for FaRPs constitute a subpopulation of LNs exhibiting GABA-immunoreactivity.

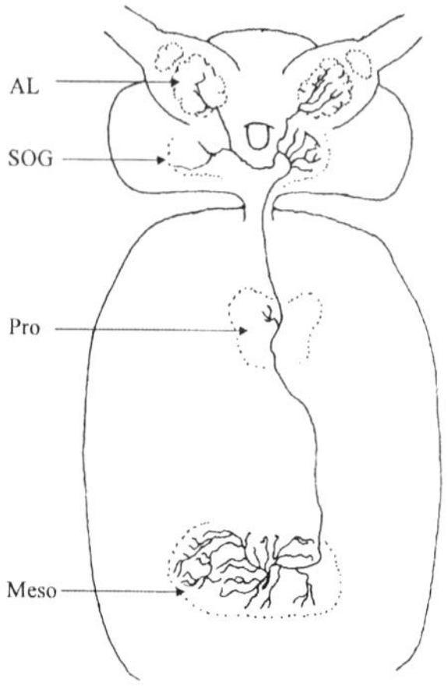

Figure 50. Schematic representation of a histamine-immunoreactive interneuron sending ascending projections from the ventral midline of the mesothoracic ganglion Meso, through the prothoracic ganglion Pro, to ramify in the suboesophageal ganglion SOG, and in both antennal lobes AL. Redrawn after Homberg and Müller [5.16].

The free radical gas nitrous oxide (NO) formed by NO synthase (NOS) acts as a signal molecule in the nervous system of invertebrates and vertebrates, without synaptic involvement. It diffuses from the production site to reach and activate guanylate cyclase, eliciting the production of cyclic guanidine monophosphate (cGMP). This regulates the action of protein kinases, diesterases, and specific ion channels (see Chapter 6). The presence of NOS was confirmed in the ALs and specifically in the glomeruli, but not in the antenno-cerebral tracts, of all insect species assayed; NOS expression in efferent projections of ORNs varied from species to species. Spherical glomeruli might compartmentalize the action of NO by confining it. Alternatively, NO diffusing from ORNs could activate individual glomeruli, whereas NO arising from LNs could modulate all glomeruli.

Physiological habituation in the honeybee, as judged by cessation of proboscis extension in response to repetitive chemosensory stimulation, but not in response to single stimuli, is compromised by inhibition of NOS in the AL; such inhibition in one AL affects habituation as mediated by only the antenna of that side. This is consistent with the independence established for each AL in processing signals of the corresponding side. NOS also functions in the formation of long-term memory in the bee [5.16, and references therein].

Memory

A fascinating question is: do insects remember previous chemical stimuli and if so to what extent? The honeybee is an insect of choice as its stereotyped and conditionable behaviour of proboscis extension in response to sucrose, provides a measure of learning and memory. This proboscis extension response (PER) is conditioned to an odour either by a single pairing of the odour with sucrose as applied to proboscis and antennae, or by repeated tests in which the odour and sucrose are paired once per test. The odour serves as the conditioned stimulus (CS) and sucrose is the unconditional stimulus (US). PER conditioning is fast, consistent with classical conditioning, and sensitive to tuning. Thus, the CS must always precede the US, shorter time intervals are more effective and are optimal at 1-3 s. Furthermore, PER conditioning is inhibited by unpaired presentation of CS and US, as the former is then associated with the absence of the latter. In addition, the fact that backward pairing inhibits conditioning suggests the operation of two opposing processes, one excitatory the other inhibitory, varying in their temporal dependency on CS and US. Layers of complexity are added by second-order conditioning and by blocking. In the former, a primary CS (CS1) may transfer its conditioning to a secondary CS (CS2) by forward pairing with it in the absence of the US. Blocking of the PER occurs when CS2 is presented jointly with CS1 to which the bee had been previously conditioned. So, the rich conditioning repertoire of the bee includes conventional or classical conditioning, inhibitory learning, second-order conditioning, and blocking [5.12, and references therein].

An understanding of associative learning requires a distinction between it and nonassociative learning or sensitization as elicited by the US: in eliciting the PER associative learning serves a releaser function; in stimulating the bee it dishabituates responses that have been habituated; it also decreases the threshold for subsequent stimuli (modulatory function); and in linking to the CS it serves a reinforcing function. Only neural components involved in associative learning are relevant to the present account.

Neural Elements

As in *M. sexta*, PNs convey chemical information in tracts from the AL to the lateral protocerebral lobe (lpl) and to the mbs. This comprises 170,000 LNs or Kenyon cells densely arranged in parallel. The axon from each forms two branches serving the α- and β-lobes respectively. The dendrites from each arborize in the calyx comprising lip, collar, and basal ring neuropiles associated with olfaction, vision, mechanoreception and contact chemoreception respectively. Although recordings are not yet available from a Kenyon cell during PER conditioning, an mb-extrinsic neuron designated PE1 has proved significant in the CS pathway. Receiving input at the base of the peduncle and α-lobe of the mb and thus from many Kenyon cells, it projects to the lpl. Responsive to odours in general it does not exhibit sensitization, but forward pairing of an odour with sucrose elicits a decreased response (conditioning) for a brief period (< 10 min).

In regard to the US pathway, a significant role is assigned to VUM neurons that connect the brain to the SOG. This receives from the antenna contact chemoreceptor axons that terminate in close proximity to the motor neurons governing proboscis movements. One such neuron, the VUM mx l, is sensitive to sucrose stimulation and its neurites arborize with the CS pathway in glomeruli, lpl, lips and basal rings of the mb calyces. Activity of this neuron could substitute for the US and mediate the conditioning or associative learning observed only in forward pairing: VUM mx l does not directly trigger proboscis extension. In summary, the sites essential for associative learning are located in the AL, mb calyces, and lpl, and are connected by two distinct pathways. The first or direct pathway connects the AL to the lpl, and the second or indirect pathway connects them through the mb [5.12]. Such data may serve as a paradigm for the better understanding of memory formation of pheromones and food odours.

Apparently, *Apis* has four separate phases of memory: (1) sensory storage lasting a few seconds; (2) a phase sensitive to the disruption of neural activity and lasting 7 min; (3) a phase lasting from three to 15 min after conditioning and sensitive only to CO_2 narcosis; and (4) long term storage of the signal beginning 15 min after conditioning. Phases (2) and (3) together represent short-term memory (STM) which is primarily non-associative. Subsequent evidence indicated that STM overlies a slower formation of associative median-term memory (MTM) lasting up to 12 hr. Long-term memory (LTM) is produced by multiple learning trials and lasts a lifetime. Formation of MTM is impaired by weak electrical stimulation of the brain and by cooling to 1-5°C through small metal probes. Use of the latter indicated that formation of STM and MTM progresses in parallel, with MTM sited in the mb calyces [5.10]. The specific mechanism would entail Kenyon cells responsive to release of octopamine by neurons such as VUM mx l [5.12].

A separate study has indicated a possible role for the ALs in memory formation in *Apis*. The volume of a specific glomerulus (T4-2(1)) increased significantly in precocious foragers and other workers exhibiting elevated levels of associative learning. LTM, formed by as few as three conditioning experiences, is insensitive to the above inducers of amnesia and the mb seems sufficient for its formation [5.41].

Biogenic amines, such as dopamine, significantly affect olfactory learning by modulating the function of olfactory centres of the honeybee brain [5.41]. Such amines could bind to receptors linked to G-proteins that activate a cascade of cyclic nucleotides.

REFERENCES

5.1 Anton, S. and Homberg, U. (1999) Antennal lobe structure. In Hansson, B.S. (ed.) *Insect Olfaction*. Springer, Berlin, pp. 97-124.

5.2 Arn, H., Städler, E. and Rauscher, S. (1975) The electroantennographic detector – a selective and sensitive tool in the gas chromatographic analysis of insect pheromones. *Z. Naturforsch.* **30**, 722-725.

5.3 Arnold, G., Masson, C. and Budharugsa, S. (1985) Comparative study of the antennal lobes and their afferent pathway in the worker bee and the drone (*Apis mellifera*). *Cell Tissue Res.* **242**, 593-605.

5.4 Baker, T.C. and Haynes, K.F. (1989) Field and laboratory electroantennographic measurements of plume structure correlated with oriental fruit moth behaviour. *Physiol. Ent.* **14**, 1-12.

5.5 Bjostad, L.B. and Roelofs, W.L. (1980) An inexpensive electronic device for measuring electroantennogram responses to sex pheromone components with a voltmeter. *Physiol. Ent.* **5**, 309-314.

5.6 Boeckh, J. (1962) Elektrophysiologische Untersuchungen an einzelnen Geruchsrezeptoren auf den Antennen des Totengräbers (*Necrophorus*, Coleoptera). *Z. vergl. Physiol.* **46**, 212-248.

5.7 de Kramer, J.J. (1985) The electrical circuitry of an olfactory sensillum in *Antheraea polyphemus*. *J. Neurosci.* **5**, 2484-2493.

5.8 de Kramer, J.J., Kaissling, K-E. and Keil, T. (1984) Passive electrical properties of insect olfactory sensilla may produce the biphasic shape of the spikes. *Chem. Sens.* **8**, 289-295.

5.9 Den Otter, C.J. (1977) Single sensillum responses in the male moth *Adoxophyes orana* (F.v.R.) to female sex pheromone components and their geometrical isomers. *J. comp. Physiol.* **121**, 205-222.

5.10 Erber, J., Masuhr, T. and Menzel, R. (1980) Localization of short term memory in the brain of the bee, *Apis mellifera*. *Physiol. Ent.* **5**, 343-358.

5.11 Färbert, P., Koch, U.T., Färbert, A. and Staten, R.T. (1997) Measuring pheromone concentrations in cotton fields with the EAG method. In Cardé, R.T. and Minks, A.K. (eds.) *Insect Pheromone Research: New Directions*. Chapman and Hall, New York, pp. 347-358.

5.12 Hammer, M. and Menzel, R. (1995) Learning and memory in the honeybee. *J. Neurosci.* **15**, 1617-1630.

5.13 Hansson, B.S. and Christensen, T.A. (1999) Functional characteristics of the antennal lobe. In Hansson, B.S. (ed.) *Insect Olfaction*. Springer, Berlin, pp. 125-163.

5.14 Hildebrand, J.G., Hall, L.M. and Osmond, B.C. (1979) Distribution of binding sites for ^{125}I-labeled α-bungarotoxin in normal and deafferented antennal lobes of *Manduca sexta*. *Proc. Natl. Acad. Sci.* **76**, 499-503.

5.15 Homberg, U., Christensen, T.A. and Hildebrand, J.G. (1989) Structure and function of the deuterocerebrum in insects. *Ann. Rev. Entomol.* **34**, 477-501.

5.16 Homberg, U. and Müller, U. (1999) Neuroactive substances in the antennal lobe. In Hansson, B.S. (ed.) *Insect Olfaction*. Springer, Berlin, pp. 181-206.

5.17 Kaissling, K-E. (1971) Insect Olfaction. In Beidler L.M. (ed.) *Handbook of Sensory Physiology* vol. IV. Springer, Berlin, pp. 351-431.

5.18 Kaissling, K-E. (1986) Chemo-electrical transduction in insect olfactory receptors. *Ann. Rev. Neurosci.* **9**, 121-145.
5.19 Kaissling, K-E. and Thorson, J. (1980) Insect olfactory sensilla: Structural, chemical and electrical aspects of the functional organisation. In Satelle, D.B., Hall, L.M. and Hildebrand, J.G. (eds.) *Receptors for Neurotransmitters, Hormones and Pheromones in Insects.* Elsevier/North Holland Biomedical Press, New York, pp. 261-282.
5.20 Kandel, E.R. and Schwartz, J.H. (1981) *Principles of Neural Science.* Elsevier/North Holland, New York.
5.21 Mankin, R.W. and Mayer, M.S. (1984) The insect antenna is not a molecular sieve. *Experientia* **40**, 1251-1252.
5.22 Matsumoto, S.G. and Hildebrand, J.G. (1981) Olfactory mechanisms in the moth *Manduca sexta*: response characteristics and morphology of central neurons in the antennal lobes. *Proc. R. Soc. Lond. ser. B* **213**, 249-277.
5.23 Mayer, M.S., Mankin, R.W. and Lemire, G.F. (1984) Quantitation of the insect electroantennogram: measurement of sensillar contributions, elimination of background potentials, and relationship to olfactory sensation. *J. Insect Physiol.* **30**, 757-763.
5.24 Mitchell, B.K., Smith, J.J.B., Albert, P.J. and Whitehead, A.T. (1990) Variance: A possible coding mechanism for gustatory sensilla on the labellum of the fleshfly *Sarcophaga bullata. J. exp. Biol.* **150**, 19-36.
5.25 Mustaparta, H. (1997) Olfactory coding mechanisms for pheromone and interspecific signal information in related moth species. In Cardé, R.T. and Minks, A.K. (eds.) *Insect Pheromone Research: New Directions.* Chapman and Hall, New York, pp. 144-163.
5.26 Nagai, T. (1981) Electroantennogram response gradient on the antenna of the European corn borer, *Ostrinia nubilalis. J. Insect Physiol.* **27**, 889-894.
5.27 O'Connell, R.J. (1975) Olfactory receptor responses to sex pheromone components in the redbanded leafroller moth. *J. Gen. Physiol.* **65**, 179-205.
5.28 Oland, L.A., Orr, G. and Tolbert, L.P. (1990) Construction of a protoglomerular template by olfactory axons initiates the formation of olfactory glomeruli in the insect brain. *J. Neurosci.* **10**, 2096-2112.
5.29 Roelofs, W.L. (1984) Electroantennogram assays: Rapid and convenient screening procedures for pheromones. In Hummel, H.E. and Miller, T.A. (eds.) *Techniques in Pheromone Research.* Springer, New York, pp. 131-159.
5.30 Rospars, J.P. and Hildebrand, J.G. (1992) Anatomical identification of glomeruli in the antennal lobes of the male sphinx moth *Manduca sexta. Cell Tissue Res.* **270**, 205-227.
5.31 Roys, C. (1954) Olfactory nerve potentials as a direct measure of chemoreception in insects. *Ann. New York Acad. Sci.* **58**, 250-255.
5.32 Rumbo, E.R. and Kaissling, K-E. (1989) Temporal resolution of odour pulses by three types of pheromone receptor cells in *Antheraea polyphemus. J. comp. Physiol.* **165**, 281-291.
5.33 Salecker, I. and Malun, D. (1999) Development of olfactory glomeruli. In Hansson, B. (ed.) *Insect Olfaction.* Springer, Berlin, pp. 208-242.
5.34 Sass, H. (1978) Olfactory receptors on the antenna of *Periplaneta americana*: response constellations that encode food odors. *J. comp. Physiol.* **128**, 227-233.
5.35 Schmidt, R.F. (1978) *Fundamentals of Sensory Physiology.* Springer, New York.

5.36 Schneider, D. (1957) Elektrophysiolgische Unterschungen von Chemo- und Mechanorezeptoren der Antenne des Seidenspinners *Bombyx mori. L. Z. Vergl. Physiol.* **40**, 8-41.
5.37 Schneiderman, A.M., Matsumoto, S.G. and Hildebrand, J.G. (1982) Trans-sexually grafted antennae influence development of sexually dimorphic neurones in moth brain. *Nature* **298**, 844-846.
5.38 Schoonhoven, L.M. (1987) What makes a caterpillar eat? The sensory code underlying feeding behavior. In Chapman, R.F., Bernays, E.A. and Stoffolano, J.G. (eds.) *Perspectives in Chemoreception and Behavior.* Springer, New York, pp. 69-97.
5.39 Schoonhoven, L.M. and Blom, F. (1988) Chemoreception and feeding behaviour in a caterpillar: towards a model of brain functioning in insects. *Entomol. Exp. Appl.* **49**, 123-129.
5.40 Selzer, R. (1981) The processing of a complex food odor by antennal olfactory receptors of *Periplaneta americana. J. comp. Physiol.* **144**, 509-519.
5.41 Sigg, D., Thompson, C.M. and Mercer, A.R. (1997) Activity dependent changes to the brain and behavior of the honeybee, *Apis mellifera* (L.). *J. Neurosci.* **17**, 7148-7156.
5.42 Steinbrecht, R.A. (1982) Electrophysiological assay of synthetic and natural sex pheromones in the African armyworm moth, *Spodoptera exempta. Entomol. Exp. Appl.* **32**, 13-22.
5.43 Stopfer, M., Wehr, M., Macleod, K. and Laurent, G. (1999) Neural dynamics, oscillatory synchronisation, and odour codes. In Hansson, B.S. (ed.) *Insect Olfaction.* Springer, Berlin, pp. 163-180.
5.44 Van der Pers, J.N.C. and Minks, A.K. (1997) Measuring pheromone dispersion in the field with the single sensillum recording technique. In Cardé, R.T. and Minks, A.K. (eds.) *Insect Pheromone Research: New Directions.* Chapman and Hall, New York, pp. 359-371.
5.45 Wadhams, L.I. (1982) Coupled gas chromatography-single cell recording: A new technique for use in the analysis of insect pheromones. *Z. Naturforsch.* **376**, 947-952.
5.46 Zufall, F., Stengl, M., Franke, C., Hildebrand, J.G. and Hatt, H. (1991) Ionic currents of cultured olfactory receptor neurons from antennae of male *Manduca sexta. J. Neurosci.* **11**, 956-965.

CHAPTER 6

BIOCHEMISTRY OF CHEMORECEPTION

There is some ambiguity in terminology affecting a key constituent of the biochemical chain of events that concludes with olfactory perception. The term sensillum, or chemosensory organ, is frequently used interchangeably with the term sensory receptor. In pharmacology, a receptor entails: (1) a macromolecule bearing sites having chemorecognitive properties for a specific natural endogenous molecule or for specific drugs; (2) site specificity on the receptor macromolecule for a particular endogenous molecule being genetically determined, with the receptor macromolecule having a genetically determined function; (3) binding of an agonist, whether an endogenous molecule or drug, causing a specific perturbation or change in state of the receptor macromolecule or its immediate environment or both, initiates a chain of events leading to a response; (4) initiation of a response by binding at a receptor site is not depending on the making or breaking of covalent bonds in the agonist [6.14]. It seems appropriate to reserve the use of the term receptor to such macromolecules, and to examine the role of such receptors in order to fully understand the chain of primary events in chemoreception. As it seems that all known receptors, whether for hormones, neurotransmitters, opioids, or for drugs, are sited in membranes, a brief account of membrane structure and function is relevant.

MEMBRANE STRUCTURE

The mosaic model, proposed in the mid 1960's by Singer and Wallach, involves globular protein embedded in and crossing a lipid bilayer core. Thus, the polar groups of lipid and protein are in direct contact with the aqueous surroundings, and the non-polar residues of both are sequestered in the heart of the structure away from the water (Fig. 51). This model is attractive in that it offers a potential route through the lipid bilayer by way of the transmembrane protein. Also, the model provides for only a fraction of its protein being bound to the membrane. This is supported by the observation that variations in ionic strength sufficient to disrupt ionic associations do not, in general, extensively dissociate membrane proteins.

The fluid-mosaic model, proposed in 1972, stressed dynamic aspects of membrane structure, and visualized globular proteins as 'icebergs' floating in a 'sea' of lipid bilayer, in addition to the globular extrinsic proteins already discussed. Mobility of globular proteins arose from the lipid bilayer, the hydrocarbon chains of which are largely highly mobile at physiological temperatures. At lower temperatures they exist as gels containing crystalline hydrocarbon regions (Fig. 51) [6.16, and references therein].

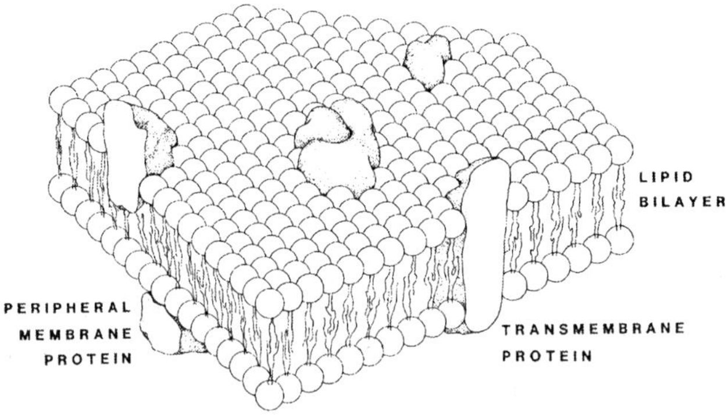

Figure 51. Diagrammatic representation of the mosaic model of membrane structure, indicating globular, integral proteins in a lipid matrix, of which the hydrophilic levels face the aqueous environment, and the hydrophobic tails are sequestered within the membrane. Redrawn after Singer and Nicholson [6.18].

RECEPTORS

As all receptors are macromolecules, with the common function of receiving a signal and transducing it to a form distinguishable by a receiving cell, they should have features in common. This has promoted common investigative approaches of which the most common is the use of a labelled form of the stimulus, or an analogue whether agonistic or antagonistic, as a probe to bind to the receptor. This distinguishes the receptor from other macromolecules in the membrane. Another common theme is reconstitution of the features of the functional receptor system that were lost during cellular disruption designed to free the receptor.

Receptor Binding Studies

A labelled probe, radiolabelled or fluorescent, may distinguish the receptor from other macromolecules. This requires the distinction of true receptor characteristics (specific binding sites) from cellular binding sites that do not reflect receptor properties (non-specific or non-receptor binding sites). Binding by a true receptor should exhibit at least the following properties: (1) high affinity for the ligand; (2) saturability; (3) reversibility; (4) tissue distribution consistent with organs on which the ligand is known to act; (5) relative binding affinities to ligand analogues consistent with their relative biological activities; and (6) a binding isotherm readily related to the concentration range over which the ligand is known to act [6.13, and references therein]. It may be said that few of any of the early investigations of an insect olfactory receptor were in a position to meet these requirements.

Specific Binding

Experimentally, binding constants are determined by incubating fixing concentrations of receptors (usually as tissue homogenates) with known concentrations of labelled ligand and measuring, after equilibrium, the specifically bound ligand concentrations. This is done by separating free ligand from the mixture by filtration or centrifugation, followed by counting the bound fraction in an automatic scintillation counter. Non-specific binding (NSB) is ascertained in the presence of a large excess (usually more than 50-fold) of unlabelled ('cold') hormone. It is presupposed that the excess 'cold' competes effectively at the true receptor site and displaces the labelled or 'hot' ligand. But some tracer is usually bound, perhaps to glassware or non-receptor protein, such that the 'cold' cannot displace it even at very high concentrations; this is the amount non-specifically bound. Subtracting NSB from total bound gives the amount specifically bound (SB). This distinction is very relevant to studies on proteins binding such hydrophobic molecules as pheromones) [6.3, 6.13, and references in both].

Receptor Families

Essentially, receptors may be classified in the following three molecular families: guanine nucleotide-binding protein (G-protein) coupled receptors; ligand-gated ion channels; and uptake transporters.

G-protein Coupled Receptors

These comprise monomeric proteins each containing seven transmembrane, hydrophobic regions. Ligand binding induces a conformational change in the receptor eliciting interaction between it and a guanosine triphosphate (GTP) - activated subunit of the G-protein in the membrane. The subunit activates an effector such as: adenylate cyclase, cyclic guanosine monophosphate (cGMP), phosphodiesterase, phospholipase C or A2; or interacts directly with an ion channel. The effector molecule modifies the synthesis of a second messenger such as cyclic adenosine monophosphate or cAMP, Ca^{++}, or ion balance of the cell, that either phosphorylates voltage-gated ion channels or directly alters conductance by the cell. An inevitable consequence of this rather elaborate pathway is a relatively slow response, in the region of several hundred ms (Fig. 52). G-protein coupled receptors cloned so far exceed 200 in number, with similarities consistent with a supergene family representing one of the largest in the mammalian genome.

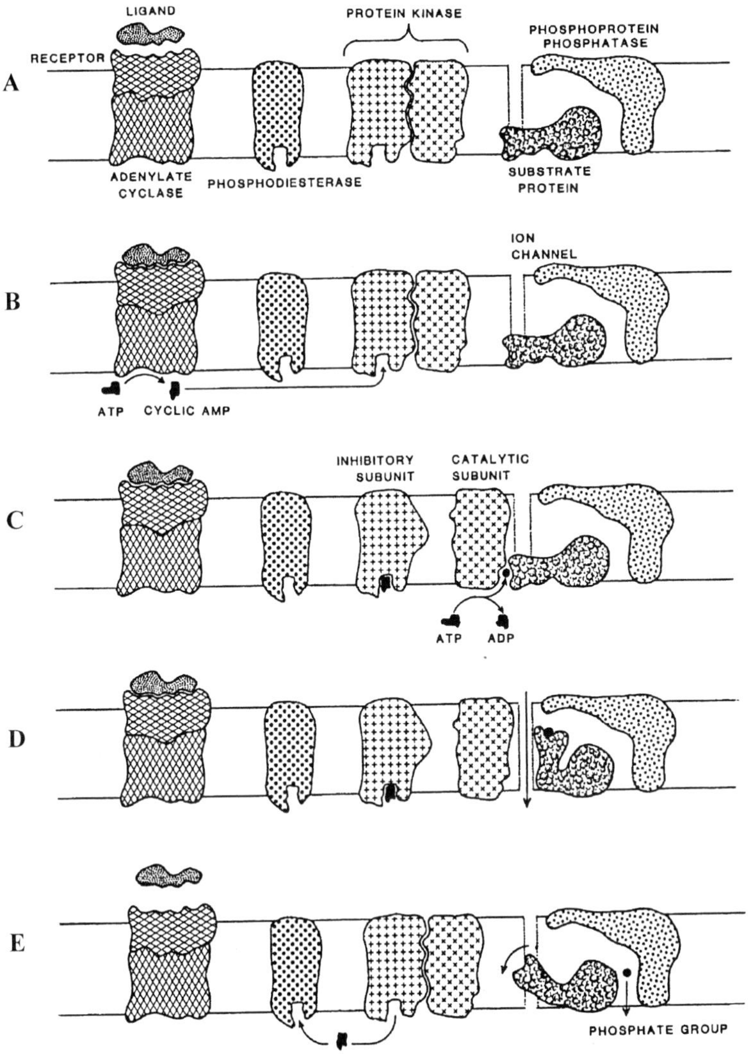

Figure 52. Generalized representation of an amplification system involving cyclic nucleotides. A) Relevant membrane components are: the receptor attached to adenylate cyclase; phosphodiesterase; a protein kinase comprising an inhibitor and catalytic subunit; a substrate protein blocking the ion channel; and a phosphoprotein phosphatase. Essentially: B) ligand-receptor binding activates adenylate cyclase to convert ATP into cAMP that binds to the inhibitory subunit of the protein kinase; C) it dissociates from the catalytic unit that consequently modifies ATP to ADP with release of one phosphate that binds to the substrate protein; D) such binding induces a conformational change that opens the gate allowing an ion flow; E) the phosphoprotein phosphatase frees the phosphate group returning the substrate protein to its original configuration, thus closing the ion gate.

Ligand-Gated Ion Channels

These are multimeric proteins assembled to form an ion channel. Ligand binding to one or several subunits opens the channel rapidly allowing ion flow that alters cell conductance. The number of subunits ranges from four to five with at least two identical. All subunits share sufficient sequence similarities with each other and with other ligand-gated ion channels, of which the best understood is that for acetylcholine, to form a supergene family. A distinguishing family feature is the rapid response that may be as fast as a fraction of a ms.

Uptake Transporters

In order to terminate neurotransmission some receptors sequester neurotransmitter from the synapse and transport it by a sodium-dependent mechanism across the neuronal or glial membrane. Most receptors serving as uptake transporters contain 12 transmembrane domains, share sequence similarities, but also exhibit less complexity as compared with the previously mentioned two supergene families [6.6, and references therein].

APPROACHES TO AN INSECT OLFACTORY RECEPTOR

A general model of the chain of primary events in insect chemoreception is as follows. Stimulus molecules contact the sensilla, penetrate one or more pores, and traverse the pore tubules to contact the dendritic surface of a sensory cell. Here, the stimulus molecule binds to a receptor molecule on the membrane inducing a conformational change that, directly or indirectly, opens channels through which ions diffuse.

It is possible on fundamental grounds to estimate of the weight of receptor in an insect sensillum. An idealized olfactory peg may be visualized as containing 50 dendritic branches, 5 μm and 0.1 μm in average length and width respectively. This gives a total dendritic surface area (2 rl x 50) of 157 μm^2. Assuming a large receptor density of 50,000 sites/μm^2, as in the acetylcholine receptor, gives 7.85×10^6 receptors. Assuming a molecular size of 40 kDa, also based on the acetylcholine receptor, gives an estimated receptor weight of 5.3×10^{-13}g/sensillum. In a species with 1,000 sensilla/antenna, the maximum yield from 1,000 specimens would therefore be 1.06×10^{-6} or 1 μg. Obviously, this estimate is based on certain assumptions but is probably a reasonable guide, at least to the order of magnitude of the likely yield. Against this framework, the prospect of purifying and characterizing a true olfactory receptor from an insect was viewed as daunting [6.15].

G-PROTEINS AND THE INSECT ANTENNA

Following significant progress with the isolation of vertebrate olfactory receptors [6.1], membrane-rich preparations from locust nervous tissue, and antennae of locust, the American cockroach *Periplaneta americana* and the moth *Antheraea polyphemus* were probed for the presence of G-proteins. The techniques were ADP-ribosylation as catalyzed by cholera toxin or by pertussis toxin, and by reactivity to an antiserum sensitive to a common sequence of the α-subunit of all G-proteins. Essentially, substrates were located for the toxins and the principal immunoreactivity resided in a 40 kDa entity, corresponding to the mobility of a G_{α}-subunit. Such preliminary data were relevant to, but did not establish, a role for G-proteins in insect olfaction [6.5].

Subsequently, crucial evidence indicated a role for second messengers in olfactory transduction in vertebrates and in insects. Stimulation of isolated olfactory cilia of rat by isomenthone, and of antennal homogenates of *P. americana* by the sex pheromone periplanone B, resulted in tissue-specific, dose-dependent, transient accumulations of only cAMP, and only inositol 1,4,5-triphosphate (IP_3) respectively. Furthermore, the time-course of recovery was dose-dependent, and the second messenger cascade in both systems seemed to be GTP-dependent; respective second messenger accumulation peaked in both systems at about 50 ms post-stimulation, consistent with published values for respective latencies [6.4].

In regard to the *P. americana*, coapplication of pheromone and GTP elicited a synergistic increase in IP_3 levels that was consistent with the involvement of G-proteins. This was supported by inhibition of IP_3 formation by the analogue GDP-β-S, established as an inhibitor of G protein activation. In contrast, IP_3 was rapidly induced by the hydrolyzing analogue GTP-γ-S, in the absence of pheromone [6.4].

This pathway may also mediate perception of general odours and plant volatiles, as IP_3-receptor antiserum labelled all ORNs of *Bombyx mori* and *A. pernyi*, and not merely the pheromone-sensitive ones. However, as axonal membranes, and apical microvilli of ancillary and epidermal cells were also labelled, such data are inconclusive. In *Drosophila melanogaster*, mRNA for the IP_3 receptor is present in large concentrations in the antenna and optic lobes. Mutations in the gene for the retinal degeneration protein B (*rdg*B), a Ca^{++} binding protein, delayed the recovery of olfactory sensitivity in the maxillary palp [6.19, and references therein].

Phospholipase C (PLC) catalyzes the formation of IP_3, a potential second messenger in the cascade. In *A. polyphemus*, PLC is present in much higher concentrations in antennae as compared with nerve and muscle tissue. GTP significantly increased PLC activity, and in the presence of odourant (2 μM bornyl acetate) this effect was double that of GTP alone. Attention was drawn to the highest IP_3 concentration occurring at the shortest measurable time after stimulation, followed by a rapid return to base level [6.2].

In 1990, a study of second messengers in male antennal extracts of *A. polyphemus*, in response to the stimulation by the female pheromone, confirmed that cAMP levels were unaffected. Although cGMP levels were elevated, this messenger was excluded for a primary role in transduction as: (1) increased cGMP levels elicited by pheromone were unrelated to amplitudes of the receptor potential; (2) such increases persisted for 10 min in whole insects and for 30 min in severed antennae, but the receptor potential recovers to half its resting value within 5 s after stimulus termination; (3) most tellingly, there was no increase in cGMP levels in sensory hairs containing dendrites of the ORNs; and (4) cGMP was not detected in isolated olfactory sensilla. Accordingly, the elevated cGMP levels observed in stimulated antennae were assigned to the somata of ORNs and/or to auxiliary cells. As cGMP was down-regulated at least 10 min after stimulation, and as such time intervals were also required for recovery of the electrophysiological response after adaptation, a correlation was proposed by which cGMP could affect the readiness of the ORN's subsequent response. It was emphasized that almost all isolated antennae exhibited a secondary cGMP increase five or more min after stimulation [6.25].

Another perspective was presented in 1992 by Hildebrand and colleagues [6.20] by drawing attention to: (1) elevated levels of IP_3 in insect ORNs within ms of stimulation by pheromones although cAMP levels were not increased; (2) extruded dendrites of moth ORNs contain cation channels that are pheromone dependent and mediated by second messengers such as diacylglicerol (DAG) and cGMP; (3) dendritic segments also contain cation channels that are second-messenger-governed (Ca^{++}, cGMP); (4) cultured ORNs from *Manduca sexta* contain both Ca^{++}-dependent cation channels and protein kinase C-dependent cation channels. These patch clamp data represent opening of the cation channels in response to pheromone diffusing from the pipette tip, and before it was puffed on to the cell soma ≥ 20 μm away. Under these conditions, latencies of < 100 ms were recorded [6.20]. Collectively, the available data led to a model of transduction entailing: G-protein activation followed by PLC stimulation; and by transient generation of IP_3, and DAG acting via protein kinases, as second messengers; followed by transient increase of Ca^{++} levels in the outer dendrite that opens the cation channels eliciting the receptor potential (Fig. 53).

Pheromone-Binding Proteins

One of the first investigations of an insect olfactory receptor was that in which a radiolabelled volatile emission of the female silk moth *A. pernyi*, was electrophoresed, apparently without allowing for NSB, with an eluant of washed male antennae. Radioactivity was confined to a male-specific protein, proposed as a sex pheromone [6.24]. A corresponding approach with *A. polyphemus* and its pheromone-bound protein species (15 kDa), designated the pheromone-binding protein (PBP), from only the male. In addition, the same pheromone bound to a protein of similar molecular size in antennal homogenates of males of the following moths: *A. pernyi*, *Hyalophora cecropia* and *M. sexta*. Further, the hydrophilic, vertebrate protein, bovine serum albumin (BSA), routinely used as a blocker of NSB in receptor assays, had an affinity only 10-fold weaker than PBP for the pheromone.

However, it was claimed that a very high concentration (ca 20 mM) of the PBP would result in most pheromone present being bound [6. 21].

Some subsequent investigations focussed on esterases of *A. polyphemus*, one present in only the male antenna, the other in cuticular tissues. Both degraded the pheromone but the latter was five-fold more potent than the former. Isolated from s. trichodea, the antennal esterase was estimated as capable of degrading pheromone at a rate exceeding 27 molecules/s. The pheromone half-life was estimated as 15 ms for pheromone concentrations within the range of 1 to 5 x 10^4 molecules/sensillum; the concentration of this esterase was estimated as 1 x 10^{-7}M. This study promoted the concept of the PBP serving as a carrier to convey a hydrophobic pheromone through the hydrophilic receptor lymph to the dendrite-bound receptor. Pheromone dissociation from this would be followed by inactivation by the esterase [6.23].

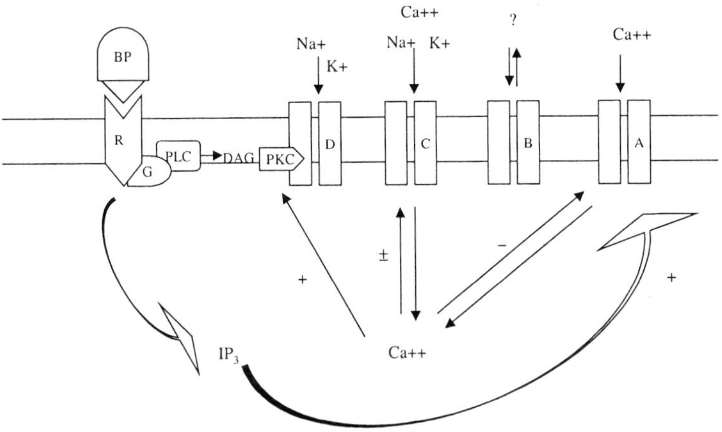

Figure 53. Schematic representation of the proposed, specific, second messenger cascade triggered by the binding protein-pheromone complex (BP), binding to a receptor (R), of the G-protein family (G), activating phospholipase C (PLC), to generate inositol triphosphate (IP_3) and diaglycerol (DAG). IP_3 elicits a Ca^{++} influx (A), that acts either through Ca^{++}-dependent, non-specific ion currents (C), or through protein kinase C (PKC)-dependent currents (D); ion flow through channel B is undetermined. Redrawn after Stengl et al. *[6.19].*

Localization of a PBP

A polyclonal antibody raised against PBP, isolated from antennal side branches of *A. polyphemus*, recognized that PBP and cross-reacted with PBPs that occur in both sexes of the gamma moth *Autographa gamma* and *B. mori*. In the latter species neither s. basiconica nor s. coeloconica were immunostained, but in s. trichodea the sensillum lymph was heavily stained. Of the auxiliary cells, the tormogen and trichogen cells have a similar ultrastructure and exhibited similar degrees of staining of electron-dense granules, especially near their apical cell membranes that border the sensillum-lymph cavity. In the receptor cells, neither the endoplasmic reticulum, Golgi apparatus, nor the cytoplasm of the outer dendritic and ciliary segments were labelled. Granules, of a different size from those in the tormogen and trichogen cells, were labelled in the inner dendritic segments and somata [6.17].

Ultimately, two principal hypotheses emerged linking PBP and esterases in transduction. One claimed that pheromone is transported through the pore tubules to the neuronal receptors. After binding to these the pheromone is quickly inactivated by binding to the PBP and metabolized by an esterase. The other claimed that the PBP transports the pheromone to the membrane-bound receptors protecting them from aggressive degradation by the sensillar esterase. After transduction, and to enable the moth to distinguish another pulse of pheromone, the pheromone is susceptible to the esterase. The combined action of the aggressive esterase and a low affinity carrier PBP could enable the regulation over six orders of magnitude of the amount of pheromone reaching the receptor [6.21].

Isolation and Characterization of a PBP

The isolation and characterization of insect PBP was subsequently pursued using incubations of male sensillar proteins with radiolabelled and photoaffinity labelled pheromone of *A. polyphemus*.

A second extracellular protein in only the male antenna of the same species is a pheromone esterase (55 kDa), 10^4 less concentrated then the PBP or equivalent to 0.01 % total soluble protein of a sensillum. Aggressively active against the acetate constituent of the pheromone it engenders a pheromone half-life of 15 ms equivalent to degrading 27 pheromone molecules/s [6.22 and references therein]. However, it also occurs in cuticular tissues in a form five-fold more potent than that in the antenna [6.23].

Sequencing PBP

Genomic DNA was extracted and digested with restriction enzymes. The resulting fragments were cloned, and screened using a radiolabelled probe (1.2 kb) from cloned *A. pernyi* PBP (APR-1). The possibility of a diversity of PBPs was specifically addressed by hybridizing genomic DNA fragments from different restriction enzymes, with labelled probe DNA. Then hybridized fractions were used to construct a partial genomic library in a vector which was rescreened. Hybridizing clones were subcloned, transformed into another vector, and purified.

The complete nucleotide sequence was derived, but of particular interest was the evidence for a second gene with a EcoRI restriction site; the APR-1 gene did not contain one. Designated APR-2, this was clearly a separate gene but with such a strong similarity (85 %) to APR-1 as to be homologous. Significant inclusions were represented by: a section of 21 amino acids following the initiator methionine and consistent with a signal peptide; and similarities of size, cysteines and hydrophobicity with a PBP. When compared with PBP nucleotide sequences from *A. polyphemus* and *M. sexta*, six cysteines, including three disulphide linkages, were conserved at identical positions in all four proteins, and APR-2 homology with *M. sexta* (70 %) was greater than that of APR-1 (64 %). Finally, a comparison of sequence substitutions between APR-1 and APR-2 revealed that these occurred in clusters and in hydrophobic regions. Such data established unequivocally the existence of PBP-microdiversity within a species and, given its expression within hydrophobic domains, strongly suggested an association between the existence of two PBPs and the pheromone constituents, aldehyde and acetate respectively. Accordingly, the role of PBPs might incorporate a discriminatory function [6.9].

Probing Cloned PBPs

In a further significant step the cDNA encoding a PBP from *A. pernyi* was integrated into the genome of the *Autographa californica* nuclear polyhedrosis virus (AcNPV) such that the protein was secreted into the culture medium (1-2 mg/l) by the baculovirus-infected cells. This rate of expression overcame the difficulties outlined above regarding the paucity of receptor-related material in insect antennal extracts [6.10]. Subsequently, most (>100 mg/l) of the recombinant form of the PBP (rPBP) was discovered in an insoluble, denatured state in inclusion bodies. Solubilization of these followed by reduction and unfolding in the reducing agent dithiothreitol (DTT), and by refolding in a cysteine-cysteine mixture, afforded a crude protein that was purified by isoelectric focussing [6.9].

Thereafter, photoaffinity- and radiolabelled-analogues of the tetradecadienyl and hexadecadienyl acetate constituents respectively of the *A. polyphemus* pheromone were used to probe the rPBP of that species (APO-3 PBP). Conditions were claimed to optimize covalent attachment of the label without NSB, although such binding does not seem to have been experimentally quantified. When the principal radiolabelled moiety of the bound rPBP was microsequenced, the Asp^{39}-Lys^{58} region contained the covalently bound site. Significantly, this stretch corresponds to the principal hydrophobic region represented in all antennal binding proteins sequenced to date, and configures as a helix-sheet-helix pattern. Strongly conserved in general odourant binding proteins (GOBPs), that are presumed to recognize and transport common odourants such as plant volatiles, this region is rather more variable in PBPs [6.8]. This might facilitate discrimination of different constituents of the pheromone bouquet.

The purified region of the rPBP that bound to the hexadecadienyl probe was readily cleaved to: a radiolabelled, hydrophilic and helical region appropriate for binding to the acetyl functional group of the pheromone; and a hydrophobic β-sheet region providing a domain for binding to the lipophilic, alkene chain of the pheromone backbone. Corresponding use of the tetrahexadecadienyl probe indicated that the major labelled site was Asp^{21}-Lys^{38} or adjoining (Asp^{39}-Lys^{58}) for the 6*E*,11*Z* probe; minor sites for the 4*E*,9*Z* probe were Asp^{39}-Lys^{58} and Leu^{125}-Val^{142}. Such data prompted a model for pheromone binding by which the acetate carbonyl forms a hydrogen bond with the protonated guanidium group of Arg^{46}, fixing the pheromone head in the helical, hydrophilic stretch of the protein, and linking the pheromone backbone with the hydrophobic domain of the site. The absence of two methylene groups renders, the 4*E*,9*Z* pheromone 2.6Å shorter, allowing two possible binding interactions: binding as before to the Arg^{46} but this would not allow optimal interaction by the hydrocarbon skeleton with the site; or hydrogen binding with Asp^{32} [6.8, 6.14, and references in both].

Improved Binding Assay

The generation of recombinant protein solved the constraint of a paucity of protein. A further difficulty is designing a receptor-binding assay to deal with the extreme hydrophobicity of pheromone molecules, with attendant extensive NSB to vials and tubes. The use of 'helpers' such as BSA has overcome this problem in other studies but the use of neither albumins, polyethylene glycol nor silanization protocols proved satisfactory with pheromones. In the event, coating reaction tubes with 1-dodecanol solved the problem by reversing the functional property of the plastic from hydrophobic to amphiphilic. Apparently, 1-dodecanol coated the tube with the hydroxy functional group extending into the aqueous phase. Furthermore, introducing PBP removed radiolabelled pheromone from the mix, apparently through binding of the PBP-pheromone complex to the amphiphilic coating. Such removal of bound pheromone provides a satisfactory footing for a binding assay, with bound material quantifiable either through displacement from PBP by a competitor, or by difference between counts in solution, before and after pheromone introduction (total binding). In conjunction with the use of recombinant protein, this step opens the way to satisfactory binding studies with a range of hydrophobic ligand-receptor interactions. In practice 1-dodecanol gave more reproducible results, and this coating was employed to quantify pheromone binding affinities with three recombinant PBP's, APR-1 and –2 from *Antheraea pernyi*, and APO l-3 from *A. polyphemus* [6.14].

This may be the first such investigation in which the time course, temperature-dependence, saturability, competitive displacement, and Scatchard analysis were all achieved. Binding was slightly higher at 4° than at 25° C; achieved half-maximal and maximal values after 15 min and 45 min respectively; was fully saturable; and the Scatchard plot with APO-3 indicated a single binding site with K_D (dissociation constant) values of 6.64 and 21.0 μM for 6*E*,11*Z*-16:Ac and 4*E*,9*Z*-14:Ac respectively. Corresponding results for APR-1 and –2, the rPBPs from *A. pernyi* were: for 6*E*,11*Z*-16:Ac, K_D of 1.83 and 11.2 μM respectively; and for 4*E*,9*Z*-14:Ac, 29.4 and 3.75 μM respectively. In other words, each rPBP might distinguish between each of the two pheromones indicating discriminatory properties for PBPs [6.14, and references therein] whether as primary molecular filters, carriers, or as scavengers. Nevertheless, the significance of PBPs in transduction is not universally accepted.

A Different Perspective

In the first instance, dendrites of sensilla trichodea of *A. polyphemus* were perfused with solutions of pheromone in Ringer either with or without PBP. The principal result was that addition of either PBP or BSA shifted the electrophysiological dose-response curve some two decades towards lower pheromone concentrations. The proposed explanation was that both proteins facilitated complete solubilization of the pheromone in the electrode, by decreasing its binding to the glass wall [6.18, and references therein]. As mentioned above, BSA is routinely incorporated in many binding assays to block ligand-glass non-specific interactions.

In the second instance, a primary role for PBP in transduction was inconsistent with single-channel records from ORNs of the male *M. sexta*. Such cultures should, of course, be free of PBPs as extracellular entities, although the remote possibility that PBPs were produced *in vitro*, and were also resistant to removal by superfusion, could not be ruled out. This meticulous study merits detailed consideration. Cultured ORNs were verified as authentic, female pheromone (bombykal)-responsive neurons by immunoreactivity with monoclonal male-olfactory-specific antibody (MOSA). Such cells reacted specifically to bombykal or to extracts of the female pheromone gland (calibrated in bombykal equivalents). Concentrations of 1 ng/ml in patch clamp experiments in the cell-attached configuration (to prevent loss of putative second messengers), elicited a large, inward, depolarizing current of 35-60 pA, that was not elicited by solvent alone, nor by the monoterpenoid plant volatile, citral. Coapplication of PBP was not a prerequisite for such pheromone-specific reactions. Furthermore, PBP and BSA could interchangeably and in high concentrations decrease pheromone concentrations, as judged by reactivation of cation channels. Such data indicate that PBP is neither necessary nor sufficient to elicit receptor-mediated responses, and rule it out as essential for presenting the pheromone to the receptor [6.20].

Scarab Beetles

Unusually, female beetles respond behaviourally to their own sex pheromone leading to the expectation that males and females would exhibit comparable receptor features. This was fulfilled in antenna-specific proteins with a high degree of conservation within N-terminal sequences expressed in various scarab species. In respect of putative PBPs, comparing the 20 bases of the N-terminal sequences indicated low homologies between species employing unrelated structures. Such a comparison would not be feasible for Lepidoptera as, with the exception of disparlure, all their pheromones are long chain hydrophobic molecules.

Furthermore, scarab pheromones provided a test of the 'filtering' capacities of a PBP. The Osaka beetle *Anomala osakana* produces and responds only to the (S)-stereoisomer of japonilure; the Japanese beetle *Popillia japonica* produces and responds to the (R)-stereoisomer that completely inhibits the efficacy of the (S)stereoisomer. The PBP isolated from *A. osakana* bound both enantiomers and apparently with a similar affinity, thus denying a discriminatory role of this entity. However, it was recognized that the binding of each enantiomer to the PBP could elicit a bound complex with different respective conformations, appropriate to bind to different and specific receptor neurons. Nevertheless, the concept of discrimination by the PBP was not sustained in that system [6.11].

General Odourant Binding Proteins and Drosophila

The amenability of *Drosophila melanogaster* to genetic analysis, indicated this as a model species for the investigation of genes expressing putative olfactory receptor proteins. The olfactory sensilla are concentrated on the third antennal segment and on the maxillary palp; the former are represented by s. trichodea, s. coeloconica, large and small s. basiconica, each with a defined odour spectrum and a particular zonal distribution; maxillary palp sensilla are some 60 s. basiconica, each with two ORNs comprising six response spectra, widely distributed on the palp surface. Six thousand antennal sensilla (unsegregated) were used as a source of RNA to isolate antennal-specific cDNAs by subtractive hybridization. Essentially RNA from heads devoid of antennae and mouth parts was used to create one cDNA library; RNA from only the third antennal segments served to create another; and the third was derived by subtraction to yield the antennal library. The antennal-specific cDNA clones resolved into five classes (B, C, D, E and F); members of each hybridized with each other but with clones of another class. One representative of B corresponded to the OS9 gene identified in an earlier study of olfactory-specific genes. Putative full-length DNAs were derived from classes C, D, and F and hybridization indicated the presence of four genes: OS-E and OS-F with sequence similarity to PBPs, and OS-C and OS-D as novel genes with no similarity to any known proteins.

Specifically, OS-F exhibited 29 % identity with APR-1, the PBP from *A. pernyi* and OS-F contained six cysteine residues each aligning with corresponding residues in APR-1. OS-F would encode a 14.1 kDa protein which is relatively close to the 15.8 kDa species reported as APR-1. OS-E exhibits 72 % identity to OS-F, 26 % identity to APR-1, and would encode a protein of 14.4 kDa. In common with OS-F and APR-1, its six cysteine residues align with those of APR-2 and APR-10 from *A. pernyi*; the PBP from *A. polyphemus*; PBPA, PBPB, GOBP1 and GOBP2 from *M. sexta*; and with the B1 and B2 proteins from the tubular male accessory sex glands of the yellow mealworm *Tenebrio molitor*. The inclusion of the latter means that the presumed PBP/GOBP family extends beyond the olfactory system, indicating a rather general transporter function for such proteins [6.12].

It was possible to localize by *in situ* RNA hybridization the expression of the four genes in regions of the third antennal segment. OS-F and OS-E are expressed in the ventro-lateral region that is consistent with the presence of s. trichodea. OS-D was expressed in the sacculus, a chamber lined by up to 40 s. coeloconica and fewer of other types. OS-C exhibited a similar pattern to that of OS-D, with additional patches [6.12]. Of course, none of the foregoing data establishes the binding affinity or specificity of any of the four produced proteins. OBPs belong to the lipocalin family of proteins, that includes many carrier proteins in a broad range of tissues. This is consistent with a transportation function for OBPs, but whether this occurs before or after transduction is problematical.

The situation was significantly advanced in 1999, by a screening of the sequence database of the *Drosophila* genome, for genes encoding G-protein-coupled receptors (GPCRs) having seven transmembrane domains; this was prompted by published data implicating such proteins in nematode (*Caenorhabditis elegans*), insect, and vertebrate olfaction. Essentially, the genome was searched, by a computer algorithm, for open reading frames (ORFs) in genes serving as candidates for GPCRs. Selection on the basis of similarity to *Drosophila* codon usage tables afforded 34 candidate ORFs, eight of which were eliminated as they matched genes of known function. Further selection based on the presence of coding for additional transmembrane domains, appropriate for splicing to the original ORFs, yielded 16 genes designated *dor* (*Drosophila* olfactory receptor). Thirteen of these were expressed after reverse transcriptase-polymerase chain reaction (RT-PCR) in the antenna, maxillary palp, or both.

This is a novel gene family with relatively little (20-30 %) sequence identity between subfamilies, with the greatest divergence in the extracellular loop and a transmembrane domain, and with more members expressed in the antenna than in the palp. Hybridization of some genes to RNA in adult *Drosophila* labelled only ORNs, and some were restricted to subsets of these. Furthermore, expression of *dor* genes is subject to regional regulation especially on the antennal surface. There is also evidence for: temporal regulation as the expression of one gene coincides with the growth of the antennal nerve; regulation of expression levels in various ORNs [6.7, and references therein]. Also in 1999, a separate similar study with *Drosophila* confirmed those results [6.24].

These significant advances open the way to the production by recombinant methods of ORN membrane entities suitable for receptor binding assays, and thus to the authentication of a true olfactory receptor.

REFERENCES

6.1 Anholt, R.R.H., Aebi, U. and Snyder, S.H. (1986) A partially purified preparation of isolated chemosensory cilia from the olfactory epithelium of the bullfrog, *Rana catesbiana*. *J. Neurosci.* **6**, 1962-1969.

6.2 Boekhoff, I., Strotmann, S., Raming, K., Tareilus, E. and Breer, H. (1990) Odorant-sensitive phospholipase C in insect antennae. *Cell. Sig.* **2**, 49-56.

6.3 Breer, H. (1997) Molecular mechanisms of pheromone reception in insect antennae. In Cardé, R.T. and Minks, A.K. (eds.) *Insect Pheromone Research: New Directions*. Chapman and Hall, New York, pp. 115-130.

6.4 Breer, H., Boekhoff, I. and Tareilus, E. (1990) Rapid kinetics of second messenger formation in olfactory transduction. *Nature* **345**, 65-68.

6.5 Breer, H., Raming, K. and Boekhoff, I. (1988) G-Proteins in the antennae of insects. *Naturwiss.* **75**, 627.

6.6 Civelli, O. (1995) The neuroreceptors: one key for many locks. *J. Rec. Sig. Trans. Res.* **15**, 161-172.

6.7 Clyne, P., Warr, C.G., Freeman, M.R., Lessing, D., Kim, J. and Carlson, J.R. (1999) A novel family of divergent seven-transmembrane proteins: candidate odorant receptors in *Drosophila*. *Neuron* **22**, 327-338.

6.8 Du, G., Ng, C.-S. and Prestwich, G.D. (1994) Odorant binding by a pheromone-binding protein: Active site mapping by photoaffinity labeling. *Biochemistry* **33**, 4812-4819.

6.9 Krieger, J., Raming, K. and Breer, H. (1991) Cloning of genomic and complementary DNA encoding insect pheromone-binding proteins: evidence for microdiversity. *Biochim. Biophys. Acta* **1088**, 277-284.

6.10 Krieger, J., Raming, K., Prestwich, G.D., Frith, D., Stabel, S. and Breer, H. (1992) Expression of a pheromone binding protein in insect cells using a baculovirus vector. *Eur. J. Biochem.* **203**, 161-166.

6.11 Leal, W.S. (1999) Scarab beetles. In Hardie, J. and Minks, A.K. (eds.) *Pheromones of Non-Lepidopteran Insects Associated with Agricultural Plants*. CABI, Wallingford, pp. 51-68.

6.12 McKenna, M.P., Hekmat-Scafe, D.S., Gaines, P. and Carlson, J.R. (1994) Putative *Drosophila* pheromone-binding proteins expressed in a subregion of the olfactory system. *J. Biol. Chem.* **269**, 16340-16347.

6.13 O'Brien, R.D. (ed.) (1980) *The Receptors: A Comprehensive Treatise*. Plenum Press, New York.

6.14 Prestwich, G.D. and Du, G. (1997) Pheromone-binding proteins, pheromone recognition, and signal transduction in moth olfaction. In Cardé, R.T. and Minks, A.K. (eds.) *Insect Pheromone Research: New Directions*. Chapman and Hall, New York, pp. 131-143.

6.15 Ryan, M.F. and Daly, P.J. (1978) Plant constituents and elucidation of primary elements of insect chemoreception. *Entomol. Exp. Appl.* **24**, 663-665.

6.16 Singer, S.J. and Nicolson, G.L. (1972) The fluid mosaic model of the structure of cell membranes. *Science* **175**, 720-731.

6.17 Steinbrecht, R.A., Ozaki, M. and Ziegelberger, G. (1992) Immunocytological localization of pheromone-binding protein in moth antennae. *Cell Tissue Res.* **270**, 287-302.

6.18 Stengl, M., Hatt, H. and Breer, H. 1992. Peripheral processes in insect olfaction. *Ann. Rev. Physiol.* **54**, 665-681.

6.19 Stengl, M., Ziegelberger, G., Boekhoff, I. and Krieger, J. (1999) Perireceptor events and transduction mechanisms in insect olfaction. In Hansson, B. (ed.) *Insect Olfaction.* Springer, Berlin, pp. 49-66.

6.20 Stengl, M., Zufall, F., Hatt, H. and Hildebrand, J.G. (1992) Olfactory receptor neurons from antennae of developing male *Manduca sexta* respond to components of the species-specific sex pheromone *in vitro. J. Neurosci.* **12**, 2523-2531.

6.21 Van den Berg, M.J. and Ziegelberger, G. (1991) On the function of the pheromone binding protein in the olfactory hairs of *Antheraea polyphemus. J. Insect Physiol.* **37**, 79-85.

6.22 Vogt, R.G. and Riddiford, L.M. (1981) Pheromone binding and inactivation by moth antennae. *Nature* **293**, 161-163.

6.23 Vogt, R.G., Riddiford, L.M. and Prestwich, G.D. (1985) Kinetic properties of a sex pheromone-degrading enzyme: The sensillar esterase of *Antheraea polyphemus. Proc. Natl. Acad. Sci.* **82**, 8827-8831.

6.24 Vosshall, L.B., Amrein, H., Morozov, P.S., Rzhetsky, A. and Axel, R. (1999) A spatial map of olfactory receptor expression in the *Drosophila* antenna. *Cell* **96**, 725-736.

6.25 Ziegelberger, G., Van den Berg, M.J., Kaissling, K.-E., Klumpp, S. and Schultz, J.E. (1990) Cyclic GMP levels and guanylate cyclase activity in pheromone-sensitive antennae of the silk moths *Antheraea polyphemus* and *Bombyx mori. J. Neurosci.* **10**, 1217-1225.

PART II

APPLIED ASPECTS

CHAPTER 7

PLANT CHEMICALS IN PEST CONTROL

At least two thousand plant species are known to contain toxic principles effective against insects [7.1, 7.14, 7.23, 7.24, 7.25, 7.26, 7.40], and terrestrial plants, encountered by more insects and microorganisms than aquatic ones, produce a broader range of defensive compounds [7.33]; several-fold that number of active compounds have been isolated and identified. The enhanced power of current analytical techniques, referred to in Chapter 2, might prompt a reanalysis of some species for potent, novel, compounds present in very small quantities. The very large number of compounds already reported as actives necessitates the use of a threshold level of activity as a basis for relevance and inclusion here. Given that many commercially-produced insecticides are active at levels of 1-10 ppm and that a 10-fold improvement on activity of lead compounds is not unusual, then a threshold of approximately 100 ppm seems reasonable. Nevertheless, given that 75 % of the planet's food is produced by Third World growers, not all of whom can afford commercial insecticides, then plant species affording usable active extracts also merit inclusion.

NATURAL INSECTICIDES AND GROWTH INHIBITORS

These two effects are considered together as many compounds elicit both.

Rotenoids

Leguminous plants containing rotenone (Fig. 54) have long been used as fish poison: macerated plants thrown into water paralyze the fish that subsequently float to the surface. In 1848, Oxley suggested tuba root as effective against leaf-eating caterpillars, and in 1877 it was reportedly used in China as an insecticide. The active chemical ingredient isolated in 1982 was named nicoulene. The name rotenone was given in 1902 by Nagai to an identical compound isolated from *Derris*. In South America and in Africa such closely related plant genera as *Lonchocarpus*, *Tephrosia*, and *Mundulea* were similarly used.

Rotenone has low vertebrate toxicity as oral administration to rat, guinea pig and chicken, eliciting LD_{50} (dose lethal to 50%) values of 132, 200 and 996 ppm respectively. LD_{50} values for various insect species are: milkweed bug *Oncopeltus fasciatus* 25 ppm; American cockroach *Periplaneta americana* 2,000 ppm; Japanese beetle *Popillia japonica* 22 ppm all by topical application; honeybee *Apis mellifera*, 3 ppm, by ingestion. By the early 1950's, more than 7 million pounds of Leguminosae roots (*Derris*, *Lonchocarpus* and *Tephrosia* spp.) were imported annually to the U.S.A. As recently as 1975 about 1.5 million pounds (ca 680 thousand kg) were used in the U.S.A. for pest control in the home and garden, and to control animal ectoparasites. Although active at low concentrations, the poor water solubility of rotenone is a factor limiting its usefulness in crop protection [7.12, and references therein].

The active principles of derris are extracted by organic solvents from dried, ground-up roots. Alternatively, the still popular "derris dust" is made by mixing the ground roots with a clay diluent. Rotenone itself forms a yellow laevorotatory crystals that dissolve in many organic solvents but are insoluble in water. Rotenoid insecticides serve to control leaf-chewing beetles and caterpillars in the garden, where toxic residues are contraindicated; degraded by light and air they do not leave residues [7.12]. Further, rotenoids from *Tephrosia* spp. incorporated with equal amounts of pyrethrins into mosquito coils were the most efficacious in terms of mortality, deterrency and feeding inhibition of a series of combinations assessed against mosquitoes in Tanzania [7.46]. Rotenone is not a nerve poison but elicits slow paralysis probably by inhibiting electron transport in mitochondria, in turn blocking nerve conduction. In isolated mitochondria, it does not affect succinate dehydrogenase, ubiquinone or the cytochromes, so it is presumed to disrupt electron transport in the region of flavin macronucleotide (FMN) [7.20].

Tobacco Alkaloids

Three hundred years ago tobacco extract was used as a plant spray in parts of Europe. Throughout the 18th century, tobacco was employed as an insecticide as an aqueous extract, or as a dust. For example, water extracts of tobacco were used to control plum curculio *Conotrachelus nenuphar*, on nectarine trees in 1746 [7.41].

Nicotine has been isolated from at least 18 species of *Nicotiana*, of which *N. tabacum* and *N. rustica* are the most common. *N. rustica*, containing up to 18 % nicotine compared to almost 6 % for *N. tabacum*, is cultivated specifically for extraction of insecticidal nicotine. LD_{50} values for nicotine topically applied to common insects are: *P. americana*, 650 ppm; squash bug *Anasa tristis*, 350 ppm; silk worm *Bombyx mori*, 4 ppm; and *A. mellifera*, 315 ppm. Nicotine is highly toxic to mammals: by oral administration, LD_{50} values of 50-60 ppm and 24 ppm against rat and mouse respectively; by dermal application, the LD_{50} value was 50 ppm against rabbit. Accidental deaths, especially following greenhouse fumigation, and suicides led to constraints on pesticidal nicotine usage in the U.S.A. [7.26].

Nornicotine is also found in *N. tabacum* and in many other *Nicotiana* species, especially *N. glutinosa* and *N. sylvestris*, and in the Australian plant, *Duboisia hopwoodii.* Insecticidal activity comparable to that of nicotine is exhibited by the bipyridyl derivative neonicotine or anabasine. Extracts of the weed *Anabasis aphylla,* that grows in Asia and Northern Africa, contain anabasine sulphate. Tested against bean aphid *Aphis fabae,* 0.12 ppm killed 85 % as compared with 1 ppm nicotine sulphate killing 68 %. Moreover, anabasine sulphate is 5-10-fold more toxic than nicotine sulphate to apple aphid *A. rumicis,* but nicotine was 2.6 times more toxic to the mosquito larvae *Culex pipiens* and *C. territans*; LD_{50} values were 90, 250 and 450 ppm for nicotine, anabasine and methyl anabasine respectively against mosquito larvae. Nevertheless, the use of insecticidal alkaloids, about 5 million pounds (ca 2.3 billion kg) p.a. until the mid-1900's, has dropped to about 1.25 million pounds (ca 560 thousand kg) of nicotine sulphate and 150,000 pounds (68 thousand kg) of nicotine alkaloid because of high production costs, disagreeable odour, extreme toxicity to mammals, and limited insecticidal activity. Anabasine was in commercial use in the USSR as recently as 1982. Nicotine disrupts neuronal transmission apparently by mimicking some properties of acetylcholine in binding to the acetylcholine receptor [7.38, and references therein].

Other Alkaloids

Many phytoalkaloids (about 300 known) are dissolved as cations in plant sap and following its evaporation, they react with available organic acid to form salts. These are deposited subsequently in vacuoles rather than in the protoplasm. Moreover, they tend to accumulate in peripheral parts that can be shed (barks, leaves, fruits) (for general properties of alkaloids see Chapter 2).

The stems and roots of *Ryania speciosa* (Flacourtiaceae) from South America afford an alkaloid ryanodine, (Fig. 54) less potent but more stable than rotenone and nicotine; additional ryanoids have been identified all apparently affecting the channel releasing Ca^{++} into muscle [7.41]. Methylxanthines including caffeine and theophylline, occur in the berries, seeds, and leaves of tea, coffee, cocoa, kola, and other plants [7.48]. The synthetic compound 3-isobutylmethylxanthine killed mosquito larvae, *Culex* sp., with an ED_{50} (medium effective dose) of 7 ppm. In contrast, confused flour beetles *Tribolium confusum,* and red flour beetles *T. castaneum,* were unaffected by doses of 3-isobutylmethylxanthine up to 30,000 ppm [7.48].

Venoms from ants of the genera *Monomorium* and *Solenopsis* are toxic to termites and to eight other species of arthropod pests. The active constituents are 2,5-disubstituted pyrrolines that might serve as leads to new insecticides [7.3].

An extensive study of more than 70 alkaloids from most structural types described the toxicity to *A. mellifera* of 11 alkaloids, with supplemental observations on feeding deterrency that will be reported later in this Chapter. The threshold for toxic effects was set at an LD_{50} of 0.01 % or 100 ppm, below which only one alkaloid, the isoquinoline berberine was rated, at 30 ppm. Receptor-binding and other assays indicated that berberine significantly intercalated DNA, strongly inhibited protein synthesis (>30 % at 1mM), bound strongly to the alpha 1 and alpha 2 adrenergic receptors, and to the serotonin and muscarinic acetylcholine receptors [7.72].

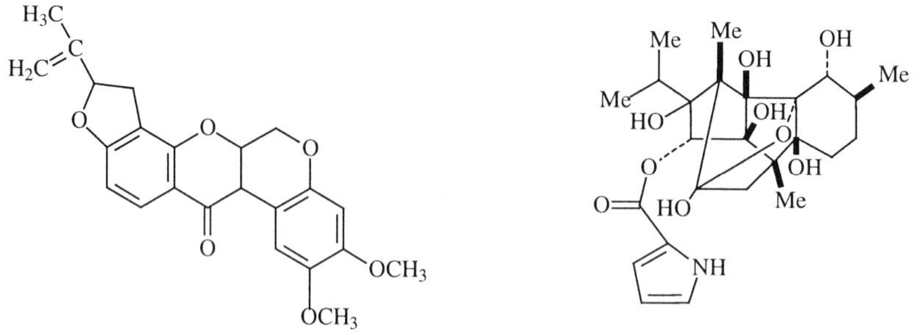

(a) (b)

Figure 54. Structures of (a) rotenone, and (b) ryanodine.

Unsaturated Isobutylamides

Highly effective insecticidal isobutylamides of unsaturated, aliphatic, straight-chain acids with 10-18 carbons, have been isolated from the families Compositae and Rutaceae. They share two properties with the pyrethrins: pungency, and rapid knockdown and kill of flying insects. In addition, they are also toxic to cockroaches and several species of chewing insect. The active compounds are pellitorines (Fig. 55) from the root of pellitory *Anacyclus pyrethrum*. In spray tests against housefly *Musca domestica*, pellitorine elicited paralysis equal to that of pyrethrum, but 50 % less mortality. One hundred ppm spilanthol, N-isobutyl-4,6-decadienamide, obtained from heads of *Spilanthes oleraceae* and *S. acmella*, immobilized mosquito larvae *Anopheles quadrimaculatus*, in 5 min and killed in 40 [7.25, and references therein].

Piperine (Fig. 55), pipercide, dihydropipercide and guineensine occur in fruit of black pepper *Piper nigrum*. Topically applied crude and purified extracts elicited 100 % mortality of rice weevil *Sitophilus oryzae*, and 95 % of cowpea weevils *Callosobruchus maculatus*, at doses of 12.5 μg/insect after 3 days. In contrast, the same dose of piperine alone elicited only 8 and 0 % mortality respectively, after 3 days. In tests against the adzuki bean weevil *Callosobruchus chinensis*, LD_{50} values (μg/insect) 48 hr after topical application were piperine (>20), pellitorine (2.0), pipercide (0.15) and pyrethrin (0.15) for males as compared with >10, 7.0, 0.25 and 0.20 respectively for females [7.43, 7.44, 7.66].

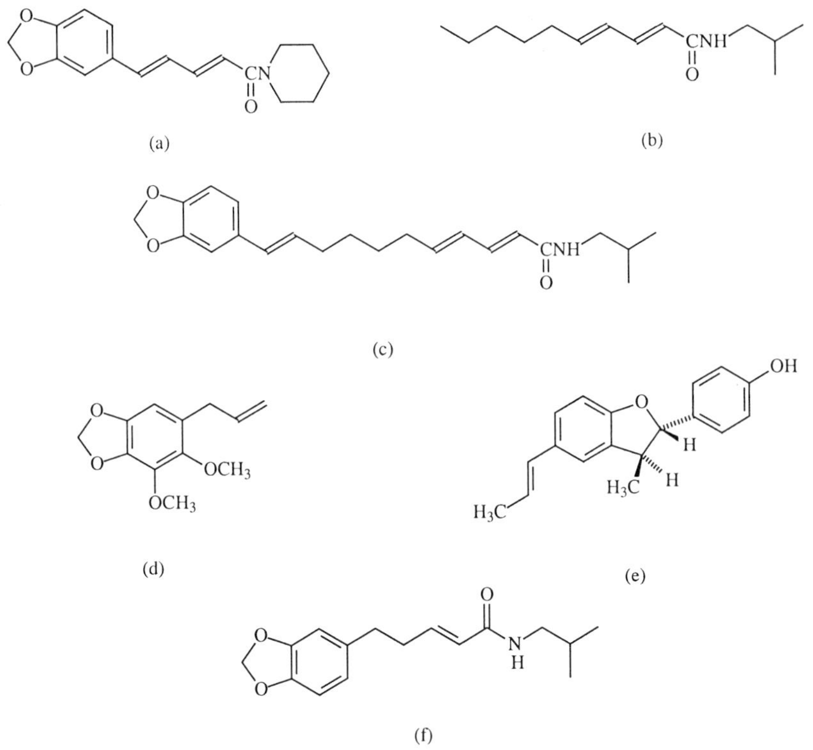

Figure 55. Structures of alkyl isobutylamides including (a) piperine, (b) pellitorine, (c) pipercide. Compiled from Miyakado et al. *[7.43]. (d) dillapiol, (e) conocarpan, and (f) piperlonguminine. Compiled from Bernard* et al. *[7.5].*

Three amides isolated from *P. nigrum*, were identified as (*E*,*E*)-N-(2-methylpropyl)-2,4 decadienamide, (*E*,*E*,*E*)-13-(1,3-benzodioxol-5-yl)-N-(2-methylpropyl)-2,4,12 tridecatrienamide, and (*E*,*E*,*E*)-11-(1,3-benzodioxol-5-yl)-N-(2-methylpropyl)-2,4,10 undecatrienamide. LD_{50} values from topical application of these three compounds against *C. maculatus* were 2.18, 0.25 and 0.84 μg/insect respectively for males, and 6.70, 1.43 and 3.88 μg/insect for females; corresponding values for piperine and pyrethrin were more than 100, and 0.08 μg/insect respectively for males; and more than 100, and 0.16 μg/insect respectively for females. Various combinations of piperacean amides were active against *C. chinensis*. For example, KT_{50} (medium knockdown time) values using 1 μg/insect were 5.5 min pipercide; 7.9 min dihydropipercide; 6.3 min guineensine; and 3.5 min for a mixture of these three compounds (1:1:1). Overall, such potencies were equivalent to one-half to one-third of the activity of pyrethrin [7.43, and references therein].

Four compounds isolated as toxicants and also as growth inhibitors from *Fagara macrophylla* and incorporated into artificial diets of first instar larvae of the pink bollworm *Pectinophora gossypiella*, and tobacco budworm *Heliothis virescens*, gave respective ED_{50} (effective dose for 50 % growth inhibition) values of: fagaramide, 440 and 350 ppm; piperlonguminine, 430 and 370 ppm; pellitorine, 15 and 270 ppm; N-isobutyl-2-*E*,4-*E*-octadienamide 70 and 600 ppm. Against the fall armyworm *Spodoptera frugiperda*, ED_{50} values were: fagaramide, 530 ppm, piperlonguminine, 500 ppm, pellitorine, 230 ppm and N-isobutyl-2-*E*,4-*E*-octadienamide, 280 ppm. These four natural isobutylamides were lethal against third-instar larvae of *C. pipiens* at concentrations of 15, 10, 5, and 15 ppm respectively [7.44].

Incorporating extracts of 14 species of American neotropical Piperaceae in the diet of the European corn borer *Ostrinia nubilalis*, indicated postdigestive toxicity. Bioassay-guided isolation using the mosquito *Aedes atropalpus* and 100 ppm doses of plant fractions afforded the monolignan dillapiol, the neolignan conocarpan, and the isobutylamide dihydropiperlonguminine (Fig. 55). Significantly, 0.1 ppm dillapiol elicited 92 % mortality of the mosquito larva and this compound also synergizes the effects of various natural insecticides including pyrethrum, azadirachtin, and tenulin at concentrations below those occurring in *P. aduncum*; 0.01 ppm conocarpan and dihydropiperlonguminine each elicited 47 % mortality (values for 0.1 ppm not given). A previous study of structure-activity relationships of insecticidal amides, associated toxicity with a methylenedioxyphenyl (MDP) moiety that also occurs in dihydropiperlonguminine (Fig. 55). Several lignans inhibit polysubstrate monooxygenase (PSMO) detoxification enzymes of herbivores [7.5, and references therein].

Terpenoids

Monoterpenes

The following 13 monoterpenes from short leaf pine *Pinus echinata*, and loblolly pine *P. taeda* tested by topical application against field-collected southern pine beetles *Dendroctonus frontalis*, gave LD_{50} values (μg/beetle) of: Δ^3-carene 0.36; D-limonene 0.47; *α*-phellandrene 0.47; DL-limonene 0.48; L-limonene 0.48; myrcene 0.55; DL-α-pinene 0.62; L-*β*-pinene 0.64; *β*-phellandrene 0.95; D-camphene 1.15; limonene dioxide 0.24. So, the limonene compounds were the most toxic [7.10].

A comprehensive study sought structure-activity relationships in the topical, fumigant and ovicidal efficacies of fourteen monoterpenes and thirty-one monoterpenoid derivatives against *M. domestica*. In topical bioassays phenols and acyclic alcohols were more potent than monocyclic and bicyclic alcohols, which was consistent with their previously reported nematicidal efficacies. In topical bioassays and ovicidally, ketones were more potent than corresponding alcohols perhaps by being less susceptible to metabolic inhibitors. Consistent with this are the higher concentrations of monoterpenoid ketones in more vulnerable, younger leaves, as compared with higher alcohol concentrations in older ones of the peppermint *Mentha piperita*, as host for the variegated cutworm *Peridroma saucia*. Topically applied, phenols were more potent than more saturated alcohols, but more saturated alcohols and ketones were more efficacious as fumigants [7.52].

In regard to derivatives acetates were more potent, topically and as fumigants, than propionates, and were more efficacious than haloacetates, topically and as ovicides, possibly due to their greater stability. However as fumigants, acetates were less effective than trifluoracetates, assignable to the greater volatility of the latter. Of all compounds bioassayed, thymol and geranyl acetate were the most efficacious, topically, and thymyl trifluoracetate, as derivative, was the most potent fumigant. As ovicides, geraniol, geranyl propionate, terpineol, carvacrol, and menthone were as potent as the standard, 20 % pyrethrins [7.52].

Sesquiterpenes

Gossypol, the predominant sesquiterpenoid in the lysigenous glands in seeds of most cotton varieties, *Gossypium* spp., is a minor terpenoid constituent in foliar glands (Fig. 56). Cotton analysis revealed an array of toxic C_{15}, C_{25} and C_{30} terpenoids. When the C_{25} terpenoids, isolated from domestic cotton *G. hirsutum* (leaves and bolls) and referred to as heliocides, were added at 4 mmoles/kg to an artificial diet, they decreased by about 75 % larval growth of *H. virescens* (Fig. 56) [7.63, and references therein]. Another sesquiterpene (5*E*)-ocimenone (Fig. 57) occurs in the tree of Mexican marigold *Tagetes minuta* (Compositae), the leaves of which are used in east Africa to repel mosquitoes and ants. At 40 ppm this compound elicits 100 % mortality in larvae of the yellow fever mosquito *Aedes aegypti* [7.37]. The heartwood powder of *Juniperus recurva*, long used as an insecticide, contains, as actives, the sesquiterpenes thujopsene and 8-cedren-13-ol. Their LD_{50} efficacies against mosquito female adults *Culex pipiens pallens*, by topical application, were 4.5 and 6.6 μg/mosquito respectively. Three sesquiterpenes (α-cedrene, 8,14-cedranoxide, and cedrol) from heartwood powder have rather low insecticidal activity with LD_{50} values of 33.5, 4.5 and 6.6 μg/mosquito respectively [7.49].

Myristicin a component of dill, celery, parsley and parsnip, is best known as a constituent of oil of nutmeg *Myristica fragrans*. Myristicin is both an insecticide and an insecticidal synergist. Crude extracts from the edible parts of parsnip *Pastinaca sativa* and synthetic myristicin were toxic to vinegar fly *Drosophila melanogaster*, yielding identical LD_{50} values of 430 μg/jar, after 1 hr exposure, and 300 μg/jar, after 24 hr. At 40 ppm, this compound elicits 100 % mortality in larvae of *A. aegypti*. Added to pyrethrum, myristicin is a synergist eliciting about a three-fold higher mortality of *D. melanogaster*. Against *M. domestica*, piperonyl butoxide enhanced the effect of pyrethroids (87 % by ratio 1:1, 36 % by ratio 1:5, and 9 % by ratio 1:10); corresponding values for myristicin with carbamate were 86 % by 1:0.2 [7.34].

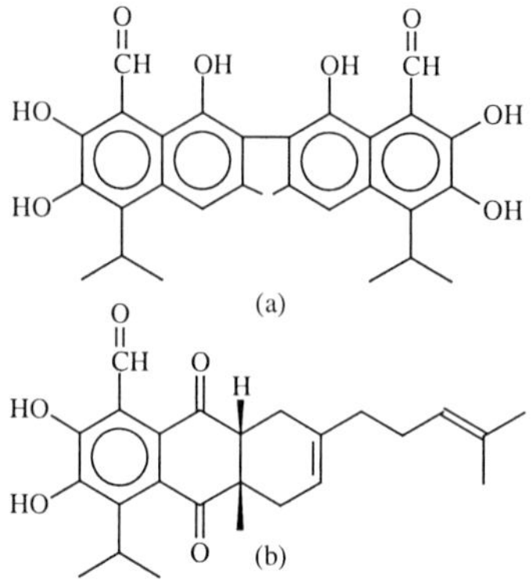

Figure 56. Structures of (a) gossypol, and (b) heliocide H_2.

Various species of Celastraceae, especially *Celastrus angulatus*, *C. gemmatus* and *Euonymus bungeanus*, used in China as natural insecticides, have yielded three new *β*-dihydroagarofuran sesquiterpenoid polyol esters (Fig. 57) [7.67].

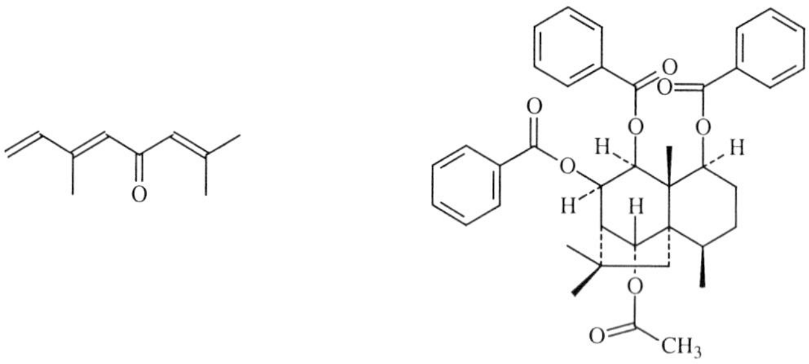

(a) (b)

*Figure 57. Structures of (a) (5*E*)-ocimenone, and (b) 6a-acetoxy-1β, 8β, 9β-tribenzoyloxy-β-dihydroagarofuran, a new sesquiterpene from seed oil of* Celastrus angulatus. *The latter structure compiled from Tu and Hu [7.67].*

Diterpenes

The best known insecticidal terpenoids are the pyrethrins (described in Chapter 1) and their use has not ended with the success of synthetic pyrethroids. Thus, Tanzania, Rwanda, Zaire and Ecuador still produce the pyrethrum flowers for export especially to the U.S.A. The continuing interest in them is such that *in vitro* culture, and genetic engineering using the chrysanthemyl diphosphate synthase gene is being attempted. Unfortunately, success in this will threaten the livelihoods of some 200,000 small farmers in East Africa [7.29].

Insecticidal diterpenes from *Isodon* spp. (Labiatae) include inflexin from leaves of *I. inflexus*, isodomedin (Fig. 58) from leaves of *I. shikokianus*, and kamebanin from *I. kameba*. Cytotoxic, all three kill lepidopteran larvae including the African armyworm *Spodoptera exempta*, with LD_{50} values of 5.4, 4.0 and 5.1 ppm respectively [7.32]. All three diterpenes also possess an exocyclic α-β-unsaturated carbonyl group.

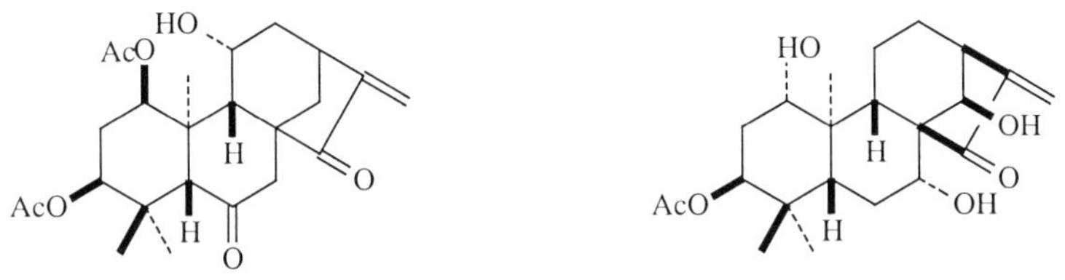

(a) (b)

Figure 58. Structures of (a) inflexin, and (b) isodomedin. Compiled from Kubo and Nakanishi [7.32].

Triterpenoids
Azadirachtin or neem, the product of the neem tree *Azadirachta indica*, a member of the mahogany family, represents for many entomologists the best hope for a commercially viable, cheap, natural insecticide. This tropical evergreen tree, also known as Indian lilac, occurs mainly in places with an annual rainfall of 400 to 800 mm and has been the subject of three international conferences [7.54, 7.55, 7.56], and many additional studies.

A limonoid (Fig. 59) azadirachtin is most concentrated in seed kernels of the fruit, that may amount to 50 kg/tree/annum, and maximally is 10 g/kg seed kernels. These contain a range of other compounds with anti-insect properties but they are overshadowed by the range of effects and high potencies of azadirachtin. Specifically, this complex of isomers, inhibits insect behaviour, metamorphosis, fecundity, fertility and fitness, all of which effects will be considered, respectively in this section.

In addition to deterring insects from landing on treated plants, neem also inhibits egg-laying by a range of pests. Antifeedant effects are long established especially against lepidopteran larvae. Feeding also decreased when azadirachtin was injected, indicating a supplemental, non-gustatory effect. Applied topically or by injection, azadirachtin retarded development, disturbed moulting, induced malformed pupae, deformed wings, and killed the adults of a range of Lepidoptera. Decreased ecdysteroid concentrations seemed to be the mode of action. Neem oil enriched with methanol extracts of neem seed kernels, applied to the desert locust

Schistocerca gregaria, did not kill but decreased flight-activity by about 50 %; such locusts could not sustain flight for more than 30 min.

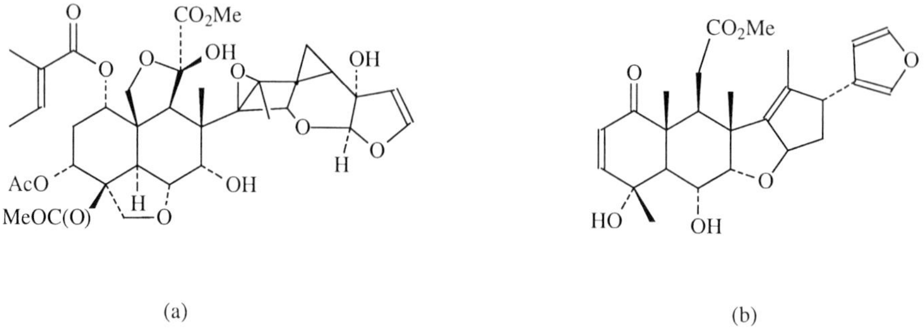

Figure 59. Structures of (a) azadirachtin, and (b) nimbandiol an open tricyclic diterpene. Compiled from Hansen et al. *[7.17].*

Topically treated females of Orthoptera, Heteroptera, Homoptera, Hymenoptera, Coleoptera, Lepidoptera and Diptera laid up to 50 % less eggs and some females became sterile. Treated females of the Mexican bean beetle *Epilachna varivestis* underwent degeneration of the ovary and laid eggs that were chorion-damaged, and infected by fungi. In treated locusts *Locusta migratoria migratorioides*, ecdysteroid concentrations were decreased, apparently through inhibition of the neuroendocrine system. Consistent data from treated *S. gregaria* showed that hormone synthesis and release by brain neurosecretory cells was delayed and decreased. Treated Homoptera were short-lived, and larvae of the Mediterranean fruit fly *Ceratitis capitata* fed an azadirachtin enriched diet gave rise to adults so short-lived that only 50 % became sexually mature. Other obvious effects include: abnormal flight, male impotence, and female failure to respond to male pheromone resulting in decreasing mating [7.54, 7.55, 7.56].

In field experiments, mortality of aphid pests of cabbage, pepper, strawberry, and lettuce ranged from 40 to 98 %, following treatment of crop plants with 1 % neem seed oil (NSO), equivalent to 20 ppm azadirachtin. Control was attributed to inhibition of adult reproduction, and failure of nymphs to moult. The field experiments gave no evidence of adverse effects on predators or parasitoids although laboratory experiments indicated moderate susceptibility [7.36].

Overall, neem and neem derivatives seem feasible as broad-spectrum: insecticides for phytophagous insects, with the added advantage that decreased feeding by larval Lepidoptera decreases adult fecundity. In addition, neem derivatives show promise in controlling stored product pests, especially through repellent effects. Although rather large concentrations are required to kill the mosquito *A. aegypti*, the effect is enhanced or synergized when used in conjunction with *Bacillus thuringiensis*. Nevertheless, the usual difficulties attending the use of many natural products also apply to azadirachtin. Specifically, environmental factors, especially low temperature and high rainfall, decrease activity such that it rarely lasts beyond 5-7 days in the field. The consequential need to repeat treatment makes it a rather uninviting prospect for the grower except for the peasant farmer with no alternative, and for whom time is not a limiting factor. A further disincentive is lack of immediate effect or knockdown as treated insects will continue to feed for some time [7.36, and references therein].

Practical experiments in the Dominican Republic sought to transform such promising research data into effective plant protection for small farmers. When they prepared water extracts from crushed neem seeds, effects in field trials were excellent. This prompted the suggestion that they should plant neem trees on marginal elements of their land [7.55, and references therein]. Such lands are typified by poor soil and unfavourable climate, but these stresses actually enhance production of secondary metabolites, although plant biomass/ha is lessened. This exemplifies a case for using marginal land to produce plants rich in useful allelochemicals.

Quassinoids

The efficacy of extracts containing quassinoids produced from wood and bark of the quassia tree *Quassia amara* (Simaroubaceae) against aphids has been known since 1939. In 1940, over one million pounds (ca 450 thousand kg) of quassia chips were used in the U.S.A., Italy consuming almost as much. Subsequently, the high cost of production and transportation, as well as the availability of more effective synthetics, virtually excluded quassia from the insecticidal market (see later for antifeedant effects).

The foregoing data, from an even wider range of plant sources than of pests, call for direct, standardized comparisons. Perhaps, surprisingly these are coming to hand only recently. Thus in 1994, *H. virescens* served in a bioassay of a range of plant-based insecticides including azadirachtin, kryocide (the synthetic form of the naturally occurring mineral cryolite), pyrethrum, rotenone, ryania, and sabadilla, from seeds of sabadilla *Schoenocaulon* spp. (Liliaceae), all incorporated at 20 ppm (except kryocide 2000 ppm) in diet. Responses were assayed in respect of survivorship, pupal weight and development. No larvae exposed to ryania or azadirachtin survived beyond the first instar; sabadilla, pyrethrum and kryocide killed 65, 60 and 43 % respectively of larvae; rotenone killed none nor did it prolong larval developmental time. In contrast, pyrethrins, sabadilla and kryocide extended this by 20, 13 and 9 days respectively which was associated with an increased number of stadia, and a significant decrease in pupal weight. Food consumption and assimilation by the fourth instar larva were significantly decreased by sabadilla, azadirachtin, kryocide and ryania. This comprehensive study could serve as a model by which the rich array of insecticidal allelochemicals might be placed in a ranked order of efficacy against, to begin with, monophagous and oligophagous test species [7.74].

Insecticidal Adjuvants

Some compounds, not in themselves toxic, are used with natural insecticides to enhance their activity. As adjuvants or synergists, they enhance pyrethrin activity by inhibiting enzymes that normally metabolize these poisons. Isobutylamides, such as sesamin from oil of sesame *Sesamum indicum* inhibit PSMOs, thus prolonging insecticidal action. These plant-derived compounds served as models for development of synthetic adjuvants such as piperonyl butoxide.

Plant Juvenoids

A search for juvenile hormone (JH) analogues (juvenoids) was stimulated by the accidental finding in Harvard University that some American paper products contained a factor that caused JH effects in the European bug *Pyrrhocoris apterus*. Bugs failed to undergo normal metamorphosis as all 5th instar larvae moulted into 6th instars, or into adult stages preserving many larval characters. This was attributable to a todomatuic acid methyl ester named juvabione present in paper towelling used to line rearing cages. The towelling source was traced to wood pulp of the American balsam fir *Abies balsamea*, [7.61, 7.70].

Although never used commercially, plant juvenoids served as lead compounds for development of JH analogues such as methoprene with low acute mammalian toxicity, $LD_{50} > 34{,}600$ ppm against rat (oral method). Methoprene and hydroprene elicited respective ID_{50} values (50 % inhibition of metamorphosis) of: *A. aegypti*, 0.00025, and 0.0078; *M. domestica*, 0.0026 and 18 μg/pupa; *H. virescens*, 0.77 and 0.30 ppm; yellow mealworm *Tenebrio molitor* 0.004 and 0.25 μg/pupa; and pea aphid *Acyrthosiphon pisum* 0.0054 and 0.0039 μg/pupa [7.19]. The beetle *Trogoderma inclusum* was completely controlled by 5 ppm methoprene in food medium [7.65], and the pupa of the khapra beetle *T. granarium* succumbed to JH analogues (6,7-epoxygeranyl, 3,4-methylenedioxy phenyl ethers with variations in the geranyl chain) with an LD_{50} value of 0.000005 μg/pupa [7.42].

Phytoecdysones

Ecdysones occur in fern trees of the genera *Pteridium*, *Polypodium* and *Osmunda* and ponasterone A was demonstrated as an insect moulting hormone from leaves of *Podocarpus nakaii* ("togariba-maki" in Japanese); rhizomes of the dried fir tree *Polypodium vulgare*, also contained phytoecdysone. The same amount of ecdysone (25 mg) extracted from half a ton (500 kg) of *B. mori* tissue was obtainable from as little as 25 g dried rhizomes. Cyasterone and ecdysterone, isolated from the leaves and root of east African medicinal plant *Ajuga remota*, were fed to larvae of *S. frugiperda* and *P. gossypiella*. Some grew three heads, or cuticles without shedding any old ones. Their mouths were buried so deeply inside the coverings that physically unable to eat, they starved to death [7.70, and references therein].

NATURAL REPELLENTS

Chemicals eliciting immediate behavioural responses in insects may be categorized as attractants (eliciting oriented movements towards the source), arrestants (causing aggregation), stimulants (eliciting feeding, phagostimulation, oviposition, etc.), repellents (causing oriented movements away from the source) and deterrents (inhibiting feeding or oviposition). The main application of repellents is control of insects such as mosquitoes and lice, that vector disease-causing organisms.

The first chemical repellent, introduced in 1911, was oil of citronella serving as the most widely used mosquito repellent from 1901 to 1938. It contained geraniol, with traces of citronellol and citronellal, the latter two viewed as the most effective. Other repellents against *A. aegypti* were: wood oil of murray river pine *Callitris glauca*, effective in 0-20 min; oil from huon pine *Dacrydium franklinii*, effective over 60 min; sassafras oils extracted from *Atherosperma moschatus* and *Doryphora sassafras*, effective in 40-60 min; tea oil extracted from black tea tree *Melaleuca bracteata*, 60 min [7.39]. Oils of turmeric *Curcuma longa*, sweetflag *Acorus calamus*, *A. indica*, and "Margosan-O", a neem-based insecticide, elicited increased repellency of *T. castaneum* with increasing concentrations from 200 ppm, turmeric being the most efficacious [7.28].

The situation in respect of synthetics may be considered for completeness. Prior to World War II, three synthetic repellents were used against biting insects: dimethyl phthalate discovered in 1939; indalone introduced in 1939; and Rutgers 612 (2-ethylhexane-1,3-diol) which became available in 1939. Requirements of the Allied armed forces for repellents in World War II led to the investigation of synthetic chemicals in the U.S.A., U.K. and Australia. From 11,000 chemicals tested between 1952 - 1962, 18 were effective against a wide range of insects including mosquitoes and flies. N,N-Diethyltoluamide, marketed as the repellent deet, protected against mosquitoes for 3-8 hr when applied to skin, and is still the mosquito repellent of choice; on clothing it protects against chiggers, ticks and fleas for several days or more [7.16].

Of behavioural responses, appropriate for manipulation to decrease crop losses, none has attracted more research than deterrence, especially oviposition deterrence, and feeding deterrence. Manipulating oviposition offers the exciting prospect of control prior to significant damage being inflicted on the plant, and by the use of essentially pest-specific and non-toxic compounds.

Neemrich I and II purified from non-edible neem oil deter oviposition by the potato tuber moth *Phthorimaea operculella* [7.59] and NSO but not azadirachtin inhibits oviposition by the African bollworm *Helicoverpa armigera* which is the greatest threat to the cotton industry of India [7.53]. In respect of purified compounds, the two glucosides 2-0-β-D-glucosyl cucurbitacin I and 2-0-β-D-glucosyl cucurbitacin E are the major oviposition deterrents in *Iberis amara* against *Pieris rapae* and *P. napi oleracea* [7.22]. Cardenolides, a group of C_{23} steroid derivatives having an α, β unsaturated γ-lactone (butenolide) ring, inhibit oviposition by the latter two species. K-Strophanthoside and erychroside were the most effective against each species respectively, and strophanthidin-based glycosides were more deterrent than digitoxigenin-based ones; glucose and cymarose glycones seemed necessary for activity [7.21].

ANTIFEEDANTS

As this area of investigation may be significantly influenced by the particular bioassays used, it is appropriate to begin with a consideration of these.

Bioassays

Available options include 'no-choice' bioassays or presenting only the treated material to the insect. This is not recommended as insects may ultimately feed on treated material from hunger or/and habituation, that might not occur in a more natural situation offering a choice. In a 'choice' bioassay a substrate may be treated, or untreated, i.e. only solvent or control. Such substrates may be dried leaves, artificial diets, filter paper, glass fibre, elderberry pith, agar discs, all for chewing insects. Alternatively, sucking insects will accept artificial diet in parafilm containers, and drinking insects may be presented with test substances in free-standing water. In addition to dual-choice are multiple-choice situations by which several candidate deterrents are simultaneously bioassayed [7.30, and references therein]. Inconsistencies have arisen in data derived from 'no-choice', as compared with 'choice' situations. Thus, the threshold for *S. exempta* was 1,000-fold higher for warburganal in a sucrose-agar diet presented in a 'no-choice' as compared with natural host plants in a choice situation [7.57, and references therein]. The relative insensitivity of responses to subtle behavioural changes in a 'two-choice' situation, where the insect returns to the same two food types has been considered [7.68].

Bioassay-guided identification of deterrents may be a lengthy procedure given the need to wait for feeding responses to each fraction. A significant improvement is offered by bioautography by which developed thin layer chromatography (TLC) plates are covered with a slender layer of artificial diet, and exposed to the test insect. Uneaten zones are matched with R_f values on a reference plate under ultraviolet (UV) light to indicate active fractions. These are scraped off and subjected to further purification by column chromatography followed by preparative TLC. By this method, originally used for identification of antifungal and antibacterial compounds, three active feeding deterrents, the furanocoumarins bergapten, xanthotoxin, and oxypeucedanin, were readily identified from *Skimmia japonica* (Rutaceae). Only four days were required for bioautography to derive appropriate R_f values as compared with dozens of experiments necessary in conventional bioassays. Nevertheless, a possible confounding problem would be undetected significant synergism between compounds, each with small individual activity [7.11].

Unlike the standardization employed for chemicals with insectical effects (LD_{50}), criteria for antifeedants vary, from concentrations decreasing food intake by 50, 80-100, 95 %, to those eliciting threshold protection [7.30, and references therein]. For consistency, the present account will emphasize decreases of at least 50 % at 100 ppm or less for pure compounds. Before proceeding to a consideration of these it is appropriate to consider investigations of whole plants. The corn wireworm *Melanotus communis* was exposed to plant extracts, from 78 plant species (24 families), applied to potato sections to screen for antifeedant activity. Using a cutoff rating system shortened the list to five of which the two most active were *Asclepias tuberosa* (Asclepiadaceae) and *Hedera helix* (Araliaceae); the other three were from *Monarda* sp. (Compositae), *Salvia sclera* (Labiatae), and *Santolina virens* (Compositae) [7.69].

A greater array of antifeedants have been identified from advanced as compared with primitive plants which is consistent with their richer range of secondary plant compounds. Numerically, alkaloids, terpenes and flavonoids predominate.

Alkaloids

Many alkaloids deter feeding by insects. One hundred ppm isoboldine, a feeding inhibitor in leaves of *Cocculus trilobus*, decreased feeding by the tobacco cutworm, *S. litura* by 70 %, and completely inhibited it at 200 ppm [7.47]. Effects of other alkaloids such as berberine, cocaine, chaconine, chinchornine, demissine, leptine, lepturine I, lepturine II, papaverine, sanguinarine and strychnine have been reviewed [7.71, and references therein]. A direct comparison of effects of eight plant-derived alkaloids on sucrose uptake by the blowfly *Phormia regina* in a 'no-choice' assay assessed the significance of various functional groups. The decreasing order of feeding inhibition over 24 hr (with functional group) was: (all values significantly different from control) quinine, 93% (quinoline); sparteine, 91% (quinolizidine); tomatine, 86% (steroidal glycoalkaloid); atropine, 83% (tropane); papaverine, 81% (isoquiniline); arecoline, 63% (pyridine); strychnine, 49% (indole); caffeine, 46% (purine). Deterrency was not correlated with numbers of nitrogen atoms or rings or molecular weight. Distinguishing between sensory and toxic effects as indicated by values above or below 80% deterrency respectively, led to the conclusion that quinine, sparteine, atropine and tomatine act through the sensory route. This was consistent with previous electrophysiological responsiveness by labellar sensilla to tomatine. Further, allocating sub-80% deterrency to toxicity was supported by caffeine, but not the other three compounds, eliciting paralysis of flies. As *P. regina* is non-herbivorous it would not usually contact any of the above compounds. Such deterrency supported the hypothesis that contact sensilla may be sensitive to the harmful effects of various compounds, whether they usually encounter them or not [7.71].

Gramine, the only indole alkaloid in barley and, amounting to 8 mg/g dry weight, deters feeding by the aphids *Schizaphis graminum* and *Rhopalosiphum padi*. Electronically-monitored feeding assays distinguished wave-forms representing: salivation (S); penetration (X); phloem ingestion (Ip); non-phloem ingestion (Inp). The absence of gramine elicited Ip and Inp for up to 80 and 40 min respectively, over a 3 hr period; Inp was not observed on a cultivar endogenously containing gramine in epidermis and parenchyma mesophyll cells; exogenous gramine applied to the deficient cultivar was concentrated in vascular bundles and it inhibited Ip. Apparently, the amount of gramine present and its location determine the extent of feeding by these aphid species on barley [7.75].

Dithyreanitrile, an entirely novel, sulphur-containing, indole alkaloid is an antifeedant isolated from *Dithyrea wislizenii* (Cruciferae) growing in the south western states of the U.S.A., and in northern Mexico. It is the first record of a natural product with two sulphur atoms and a nitrile group bound to the same carbon atom (Fig. 60), which is also a rarity in synthetic compounds. A proposed mode of action was enzymatic hydrolysis to release hydrogen cyanide, methane thiol and the corresponding carboxylic acid, which is not unlike that of cyanogenic glycosides. The simplicity of this molecule promoted the expectation of generating more powerful analogues for commercial use [7.51].

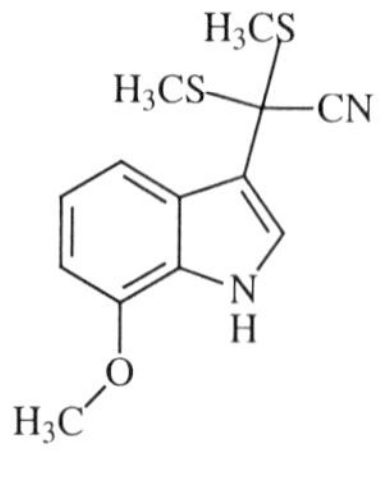

Figure 60. Structure of dithyreanitrile, a unique, sulphur-containing indole alkaloid. After Powell et al. *[7.51].*

Tritrophic interactions involving alkaloids are established for quinolizidine alkaloids (QA) by which the aphids *Aphis genistae*, *A. cytisorium* and *Macrosiphum albifrons*, store these as derived from plants of the genera *Spartium*, *Laburnum*, and *Lupinus* respectively. When batches of the carnivorous beetle *Carabus problematicus*, were allowed to feed on *M. albifrons* containing QA at 1.3 mg/g fresh weight, they were narcotized for more than 48 hr; control beetles that fed on QA-free aphids were unaffected [7.71, and references therein].

Terpenoids

Monoterpenes
Poorly represented in lower plants such as ferns, they act as deterrents in higher plants. Increasing complexity of terpenes is marked in the angiosperms and is associated with enhanced activity as antifeedants [7.8].

Sesquiterpenes
The barks of the East African genus *Warburgia* (Canellaceae), comprising only two species *W. ugandensis* and *W. stuhlmannii*, yield three sesquiterpenes with a drimane skeleton: polygodial, ugandensidial and warburganal (Fig. 61). The water-pepper *Polygonum hydropiper* (Polygonaceae) growing wild in England also affords polygodial. Of the three compounds, warburganal is by far the most active antifeedant; 0.1 ppm abolished feeding of *S. exempta* [7.31]. Polygodial loses activity when the aldehyde is reduced to a hydroxyl group or oxidized to a carbonyl [7.32] indicating that the A-ring is relatively unimportant for activity. However, such effects are dependent on the stereochemistry of the C-9 aldehyde group, i.e. only the 9 β-isomer is both antifeedant for insects and hot for human taste; the 9 α-isomer is inactive against insects and tasteless to humans, e.g. (-)-polygodial and isopolygodial respectively. Similarly, saccalutal is lethal at 0.4 ppm to fish but isosaccalutal is inactive at 10,000 ppm. However, absolute configuration is not an issue as (+)-polygodial was as potent as the natural (-)-isomer. The use of paired comparison of derivatives confirmed the significance of the C9 and indicated that the distance between the two CHO groups was shorter in the bioactives. This was consistent with the hypothesis that activity may be associated with formation of covalent bonds with primary amino groups as receptors, and that the receptor site could accommodate only bicyclic as compared with tricyclic or tetracyclic skeletons [7.9].

A separate study of derivatives, coupling whole-insect bioassays with electrophysiological recordings using *Spodoptera* spp. and *Heliothis* spp., focussed on changes to the B-ring and also emphasized the significance of the C-9, but indicated that the CHO group is more effective in the β than in the α configuration; unnatural racemic, synthetic compounds were as active as optically pure warburganal and polygodial. This study also emphasized the need to test putative antifeedants against a range of phagostimulants, including a preferred host plant [7.7].

Behavioural and electrophysiological methods also served to establish the mode of action of polygodial, warburganal and muzigadial. Essentially, these three sesquiterpenes stimulate the deterrent neuron in the medial styloconic sensillum of the cabbage white butterfly *Pieris brassicae*. In addition sensitivities of a glucosinolate cell in the medium and lateral sensilla, the amino acid receptor, and a sucrose receptor in the lateral sensillum, all decrease, some by more than 50%. Deterrency was viewed as a product of stimulation of the deterrent receptor, and inhibition of phagostimulatory receptors. Such general effects were viewed as unrelated to receptor specificity [7.58].

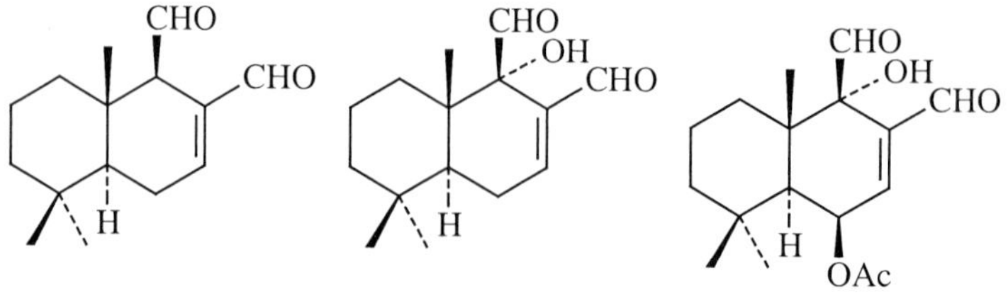

(a) (b) (c)

Figure 61. Structures of (a) polygodial, (b) warburganal, and (c) ugandensidial. Compiled from Kubo and Nakanishi [7.32].

As (-)-polygodial is the simplest of these molecules and has the broadest spectrum of activity, it was chosen for field trials against the aphid *R. padi* serving as vector for barley yellow dwarf virus. Barley sprayed with 50 g/ha polygodial afforded a less damaged crop and increased yield by 36%. It was not necessary for this antifeedant to be systemic for effectiveness against aphids, as antifeedant painted on the leaf surface was as effective as that absorbed through petiole or roots. But repeated applications were required to preclude aphid immigration, and such lack of persistence is inappropriate for commercial use [7.50].

These observations were extended to a comparison of natural (-)-polygodial and the synthetic (+) isomers in addition to a range of related synthetics in relationship to settling responses of the peach-potato aphid *Myzus persicae*. The most efficacious compounds were the natural dialdehydes (–)-polygodial, (-)-warburganal, and the synthetic enantiomer (+)-polygodial; *cis*-polygodial was inactive. As the equal potency of the natural and synthetic enantiomers was unexpected, it was confirmed by a separate experiment. Additionally, as 9α-hydroxydrimenoate and 9α-hydroxydrimenal, active against lepidopteran larvae, were inactive against aphids, this indicated different structure-activity relationships for aphids, as compared with lepidopteran larvae [7.2].

Diterpenes

The Verbenaceae, especially *Clerodendron tricotomum*, produce diterpenoid antifeedants, effective at 50-200 ppm against the *Spodoptera litura* larva, including clerodane derivatives such as: clerodendrin A and B, clerodin, caryoptin, dihydrocaryoptin, caryoptin hemiacetal, caryoptionol, 3-epicaryoptin, 3-epidihydrocaryoptin, dihydroclerodin-1, and clerodin hemiacetal. However, as clerodendrin A and B were ineffective below 5,000 ppm against the larva of *Calospilos miranda* and *O. nubilalis*, such diterpenoids seemed more potent against polyphagous as compared with monophagous insects [7.47, and references therein].

The plant family Labiatae, especially the genera *Ajuga* and *Teucrium*, produce diterpenoids also with a clerodane skeleton and having powerful antifeedant properties. Thus, ajugarins I, and II (Fig. 62), are diterpenoid antifeedants isolated from the leaves of *A. remota* resistant to *S. exempta*; 100 ppm was the minimal concentration deterring *S. littoralis* in leaf disc tests, whilst 0.001 ppm injected into *S. littoralis* elicited 100% mortality [7.30, and references therein]. But 1% ajugarin I in common with other clerodanes was ineffective against aphids [7.50].

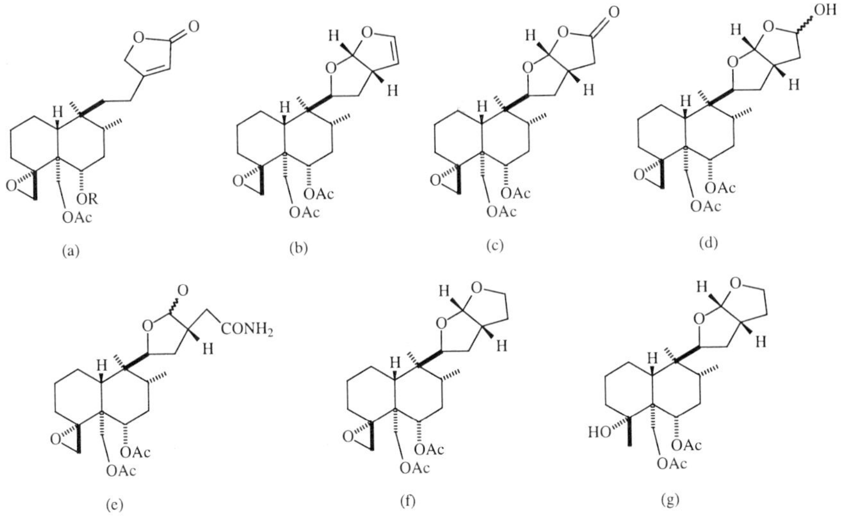

Figure 62. Structures of (a) ajugarins I (R=Ac) and II (R=H), (b) clerodin, (c) clerodin lactone, (d) clerodin hemiacetal, (e) clerodin-14-bromo hemiacetal, (f) dihydroclerodin, (g) tetrahydroclerodin. Compiled from Blaney et al. *[7.6].*

A range of natural and synthetic clerodanes have been assessed, behaviourally and electrophysiologically, as antifeedants against the final instar larva of the lepidopteran pests *S. exempta*, *S. littoralis*, *S. frugiperda*, *H. armigera* and *H. virescens*. Behaviourally, ajugarin I was most potent against oligophagous *S. exempta*; 11-(5) hydroxy-13,14-dihydroajugarin I was reasonably and consistently potent against all five species; clerodin, but not clerodin lactone, was potent against all the *Spodoptera* species, so oxidation of the side-chain at C_9 decreased activity as did reduction of the butenolide group in ajugarin. The group at C_{15} is also significant, as clerodin hemiacetal is more potent than clerodin lactone as, correspondingly, clerodin-14-bromo hemiacetal is more potent than clerodin-14-bromolactone. Other side chain changes such as to produce clerodin hydroxyamide enhanced activity. Also significant is the epoxy group as indicated by the potent dihydroclerodin compared with the very much weaker tetrahydroclerodin. Removal of the acetoxy group at C_6 diminished potency and removal of the C_9 side chain together with the decalin ring abolished it (Fig. 62). Electrophysiologically, *S. exempta* again exhibited a 'deterrent' neuron in the median styloconic sensillum and, in both *S. littoralis* and *S. frugiperda*, 'neuron A' was stimulated by all compounds tested. The firing rate of the 'deterrent' neuron was positively and very highly significantly correlated with deterrency. As in the previous study, the firing rate of 'neuron A' when expressed as percentage of total firing during the first second of stimulation, correlated positively and significantly with behaviour. Collectively, these data highlight the significance of both the furofuran and decalin entities. Also the similar responses to structurally-unrelated compounds argue for a rather unspecific effect by deterrents on the dendritic receptor membrane rather then for a 'lock and key' effect, although the possibility of ligand conformational changes to offer shared features to a specific receptor was not overlooked [7.6].

Clerodane-related diterpenoids, isolated from *Salvia reflexa* and *Baccharis rhetinoides*, and derivatives, were bioassayed for antifeedant and repellent properties using *T. molitor*. The most active antifeedants shared an α, β, unsaturated γ -lactone system on the decalin moiety and a furan ring in the side chain; hydrogenation abolished activity. Further studies with *S. littoralis* and *H. armigera* indicated the significance of an epoxy group at the C-4 position together with C-5-methylacetoxy or C-12-acyloxy substitutions. Repellency effects were consistent with those for antifeedants [7.62].

As perhaps no other antifeedant group has been as thoroughly and systematically explored as the diterpenoids especially the clerodanes, it seems appropriate to seek general principles of activity. Structurally significant features seem to include: primarily, the furofuran and decalin rings, with the former influencing the varying responses by various species, an appropriate degree of unsaturation, and the side chain at C-9 influencing responses of styloconic sensilla; secondarily, the presence of epoxy and diacetate groups on the decalin ring.

Deterrent diterpenoids, the ginkgolides, A, B and C, have been isolated from the world's most ancient living tree *Ginkgo biloba*. Although some fractions of this tree decreased food intake of *P. brassicae* by up to 85% at 25-50 ppm, this falls short of efficacies assignable to warburganal and azadirachtin that abolish feeding at 1 ppm. Ginkgolides A, B, and C and various fractions were feeding deterrents, and electrophysiologically activated the deterrent sensillum in the medial maxillary taste hair of *P. brassicae*, and the lateral hair of *P. rapae*. The notable resistance of this tree to insects was, therefore, assigned to the interaction of a range of substances conferring broad spectrum protection [7.13, and references therein].

Triterpenes

The most significant triterpenoid is azadirachtin established as an antifeedant against a range of pests including: *P. saucia*; striped and spotted cucumber beetles

Acalymma vittata and *Diabrotica undecimpunctata* respectively; green leafhopper *Nephotettix virescens* and brown planthopper *Nilaparvata lugens*. Feeding by *S. gregaria* is abolished by concentrations as small as 0.01 ppm, but the migratory grasshopper *Melanoplus sanguinipes* is undeterred by 500 ppm. A key issue is efficacy in field use by peasant farmers of 1% NSO. Experiments with the strawberry aphid *Chaetosiphon fragaefolii* indicated firstly, that this aphid was deterred within the first hour, the response decreasing only slightly up to 48 hr; NSO also deterred *A. pisum* and an unidentified aphid *Chaetosiphon* species (probably *C. thomasi*), but not *Fimbriaphis fimbriata*, nor *M. persicae,* nor the lettuce aphid *Nasonovia ribisnigri*. Secondly, neem was not repellent except to the sweet-potato whitefly *Bemisia tabaci*, as all species attempted to probe treated material. Thirdly, deterrency was not related to azadirachtin concentrations within the NSO, nor to those of di-*n*-propyl disulphide comprising 76% of neem volatiles. Given the range of types and numbers of triterpenoids associated with neem, the observed feeding deterrency was assigned to the interaction of multiple compounds. Finally, in the greenhouse, 1 and 2% NSO lost deterrency after 12 and 24 hr respectively and such loss would be even more rapid in the field where 600 ppm azadirachtin no longer deterred after 72 hr [7.35, and references therein]. Thus, neem seems significantly less effective as a feeding deterrent than as an insecticide.

The mode of action of azadirachtin, the principal constituent of NSO, was investigated by a combination of antifeedant tests, electrophysiological and receptor-binding assays. Azadirachtin was a very potent antifeedant against the polyphagous species *Spodoptera littoralis*, *S. frugiperda* and *Schistocerca gregaria*, but the oligophagous species *L. migratoria* was less affected. Azadirachtin and analogues stimulated the deterrent neuron of the medial styloconic sensillum of the lepidopteran larvae. Comparative efficacies of the analogues indicated that alterations at the C11 position and bulky substitutions in the C22, 23 region decreased and abolished activity respectively. These responses were matched by decreased binding to locust testis and Sf 9 cells (from an ovarian cell line of *S. frugiperda*), with the suggestion that the binding site is associated with cellular RNA [7.45, and references therein].

Other triterpenoids present in extracts of *A. indica* include nimbin and salannin which are photooxidized, more rapidly than azadirachtin, by UV light in the presence of oxygen to more polar, unstable intermediates [7.27]. Isonimbinolide was as potent as azadirachtin as an antifeedant against *S. littoralis*, *L. migratoria* and *S. gregaria*. Isosalanninolide was a potent antifeedant and growth inhibitor against *S. littoralis* [7.18].

The neutral fraction of the bark of the *A. indica* stem has afforded two new isomeric diterpenes, nimbonone and nimbonolone. Root bark has afforded two new tricyclic diterpenes, nimolinin and nimbandiol (Fig. 59), and neem fruit has afforded a new tetracyclic terpenoid, azadirol. Overall, the triterpenoids of neem comprise two principal categories, triterpenoids and meliacins (limonoids or tetranortriterpenoids) [7.17]. Given that the bioactivities of many of these compounds, the rings of which may be intact or open, are as yet unestablished, this plant remains an enduringly intriguing prospect.

Other limonoids, a group of tetranortriterpenes, occurring in the Meliaceae, Cneoraceae and Rutaceae, exhibit antifeedant activity against: *S. frugiperda*; cotton bollworm *Helicoverpa zea*; African sugarcane borer *Eldana saccharina*; cowpea borer *Maruca testulalis*; and Colorado potato beetle *Leptinotarsa decemlineata*. As limonin is readily available in large quantities by extraction from citrus (Rutaceae) seeds, it proved an attractive lead for structure-activity relationships. This compound also contains polar functional groups appropriate both for receptor-binding and structural modification. Ten limonoids derived synthetically and systematically from limonin (Fig. 63) were investigated as antifeedants against the fourth instar larva of *L. decemlineata* in no-choice leaf disc assays at doses of 1, 3, 10, 32 and 100 μg/cm^2. Modification of the ketone group at C_7 to an acetate or an alcohol significantly increased antifeedant activity but only at the highest concentration (100 μg/cm^2). In contrast, reducing both the 7-keto and 14-15 epoxide abolished activity at doses of 10 μg and lower, and significantly decreased it at 31,7 μg/disc. Furthermore, de-epoxidation of limonol to deoxylimonol substantially decreased activity at lower doses i.e. those of greatest interest. Correspondingly, de-epoxidation of limonin to deoxylimonin abolished activity and at the largest concentration (Fig. 63). A possible mode of action by the epoxy group in receptor alkylation was considered but not sustained as some activity persisted in deoxy analogues with a 7-hydroxyl. However, this would not preclude alkylation if the hydroxyl has an ancillary role. An absolute requirement was the furan ring, as its hydrogenation abolished activity at all concentrations [7.4].

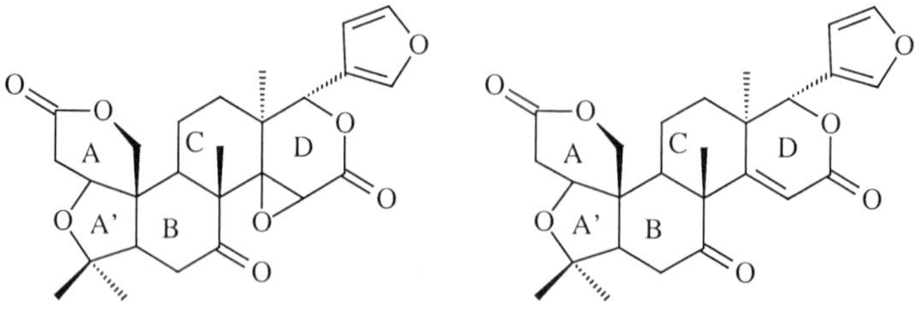

(a) (b)

Figure 63. Structures of (a) limonin, and (b) deoxylimonin. Compiled from Bentley et al.*[7.4].*

The pentacyclic limonoid harrisonin, isolated from root bark of the East African shrub *Harrisonia abyssinica* inhibited feeding activity by *S. exempta* at the 20 ppm level in leaf disc tests. One hundred ppm xylomolin, from unripe fruit of *Xylocarpus moluccensis*, decreased such feeding by 50% [7.32] (Fig. 64).

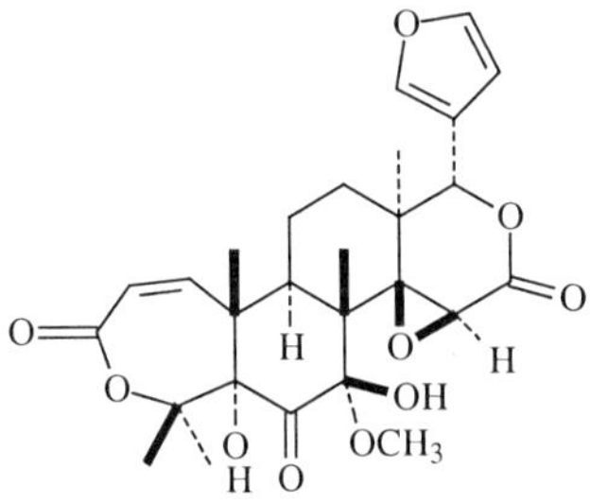

(a)

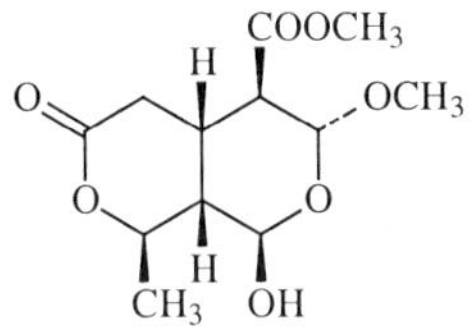

(b)

Figure 64. Structures of (a) harrisonin, and (b) xylomolin. Compiled from Kubo and Nakanishi [7.32].

Flavonoids

This important family includes: flavonols, hydrochalcones, chalcones, flavanones, leucoanthocyanidins, flavones, anthocyanidins, coumarins, flavononols and may be considered in conjunction with tannins and lignans. Differences between individual compounds such as quercetin (less saturated at C_3 and a more potent antifeedant than catechin (less oxidized) suggested a framework for structure-activity relationships. This was systematically explored in a series of 21 chalcones, flavanes, and flavanones, identified or derived from the genera *Lonchocarpus* and *Tephrosia* (Leguminosae), and subjected to both behavioural and electrophysiological bioassays. The compounds represented a homogenous batch of C-15 derivatives, all with one isoprenyl group, viewed as intermediates in the biosynthesis of rotenoids. In a dual-choice bioassay using glass-fibre discs, antifeedant effects were strongest against the larva of the oligophagous *S. exempta* (that feeds mainly on Gramineae) as compared with that of the polyphagous *S. littoralis*. Flavanones were more active than chalcones. This pattern was confirmed in leaf disc assays except that antifeedant activity was generally less. Structure-activity relationships were indicated when the same substitutions in both chalcones and flavanones (Fig. 65) elicited deterrence in both species. Generally, a methoxy substitution at R_2 was more potent than a hydroxy, and unsaturation of the dimethyl chromene ring of chalcone and flavanone was also significant as reduced analogues were less potent.

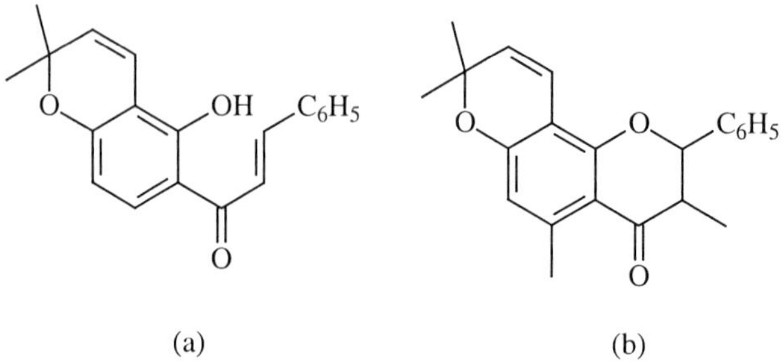

Figure 65. Structures of (a) a chalcone, and (b) a flavanone, both active as antifeedants. Compiled from Simmonds et al. *[7.60].*

Electrophysiologically, activity of the medial maxillary styloconic sensilla gave the best correlation with antifeedant activity. Each sensillum has four neurons of which A, in oligophagous *S. exempta*, was clearly the deterrent neuron by responding to most deterrents. In contrast, in polyphagous *S. littoralis*, three neurons (A, C, and D) responded and, surprisingly, more deterrency elicited less total firing. However, firing by neuron A during the first second of activity, as a percentage of total firing, was positively and significantly correlated with deterrency. Therefore, such deterrents are perceived through an alteration in firing ratios [7.60].

Quassinoids (considered above as insecticides) were investigated as a series for antifeedant activity against the aphid *M. persicae*. Of the active compounds, isobrucein A, isobrucein B, brucein B, brucein C, glaucarubinone and quassin, only the latter was an effective antifeedant (down to 0.05 %), without eliciting phytotoxicity [7.15].

WHOLE PLANTS

The foregoing account reflects a long-standing preoccupation of the developed world with a single or "magic bullet" solution for insect pest and other problems. As such it could be seen in lineage to such concepts as - "one pheromone, one compound"; "one neuron, one neurotransmitter". Whether the philosophy of the Third World has a propensity for corresponding dogmas is unclear, but what is beyond dispute is that it can rarely afford the products. Third World peasant farmers perforce rely on whole plants or crude extracts rather than on pure single compounds, and as their approach is not without success it merits scrutiny.

No civilization has a record to match China's in the use plants for man's benefit. Specifically, the *Classical Pharmacopoeia of the Heavenly Husbandman* referred to *Melia azedarach* (related to the neem tree) as an anthelminthic and active against lice and fleas as early as 25-220 AD. Subsequent dynasties gave rise to publications detailing herbal pesticides against lice, fleas and mosquitoes. In general, the use of pesticidal plants is an extension of China's unique repertoire of knowledge of indigenous medicinal plants. Secondly, China's isolation from the West, especially from the 1950's, stimulated a general self-reliance, part of which was a concerted research, development, and social programme to eliminate the four most destructive pests (flies, mosquitoes, fleas and rats) by use of indigenous pesticides. Although that programme was short-lived (four yr), not least because small advances were exaggerated leading to depletion of useful plants, and because of the absence of scientific rigour, it generated a checklist of some 267 plant species with pesticidal or germicidal properties. Three well-documented plants merit detailed consideration. *Tripterygium wilfordii* (Celastraceae) as a sprayed aqueous extract of root gave at least 80 % control of a range of caterpillars, moths and beetles; a 1:20 preparation gave 100 % kill of mosquito larvae, and 1:10 decreased rust spore germination on wheat by 70-80 %. *M. azedarach* (Meliaceae) is prepared as ground seeds in 1:3 aqueous extract or as a boiled water extract of leaves and twigs, followed by filtering and addition of soap and kerosene and made up 1:10. A water spray killed rice borers, cotton aphids and armyworms inter alia. Dried leaves protected stored produce. Powdered seed and bark in water inhibited wheat rust spore germination by 70 %. *Stellera chamaejasme* (Thymeleaceae), long established as medicinal and poisonous, is used as powdered root at 45 kg/ha in topsoil against soil pests, and as a 1:40 aqueous spray against caterpillars and mustard beetles [7.73, and references therein].

Such deployment of whole plants or crude extracts is of course not confined to China but extends through the underdeveloped countries of Asia, Africa, and tropical America. Specific uses include insecticidal plants, mixtures, baits and traps, and protection of stored produce. For example, *Minthostachys glabrescens* (Labiatae) is one of 12 species of this genus occurring from Venezuela to Argentina. Colloquially known as muna it grows in the high Andes and is harvested in summer when concentrations of essential oils are highest. Potatoes are stored in boxes, silos, pits or cylinders made with or lined by muna twigs that, in replicated trials, decreased germination losses after 60 and 120 days by 74 and 51% respectively, and decreased the incidence of rotten tubers after 120 days by 56% [7.64].

Many of these plants protect against several types of pest and it is surely significant that the better-documented Chinese plants are potent against multiple targets ranging from fungi to insects.

Accordingly, a significant contribution of the developed countries to the Third World would be the installation of production, including cell culture, facilities to generate standardized extracts of plants, rigorously established as multipotent, for indigenous use. Such an approach might serve at least as well as the standard one of giving either bilateral or World Bank loans directly to governments for projects that are not always completed. The frittering of these monies combined with the inexorable debt repayments have contributed to the pauperization all but the ruling elites of many Third World countries. A supplemental approach by the West would be development and dissemination of pest-resistant cultivars.

REFERENCES

7.1 Arnason, J.T., Philogène, B.J.R. and Morand, P. (1989) *Insecticides of Plant Origin.* ACS Symposium Series 387, American Chemical Society, Washington, DC.

7.2 Asakawa, Y., Dawson, G.W., Griffiths, D.C., Lallemand, J.-Y., Ley, S.V., Mori, K., Mudd, A., Pezechk-Leclaire, M., Pickett, J.A., Watanabe, H., Woodcock, C.M. and Zhong-Ning, Z. (1988) Activity of drimane antifeedants and related compounds against aphids, and comparative biological effects and chemical reactivity of (-)-and (+)-polygodial. *J. Chem. Ecol.* **14**, 1845-1855.

7.3 Bacos, D., Basselier, J.J., Celerier, J.P., Lange, C., Marx, E., Lhommet, G., Escoubas, P., Lemaire, M. and Clement, J. L. (1988) Ant venom alkaloids from *Monomorium* species: natural insecticides *Tetrah. Lett.* **29**, 3061-3064.

7.4 Bentley, M.D., Rajab, M.S., Alford, A.R., Mendel, M.J. and Hassanali, A. (1988) Structure-activity studies of modified citrus limonoids as antifeedants for Colorado potato beetle larvae, *Leptinotarsa decemlineata. Entomol. Exp. Appl.* **49**,189-193.

7.5 Bernard, C.B., Krishnamurty, H.G., Chauret, D., Durst, T., Philogène, B.J.R., Sánchez-Vindas, P., Hasbun, C., Poveda, L., San Román, L. and Arnason, J.T. (1995) Insecticidal defenses of Piperaceae from the neotropics. *J. Chem. Ecol.* **21**, 801-814.

7.6 Blaney, W.M., Simmonds, M.S.J., Ley, S.V. and Jones, P.S. (1988) Insect antifeedants: a behavioural and electrophysiological investigation of natural and synthetically derived clerodane diterpenoids. *Entomol. Exp. Appl.* **46**, 267-274.

7.7 Blaney, W.M., Simmonds, M.S.J., Ley, S.V. and Katz, R.B. (1987) An electrophysiological and behavioural study of insect antifeedant properties of natural and synthetic drimane-related compounds. *Physiol. Entomol.* **12**, 281-291.

7.8 Brattsten, L.B. (1983) Cytochrome P-450 involvement in the interactions between plant terpenes and insect herbivores. In Hedin, P.A. (ed.) *Plant Resistance to Insects.* ACS Symposium Series 208, American Chemical Society, Washington, DC, pp. 173-195.

7.9 Caprioli, V., Cimino, G., Colle, R., Gavagnin, M., Sodano, G. and Spinella, A. (1987) Insect antifeedant activity and hot taste for humans of selected natural and synthetic 1,4-dialdehydes. *J. Nat. Prod.* **50**, 146-151.

7.10 Coyne, J.F. and Lott, L.H. (1976) Toxicity of substance in pine oleoresin to Southern pine beetle. *J. Ga. Entomol. Soc.* **11**, 301-305.

7.11 Escoubas, P., Fukushi, Y., Lajide, L. and Mizutani, J. 1992. A new method for fast isolation of insect antifeedant compounds from complex mixtures. *J. Chem. Ecol.* **18**, 1819-1832.

7.12 Fukami, H. and Nakajima, M. 1971. Rotenone and the rotenoids. In Jacobson, M. and Crosby, D.G. (eds.) *Naturally Occurring Insecticides.* Dekker, New York. pp. 71-97.

7.13 Fu-Shun, Y., Evans, K.A., Stevens, L.H. van Beek, T.A. and Schoonhoven, L.M. (1990) Deterrents extracted from the leaves of *Ginkgo biloba:* effects on feeding and contact chemoreceptors. *Entomol. Exp. Appl.* **54**, 57-64.

7.14 Grainge, M. and Ahmed, S. (1988) *Handbook of Plants with Pest-Control Properties.* Wiley and Sons, New York.

7.15 Griffiths, D.C., Pickett, J.A., Smart, L.E. and Woodcock, C.M. (1989) Use of insect antifeedants against aphid vectors of plant virus disease. *Pestic. Sci.* **27**, 269-276.

7.16 Hall, S.A., Green, N. and Beroza, M. (1957) Insect repellents and attractants. *J. Agric. Fd. Chem.* **5**, 663-669.

7.17 Hansen, D., Cuomo, J., Khan, M., Gallagher, R.T. and Ellenberger, W.P. (1994) Advances in neem and azadarichtin chemistry and bioactivity. In Hedin, P.A., Menn, J.J. and Hollingworth, R.M. (eds.) *Natural and Engineered Pest Management Agents.* ACS Symposium Series 551, American Chemical Society, Washington, DC, pp. 103-129.

7.18 Hassanali, A. and Bentley, M.D. (1987) Comparison of the insect antifeedant activities of some limonoids. In Schmutterer, H. and Ascher, K.R.S. (eds) *Natural Pesticides from the Neem Tree (*Azadirachta indica *A. Juss) and Other Tropical Plants.* Proc. 3rd Int. Neem Conf. German Agency for Technical Cooperation (GTZ) Eschborn, Germany, pp. 683-689.

7.19 Henrick, C.A., Willy, W.E., Garcia, B.A. and Staal, G.B. (1975) Insect juvenile hormone activity of the stereoisomers of ethyl 3, 7, 11-trimethyl –2, 4 dodecadienoate. *J. Agric. Fd. Chem.* **23**, 396-400.

7.20 Horgan, D.J., Ohno, H., Singer, T.P. and Casida, J.E. (1968) Studies on the respiratory chain-linked reduced nicotinamide adenine dinucleotide dehydrogenase. *J. Biol. Chem.* **243**, 5967-5976.

7.21 Huang, X. and Renwick, J.A.A. (1994) Cardenolides as oviposition deterrents to two *Pieris* species: Structure-activity relationships. *J. Chem. Ecol.* **20**, 1039-1051.

7.22 Huang, X., Renwick, J.A.A. and Sachdev-Gupta, K. (1993) Oviposition stimulants and deterrents regulating differential acceptance of *Iberis amara* by *Pieris rapae* and *P. napi oleracea. J. Chem. Ecol* **19**, 1645-1663.

7.23 Jacobson, M. (1958) *Insecticides from plants. A review of the literature, 1941-1953.* Handbook No. 154, U.S. Dept. of Agriculture, Washington, DC.

7.24 Jacobson, M. (1975) *Insecticides from plants. A review of the literature, 1954-1971.* Handbook No. 461, U.S. Dept. of Agriculture, Washington, DC.

7.25 Jacobson, M. (1982) Plants, insects, and man-their interrelationships. *Econ. Bot.* **36**, 346-354.

7.26 Jacobson, M. and Crosby, D.G. (eds.) (1971) *Naturally Occurring Insecticides.* Dekker, New York.

7.27 Jarvis, A.P., Johnson, S., Morgan, E.D., Simmonds, M.S.J. and Blaney, W.M. (1997) Photooxidation of nimbin and salannin, tetranortriterpenoids from the neem tree (*Azadirachta indica). J. Chem. Ecol.* **23**, 2841-2860.

7.28 Jilani, G., Saxena, R.C. and Rueda, B.P. (1988) Repellent and growth-inhibiting effects of turmeric oil, sweetflag oil, neem oil, and "Margosan-O" on red flour beetle (Coleoptera: Tenebrionidae). *J. Econ. Entomol.* **81**, 1226-1230.

7.29 Jovetic, S. (1994) Natural pyrethrins and biotechnological alternatives. *Biotech. Dev. Mon.* **21**, 12-13.

7.30 Kubo, I. (1991) Screening techniques for plant-insect interactions. In Hostettman, K. (ed.) *Methods in Plant Biochemistry* Vol. 6 *Assays for Bioactivity.* Academic Press, New York, pp.179-193.

7.31 Kubo, I., Lee, Y-W., Pettei, M., Pilkiewicz, F. and Nakanishi, K. (1976) Potent army worm antifeedants from the East African *Warburgia* plants. *J.C.S. Chem. Comm.* 1013-1014.

7.32 Kubo, I. and Nakanishi, K. (1977) Insect antifeedants and repellents from African plants. In Hedin, P.A. (ed.) *Host Plant Resistance to Pests.* ACS Symposium Series 62, American Chemical Society, Washington, DC, pp. 165-178.

7.33 Levin, D.A. (1971) Plant phenolics: an ecological perspective. *Am. Nat.* **105**, 157-181.

7.34 Lichtenstein, E.P. and Casida, J. E. (1963) Myristicin, an insecticide and synergist occurring naturally in the edible parts of parsnip. *J. Agric. Fd. Chem.* **11**, 410-415.

7.35 Lowery, D.T. and Isman, M.B. (1993) Antifeedant activity from neem, *Azadirachta indica*, to strawberry aphid, *Chaetosiphon fragaefolii. J. Chem. Ecol.* **19**, 1761-1773.

7.36 Lowery, D.T. and Isman, M.B. (1994) Effects of neem and azadirachtin on aphids and their natural enemies. In Hedin, P.A. (ed). *Bioregulators for Crop Protection and Pest Control.* ACS Symposium Series 557, American Chemical Society. Washington, DC, pp. 78-91.

7.37 Maradufu, A., Lubega, R. and Dorn, F. (1978) Isolation of (5E)-ocimenone, a mosquito larvicide from *Tagetes minuta. Lloydia* **41**, 181-183.

7.38 Martin, H. and Woodcock, D. (1983) *The Scientific Principles of Crop Protection*, Seventh Edition, Edward Arnold, London.

7.39 McCullogh, R. and Waterhouse, D. (1947) Laboratory and field tests of mosquito repellents. *Aust. Coun. Sci. Ind. Res. Bull.* **213**, 9-28.

7.40 McIndoo, N.E. (1945) *Plants of possible insecticidal value, a review of the literature up to 1941.* U.S. Bureau of Entomology and Plant Quarantine.

7.41 Metcalf, R.L. (1955) *Organic Insecticides, Their Chemistry and Mode of Action.* Interscience, New York.

7.42 Metwally, M.W. and Sehnal, F. (1973) Effects of juvenile hormone analogues on the methamorphosis of the beetles *Trogoderma granarium* and *Caryedon gonagra. Biol. Bull.* **144**, 368-382.

7.43 Miyakado, M., Nakayama, I. and Ohno, N. (1989) Insecticidal unsaturated isobutylamides: from natural products to agrochemical leads. In Arnason, J.T., Philogène, B.J.R. and Morand, P. (eds.) *Insecticides of Plant Origin.* ACS Symposium Series 387, American Chemical Society, Washington, DC, pp. 173-187.

7.44 Miyakado, M., Nakayama, I., Ohno, N. and Yoshioka, H. (1983) Structure, chemistry and actions of the piperaceae. Amides: new insecticidal constituents isolated from the pepper plant. In Whitehead, D. and Bowers, W (eds.) *Natural Products for Innovative Pest Management.* Pergamon Press, New York, pp. 369-382.

7.45 Mordue (Luntz), A.J., Simmonds, M.S.J., Ley, S.V., Blaney, W.M., Mordue, W., Nasiruddin, M. and Nisbet, A.J. (1998) Actions of azadirachtin, a plant allelochemical, against insects. *Pestic. Sci.* **54**, 277-284.

7.46 Mosha, F.W., Njau, R.J.A. and Myamba, J. Biological efficacy of new formulations of mosquito coils and a critical review of test methods. *Pyreth. Post.* 2: 47-51.

7.47 Munakata, K. (1977) Insect antifeedants of *Spodoptera litura* in plants. In Hedin, P.A. (ed.) *Host Plant Resistance to Pests.* ACS Symposium Series 62, American Chemical Society, Washington, DC, pp.185-196.

7.48 Nathanson, J.A. (1984) Caffeine and related methylxanthines: possible naturally occurring pesticide. *Science* **226**, 184-187.

7.49 Oda, J., Ando, N., Nakajima, Y. and Inouye, Y. (1977) Studies on insecticidal constituents of *Juniperus recurva*. *Agric. Biol. Chem.* **41**, 201-204.
7.50 Pickett, J.A., Dawson, G.W., Griffiths, D.C., Hassanali, A., Merritt, L.A., Mudd, A., Smith, M.C., Wadhams, L.J., Woodcock, C.M. and Zhong-ning, Z. (1987) Development of plant-derived antifeedants for crop protection. In Greenhalgh, R. and Roberts, T.R. (eds.). *Pesticide Science and Biotechnology*. Blackwell Scientific, Oxford, pp. 125-128.
7.51 Powell, R.G., Mikolajcwak, K.L., Zilkowski, B.W., Lu., H.S.M., Mantus, E.K. and Clardy, J. (1991) Dithyreanitrile: an unusual insect antifeedant from *Dithyrea wislizenii*. *Experientia* **47**, 304-306.
7.52 Rice, P.J. and Coats, J.R. (1994) Structural requirements for monoterpenoid activity against insects. In Hedin, P.A. (ed.) *Bioregulators for Plant Protection and Pest Control.* ACS Symposium Series 557, American Chemical Society, Washington, DC, pp. 92-108.
7.53 Saxena, K.M. and Rembold, H. (1984) Orientation and ovipositional responses of *Heliothis armigera* to certain neem constituents. In Schmutterer, H. and Ascher, K.R.S. (eds.) *Natural Pesticides from the Neem Tree* (Azadirachta indica *A. Juss) and other Tropical Plants.* Proc. 2nd Int. Neem Conf. German Agency for Technical Cooperation (GTZ), Echsborn, Germany.
7.54 Schmutterer, H. and Ascher, K.R.S. (1984) *Natural Pesticides from the Neem Tree* (Azadirachta indica *A. Juss) and other Tropical Plants*. Proc. 2nd Int. Neem Conf. German Agency for Technical Cooperation (GTZ). Echsborn, Germany.
7.55 Schmutterer, H. and Ascher, K.R.S. (1987) *Natural Pesticides from the Neem Tree* (Azadirachta indica *A. Juss*) *and other Tropical Plants*. Proc. 3rd Int. Neem Conf. German Agency for Technical Cooperation (GTZ). Echsborn, Germany.
7.56 Schmutterer, H. and Ascher, K.R.S. Rembold, H. (1981) *Natural Pesticides from the Neem Tree* (Azadirachta indica *A. Juss*). Proc. 1st Int. Neem Conf. German Agency for Technical Cooperation (GTZ). Echsborn, Germany.
7.57 Schoonhoven, L.M. (1982) Biological aspects of antifeedants. *Entomol. Exp. Appl.* **31**: 57-69.
7.58 Schoonhoven, L.M. and Fu-Shun, Y. (1989) Interference with normal chemoreceptor activity by some sesquiterpenoid antifeedants in an herbivorous insect *Pieris brassicae*. *J. Insect Physiol.* **35**, 725-728.
7.59 Sharma, R.N., Nagasampagi, B.A., Bhosale, A.S., Kulkarni, M. and Tungikar, V. B. (1984) "Neemrich": the concept of enriched fractions from neem for behavioural and physiological control of insects. In Schmutterer, H. and Ascher, K.R.S. (eds.) *Natural Pesticides from the Neem Tree* (Azadirachta indica *A. Juss*) *and other Tropical Plants*. Proc. 2nd Int. Neem Conf. German Agency for Technical Cooperation (GTZ). Echsborn, Germany.
7.60 Simmonds, M.S.J., Blaney, W.M., Delle Monache, F. and Marini Bettolo, G.B. (1990) Insect antifeedant activity associated with compounds isolated from species of *Lonchocarpus* and *Tephrosia*. *J. Chem. Ecol.* **16**, 365-380.
7.61 Slama, K.K. and Williams, C.M. (1966) The juvenile hormone, V. The sensitivity of the bug *Pyrrhocoris apteris* to hormonally active factor in American paper pulp. *Biol. Bull.* **130**, 235-246.
7.62 Sosa, M.E., Tonn, C.E. and Giordano, O.S. (1994) Insect antifeedant activity of clerodane diterpenoids. *J. Nat. Prod.* **57**, 1262-1265.
7.63 Stipanovic, R.D., Williams, H.J. and Smith, L.A. (1986) Cotton terpenoid inhibition of *Heliothis virescens* development. In Green, M.B. and Hedin, P.A.

(eds.) ACS Symposium Series 296. American Chemical Society, Washington, DC, pp. 79-94.

7.64 Stoll, G. (1987) *Natural Crop Protection*. Verlag J. Margraf, Langen, Germany.

7.65 Strong, R.G. and Diekman, J. (1973) Comparative effectiveness of fifteen insect growth regulators against several pests of stored products. *J. Econ. Entomol.* **66**, 1167-1173.

7.66 Su, H.C.F. (1977) Insecticidal properties of black pepper to rice weevils and cowpea weevils. *J. Econ. Entomol.* **70**, 18-21.

7.67 Tu, Y.Q. and Hu, Y.J. (1993) Structure of sesquiterpenoids from *Celastrus angulatus*. *J. Nat. Prod.* **56**, 126-129.

7.68 Usher, B.F., Bernays, E.A. and Barbehenn, R.V. (1988) Antifeedant tests with larvae of *Pseudaletia unipuncta*: variability of behavioral response. *Entomol. Exp. Appl.* **48**, 203-212.

7.69 Villani, M. and Gould, F. (1985) Screening of crude plant extracts as feeding deterrents of the wireworm *Melanotus communis*. *Entomol. Exp. Appl.* **37**, 69-75.

7.70 Williams, C.M. (1970) Hormonal interactions between plants and insects. In Sondheimer, E. and Simeone, J.B. (eds.) *Chemical Ecology*. Academic Press, New York, pp. 103-132.

7.71 Wink, M. (1998) Chemical ecology of alkaloids. In Roberts, M.F. and Wink, M. (eds.) *Alkaloids Biochemistry, Ecology, and Medicinal Applications*. Plenum Press, New York, pp. 265-300.

7.72 Wink, M., Schmeller, T. and Latz-Brüning, B. (1998) Modes of action of allelochemical alkaloids: interaction with neuroreceptors, DNA, and other molecular targets. *J. Chem. Ecol.* **24**, 1881-1937.

7.73 Yang, R.Z. and Tang, C.S. (1988) Plants used for pest control in China: A literature review. *Econ. Bot.* **42**, 376-406.

7.74 Yoshida, H.A. and Toscano, N.C. (1994) Comparative effects of selected natural insecticides on *Heliothis virescens* (Lepidoptera: Noctuidae) larvae. *J. Econ. Entomol.* **87**, 305-310.

7.75 Zúñiga, G.E., Varanda, E.M. and Corcuera, L.J. (1988) Effect of gramine on the feeding behavior of the aphids *Schizaphis graminum* and *Rhopalosiphum padi*. *Entomol. Exp. Appl.* **47**, 161-165.

CHAPTER 8

HOST PLANT RESISTANCE

A natural consequence of identifying intrinsic plant defence chemicals is a search for crop cultivars expressing them in more effective concentrations, such as decreased concentrations of attractants, enhanced concentrations of insecticides, repellents and antifeedants, or a combination of these. Such selected cultivars would then have natural resistance to the appropriate pest.

Host plant resistance (HPR) to insect attack has been defined as: the relative amount of its heritable qualities, possessed by the plant, which influence the ultimate degree of damage done by the insect [8.49]; and as the collective heritable characteristics by which a plant species, race, clone, or individual may decrease the possibility of successful utilization of a plant as a host by an insect species, race, biotype or individual [8.3]. Resistance usually is quantified by reference to susceptible cultivars of the same plant species serving as controls. The terms "host plant" and "immune" are mutually exclusive, i.e., an immune plant is non-host and any degree of host reaction less than immunity is resistance. HPR has been reviewed [8.44] and more recently [8.5, 8.61] to include physical and other aspects that are outside the scope of the present work that, by definition, focusses on only chemical aspects. Furthermore, the use of the term HPR in the present study is confined to relationships involving altered concentrations in cultivars of chemicals that are associated with plant defence. The fact that a particular plant species benefits from protection by intrinsic chemicals is not taken as adequate evidence of HPR.

CLASSIFICATION OF RESISTANCE

Resistance may be classified as follows [8.49]:

(1) Immunity: an immune cultivar is one that a specific insect will never consume or injure under any known condition;

(2) High resistance: is demonstrated by a cultivar that has qualities resulting in small damage by a specific insect under a given set of conditions;

(3) Low resistance: describes a cultivar showing less damage or infestation than the average for the crop considered;

(4) Susceptibility: a susceptible cultivar shows average or more than average damage by one insect species;

(5) High susceptibility: is expressed when much more than average damage is caused by a specific insect.

The term pseudoresistance is defined as apparent resistance in potentially susceptible host plants resulting from chance, transitory non-heritable traits, or from environmental conditions. Three types may be distinguished: (1) host evasion, as when the host may pass through the most susceptible stage quickly if insect numbers are decreased; (2) induced resistance, or temporarily-increased resistance resulting, for example, from a change in the amount of water or soil fertility, although this term has now been redefined (see below); (3) escape, through such transitory circumstances as incomplete infestation.

The foregoing definitions are largely the product of the perspective of the entomologist seeking decreased insect attack. It is also appropriate to consider the perspective of the plant physiologist to delineate intrinsic physiological constraints on plant chemical defences. These provide the upper limit within which the entomologist must operate.

CONSTRAINTS

The production of plant-defensive chemicals is determined by the interaction between the cost of the chemicals, their effectiveness in conferring defence, and the cost of herbivory [8.46]. Chemical costs comprise those of biosynthetic maintenance and turnover, in addition to the fundamental costs of the elements employed. These reflect the contribution the elements make to other vital plant activities including reproduction, photosynthesis, nutrient uptake, and physical infrastructure. Accordingly, the concentration of chemicals available for defence must be the product of a tradeoff between these competing demands. Nitrogen and carbon contribute to the synthesis of defence compounds and defence structures, and carbon further contributes as an energy source. Temperature provides an overriding influence by regulating the availability of such resources.

Nitrogen

Nitrogen is a significant constituent of plant defence chemicals, especially cyanogenic glycosides, glucosinolates, and alkaloids. Leaf nitrogen levels limit insect growth and the leaf's capacity to assimilate carbon. Nitrogen may amount to as much as 5 % of leaf matter, and is also highly motile as 20-40 % of nitrogen in forest leaves is resorbed and translocated. The beneficiaries are usually vulnerable phases such as developing fruits, younger leaves, or seeds. In turn, nitrogen-containing defence chemicals in seeds may be completely degraded during germination.

It is well established that plant growth reflects the nitrogen-content of the fertilizer that is applied. Nitrogen is frequently, but not always, the major limiting factor affecting plant growth and there may be limited scope for allocation of nitrogen to defensive chemicals. Nevertheless, temporary sources of abundant nitrogen may serve as a supplement. For example, habitats that are disturbed or semi-disturbed (old fields) may provide sudden discharges of nutrients, especially nitrogen, that enable much of the growth of many plant species. Usually, the entire quota of a plant's nitrogen stock is taken up during vegetative growth, and translocated and redistributed thereafter. Accordingly, the distribution and concentration of nitrogen-rich defensive chemicals reflects the quantity that is surplus to growth requirements [8.46].

Carbon

Carbon metabolism differs from nitrogen metabolism in one important respect. Almost all nitrogen-containing chemicals are reusable within the plant but most structural ingredients especially cellulose and lignin are not. This greater constraint on the supply of carbon for defensive purposes may be visualized by reference to the deciduous forest.

Some 80 % of the canopy of the deciduous forest develops in the spring. This drains carbohydrate reserves, as photosynthesis is inadequate to meet the needs of leaf growth, flower production, and shoot elongation. This loss seems to be made up soon afterwards. Young leaves thus produced lack the physical protection of sclerophylly and are intensively fed on by insects. In principle, the duration of this phase could be manipulated by, for example, decreasing the protein content of leaves to retard growth and decrease photosynthesis. But in either eventuality, development time would be extended which would merely increase opportunities for insect attack. Conversely, sclerophylly could be advanced but lignification and development of more rigid cell walls would retard leaf expansion. Lengthening the demand time on carbon reserves would decrease opportunities for reconstruction of the reserve.

These considerations almost certainly have wider applicability. It is commonplace in crop/insect relationships to find a period of enhanced vulnerability to insect attack coinciding with shoot/leaf elaboration. Indeed this period frequently coincides with the production of the first instar or invading larva of many species of phytophagous pests. Accordingly, it seems likely that species such as the wheat-bulb fly *Delia* (=*Leptophylemia*) *coarctata* have evolved to synchronize larval hatching with the production of such young, tender, unprotected shoots or leaves [8.56].

MECHANISMS OF RESISTANCE

Mechanisms of resistance were originally classified into three categories [8.49] to which a fourth has been added:

(1) Non-preference / Antixenosis: the latter is the term of choice as it is a plant property whereas the former is an insect one. Also it is a parallel term to antibiosis and conveys that the plant is avoided as a poor host [8.37]. Through coevolution, insects have adapted to their hosts as sources of attractants, feeding and oviposition stimulants. Antixenosis results from the lack of these stimulants, or from the presence of chemical and physical deterrents, or from an effective balance between these.

(2) Antibiosis: comprises the defensive mechanism of plants against their pests through adverse influence on growth, and survival or reproduction, by means of chemical or morphological factors;

(3) Tolerance: or the ability to withstand attack without appreciable loss of vigour or crop yield.

The foregoing three categories relate to constitutive resistance derived from factors present prior to insect attack.

(4) Induced resistance: recent work has emphasized the need to take greater account of injury-dependent, post-infestation responses of plants to insect attack. Such induced resistance, long recognized as elicited by pathogen attack, has been redefined as:

> the qualitative or quantitative enhancement of a plant's defence mechanisms against pests in response to extrinsic physical or chemical stimuli inducers [8.38].

Before considering each mechanism in detail it is important to distinguish between evidence implicating a particular chemical in plant defence, and satisfactory evidence that it is relevant to HPR. It does not necessarily follow that a compound contributing to plant defence is also a resistance factor. Establishing this requires, as a minimum, an association between concentrations of the compound in resistant and susceptible cultivars on the one hand, and degrees of resistance on the other. This distinction, frequently overlooked or avoided, is one reason why there are relatively few convincing few examples of chemical-based intraspecific resistance [8.4].

One rather rigorous protocol proposed to address this difficulty is the use of bioassay-driven fractionation to identify behaviourally-significant chemicals, coupled with collection of two kinds of data to establish the effect in planta: direct evidence for an effect of the chemical(s) on the herbivore, perhaps through incorporation in an artificial diet (although this would serve well in regard to only antixenosis and antibiosis), and indirect evidence comprising associations between allelochemical concentrations and degrees of resistance [8.4].

ANTIXENOSIS

Oviposition

Among insects that lay eggs on or near plants utilized by the progeny, resistance to oviposition is the first point at which the plant may evince resistance. Oviposition is not a simple act but may involve a catenary series of behavioural events such as: recognition and orientation to the host plant; settling; selecting the oviposition site; and laying eggs. As each stage may be influenced by different plant characteristics, resistance may derive from: (1) failure to provide a stimulus for one or more of the behavioural components; (2) provision of inhibitory stimuli; (3) an effective balance of these.

The initial orientation of the gravid female towards the host plant may occur in response to host plant volatiles. Migrating cotton boll weevils *Anthonomus grandis* are attracted to fruiting cotton and the period of maximum migration (July/August) coincides with peak production of cotton volatiles [8.22]. The carrot fly *Psila rosae* (Diptera) represents one of the rather few well-worked species evincing a clear relationship between HPR and concentrations of oviposition stimulating chemicals.

Chronologically, it was shown that oviposition was stimulated by three volatiles isolated from carrot leaves with *trans*-1,2,dimethoxy-4-propenyl benzene or *trans*-methylisoeugenol as the most active. The *cis* isomer is inactive as are eugenol, methyleugenol and *trans* and *cis*-isoeugenol, indicating considerable stereospecificity in the female sensory system. The other two actives are 4-allylanisole and anisaldehyde. Subsequently, *trans*-asarone, an analogue of *trans*-methylisoeugenol, was shown to attract flies to foliage, and to act synergistically with other ingredients in stimulating oviposition [8.64, and references therein]. However, a subsequent investigation indicated that falcarindiol (Fig. 66) could be a significant stimulus for oviposition, as judged by the criterion of presence in carrot leaf of concentrations well in excess of the estimated threshold for oviposition, 3,200 ng and 6 ng respectively. Subsequently, four-fold and statistically significant variations in egg numbers were elicited by different cultivars but *trans*-methylisoeugenol did not vary in concentration in association with their resistance [8.42].

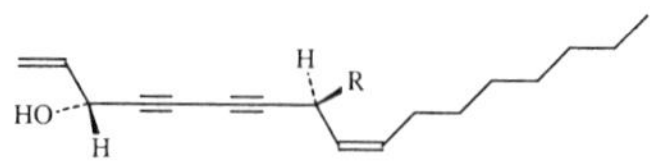

Figure 66. Structure of falcarindiol.

The use of egg traps, providing total enclosure of root but leaving foliage unaffected, abolished differential oviposition between resistant and susceptible cultivars. Modifying the egg trap to expose the root, restored the difference. Evidently, it is root not foliage that regulates nonpreference oviposition by the fly [8.41]. Furthermore, there is more falcarindiol in root that in leaf (means of 135 μg and 3 μg respectively), and the difference between resistant and susceptible roots is ca 800 μg or 133-fold larger than threshold. Clearly, the differential concentration of this compound in root will exceed in significance any differences assignable to the leaf [8.42]. Significantly, the oviposition stimulating chemicals first detected were not truly associated with resistance, providing further evidence of the need for caution in assigning allelochemicals a role in HPR merely because they modify insect behaviour.

Oviposition is not usually governed by single compounds. In Cruciferae a combination of several glucosinolates is required to elicit oviposition by *D. radicum* and *D. floralis* [8.7]. The discovery of oviposition deterrents, such as pentadecanal in rice serving to deter *Chilo suppressalis*, has emphasized the significance of balance between stimulants and inhibitors [8.59, and references therein].

Penetration and Feeding

Antixenosis may also affect orientation, biting, and maintenance of feeding by insects. Feeding inhibitors are classified as: (1) repellents; (2) suppressants of initial biting and piercing; and (3) deterrents of continued feeding [8.3]. As β-pinene is repellent to the Douglas-fir beetle *Dendroctonus pseudotsugae* and as α-pinene is an attractant, the tissue specificity shown by this insect may be related to the ratio of α- / β-pinene in the bark of Douglas fir *Pseudotsuga menziezii* [8.11]. Alkaloids and monoterpenoids are the most consistently deterrent compounds, at low concentrations, to feeding by the locust *Locusta migratoria*; grasses readily acceptable to both the locust and to grasshopper *Chorthippus parallelus* do not contain deterrents in sufficient quantity [8.6]. Steam distillates of resistant rice cultivars repelled and killed the brown planthopper (BPH) *Nilaparvata lugens* and the green leafhopper *Nephotettix virescens* respectively and significantly decreased their feeding. Furthermore, the foliar wax coating of rice, and specifically the hydrocarbon- and carbonyl-containing fractions adversely affect hopper feeding behaviour by causing the BPH to move away from preferred feeding sites [8.77].

In regard to larval feeding, the investigation of carrot resistance to *P. rosae* is again relevant by showing two distinct patterns of larval establishment. In the expected one, small egg numbers were followed by small attacks, and large egg numbers by large attacks. The converse is true in the second pattern, where small egg numbers are followed by a substantial attack, or where large egg numbers are followed by low attack levels (Fig. 67). Clearly, larval numbers in root do not simply reflect numbers of eggs laid but are affected by root resistance to larval invasion. In other words, the carrot plant has a second line of defence after taking effect on oviposition. As larval damage to root is the economically important event, then root resistance to the larva is the crucial requirement for meaningful resistance to *Psila* [8.41]. This effect was confirmed in a separate set of cultivars.

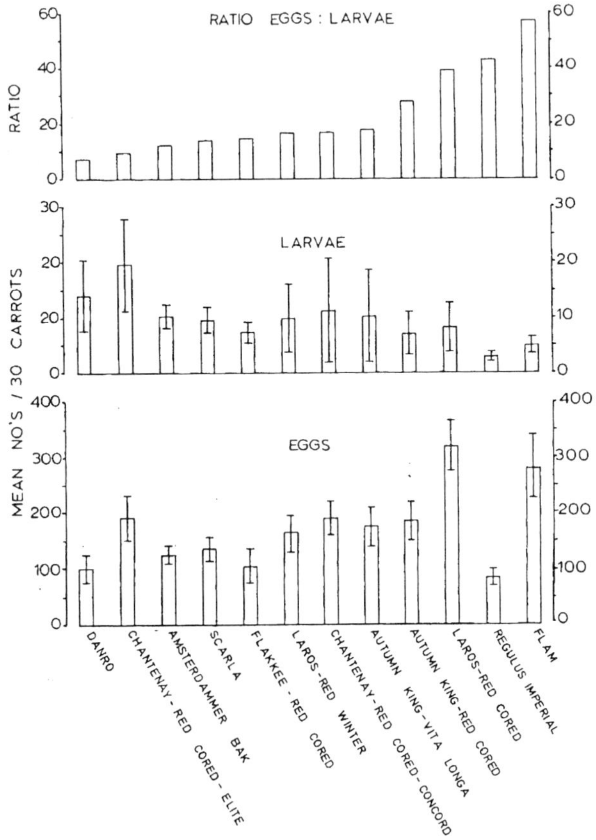

Figure 67. Resistance modes of carrot cultivars to carrot fly Psila rosae *involving antixenosis in terms of decreased numbers of eggs laid, and decreased numbers of larvae established in root. The range in egg numbers required to establish a single larva in a plant (ca 60-10), indicates the independence of root resistance to larvae. Compiled from Maki and Ryan [8.41].*

The steam-distillate and headspace vapour from roots of a resistant cultivar yielded 38 and 23 % less oil respectively than from roots of a susceptible one. The carbonyl-rich fractions contained the actives, of which the neurotoxin falcarindiol was a crucial compound by serving as: (1) the principal observed stimulus for oviposition as judged by choice bioassays and electrophysiological recordings; (2) a potent host-location stimulus for the larva and (3) as an important factor, in decreased concentrations, in increased root resistance (see above). Coevolution has transformed the role of many toxins into host-location cues but this seems a rare example of a neurotoxin eliciting antixenosis through decreased concentrations [8.42].

Trichomes and Glands

Trichomes are epidermal appendages in the form of glandular hairs and scales or peltate hairs. Glandular trichomes produce a variety of secretions including essential oils, resins, gums, and other sticky substances. They are viewed as the precursors of nectaries that evolved as insect attractants. Apparently, nectaries attracted pollinators, and extrafloral ones rich in sugars attracted ants that defended the plant against insect herbivores except, of course, some aphid species [8.65 and 8.74, and references in both].

Physical Properties

Some trichomes are nonsecretory and act as physical barriers. Their density, length, and degree of branching is positively associated with insect mortality. Furthermore, they may decrease by up to one-third oviposition by the cereal leaf beetle *Oulema melanopus* on pubescent wheat, as compared with a glabrous control. Such eggs as were laid decreased in viability with increased pubescence due, apparently, to desiccation from exposure. Survival and growth of the first instar larva also decreased [8.26].

Hooked trichomes on the bean *Phaseolus vulgaris* entrap insect herbivores such as the potato leafhopper *Empoasca fabae* and the aphids *Myzus persicae*, *Aphis fabae* and *A. craccivora*. Furthermore, trichomes piercing *E. fabae* nymphs killed sufficient of them to constitute the major mortality factor affecting this species on *P. vulgaris* in the field [8.51]. The more pubescent plants of the soybean *Glycine max* supported fewer *E. fabae*, that also were less fecund. Trichomes impede movement, and preclude access by the proboscis to the mesophage [8.39].

Chemical Properties

Allelochemical constituents of trichome secretions include, terpenes, alkaloids, flavonoids, fatty acids and alcohols. Indeed, more than 100 mono-, sesqui- and diterpenes have been isolated from glandular trichomes serving as the first line of defence of many plant species especially Solanaceae against pests and pathogens [8.28, 8.65, and references in both]; the role of such compounds in plant-insect relationships has been considered in detail in Chapter 2. Essentially, individual chemicals from each of the foregoing groups contribute to resistance, in a general sense, by their presence in trichomes. The answer to the more searching question previously raised is less clear: are variations in the concentration of such compounds systematically associated with different degrees of resistance between cultivars of a particular species? However, as it is established that cultivars vary in degrees of pubescence the concentration of allelochemicals might vary correspondingly.

One example of specific resistance involving trichomes is provided by cotton. The glandless strain was eaten in preference to the glanded one by bollworms, *Helicoverpa* and *Heliothis* species, the cotton leaf worm *Alabama argillacea*, and the beet armyworm *Spodoptera exigua*. In no-choice field tests, a glandless genotype sustained a 2.5-fold larger infestation of the lygus bug *Lygus hesperus*, than isogenic glanded plants. This effect was assigned to 2-fold greater nymphal growth and survival on the glandless lines [8.70]. In addition, glandless cotton is attacked by the following arthropod pests that rarely attack cotton: blister beetles *Epicauta* spp.; cucumber beetles *Diabrotica* spp.; grape colapsis *Maecolaspis flavida*; various cutworm species; and pillbugs *Porcellio* spp. The pigment glands anatomically characteristic of cotton seed contain gossypol demonstrated experimentally as lethal to: the cotton aphid *Aphis gossypii*; lygus bug *Lygus hesperus*; the salt marsh caterpillar *Estigmene acraea*; various thurberia boll weevils including *Anthonomus grandis thurberiae*; bollworms. Conclusively, there was a four-fold larger concentration of gossypol in glanded as compared with glandless cotton. A compound with remarkable properties, this sesquiterpene dimer is also bactericidal, viricidal, anti-tumour, and decreases the fertility of the human male. Further, as an inhibitor of acetylcholinesterase (see Chapter 2 [2.49, 2.35 respectively]), it could decrease the mobility of herbivorous insects (see later for role in antibiosis).

The role of trichomes in resistance has also been demonstrated in species and cultivars of potato, *Solanum* species, as studied by Tingey and associates. The gummy exudates immobilize and entrap arthropod pests especially soft-bodied ones. Two trichomes types exist in *Solanum* species: Type A, short with a tetralobulate gland at tip; and Type B, a simpler, longer, multicellular hair with an ovoid gland at the apex. Their exudates accumulate on mouthparts of *E. fabae*, to occlude the labium tip and encase the tarsi (Fig. 68). Type A and Type B exudates may interact. Type B exudate is very viscous and adhesive such that aphids, coated on mouthparts and antennae, try to remove it and spend less time feeding; tarsi coated with Type B exudate readily discharged, from type A trichomes, a clear viscous exudate that rapidly darkened and hardened, encasing the affected insect appendages [8.47]. Direct biochemical evidence was obtained for the action of polyphenolic oxidase (PO) and polyphenolic peroxidase (PPO) in the trichome secretions. Essentially, phenolic substrates were oxidized by the exudate, and this reaction was blocked by inhibitors that bind with the copper prosthetic group of PPO and the heme group of PO [8.55].

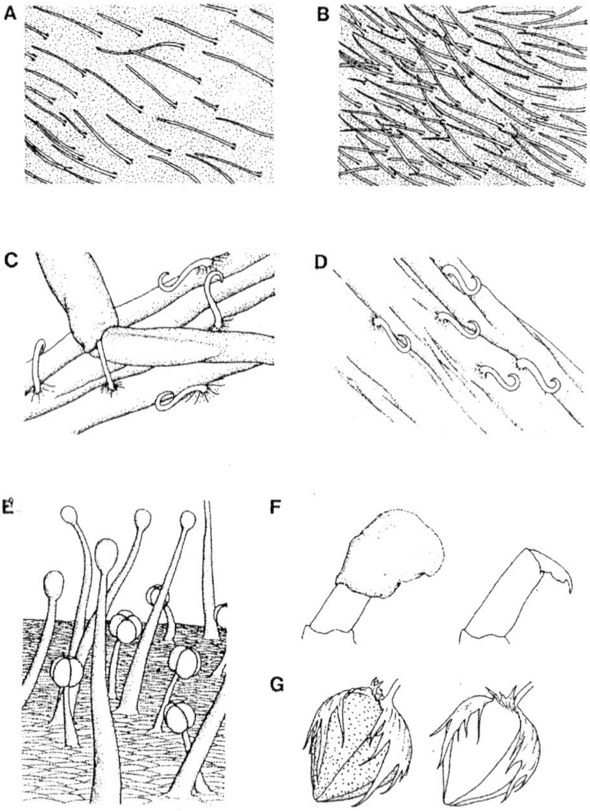

Figure 68. Schematic representation of trichome expression and effects in terms of: A) medium and B) profuse density; C) insect limb impaled by D) recurved trichomes; E) tetralobular (type A) and simple (type B) trichomes associated with Solanum *spp.; F) tarsus encased in trichome exudate, and free; G, glanded and glandless cotton bolls, of which the former produces gossypol. Redrawn after Tingey* et al. *[8.70].*

Trichomes on the resistant, wild, Bolivian potato species *Solanum berthaultii* affect *M. persicae* by delaying probing, eliciting probes of shorter duration, and by decreasing total feeding time, as compared with the susceptible Hudson cultivar of *S. tuberosum*. Experimentally removing type B droplets, and transferring them to leaves of cv. Hudson attributed the effect to type B glandular trichomes [8.71, and references therein]. Furthermore, *S. berthaultii* slowed larval development of the Colorado potato beetle *Leptinotarsa decemlineata* and only 2 % of larvae survived to adulthood, as compared with 41 % survival on *S. tuberosum* cv. Katahdin. A hybrid family presented intermediate levels of defoliation and of population density of adult *Leptinotarsa*. In the second generation, population densities of small and large larvae were decreased by 90 and 87 % respectively on the hybrid family, and by 96 and 97 % respectively on *S. berthaultii*, as compared with *S. tuberosum*. In one of the few explicit tests of the practicality of such resistance, Tingey's group showed that three to four insecticidal treatments were required for acceptable levels of control on three hybrid families compared with one on *S. berthaultii*, whereas three were insufficient for control on *S. tuberosum* [8.78]. However, trichomes also decrease population densities of predators, especially Coccinellidae, and parasitoids of aphids. Specifically, they decrease the mobility of first instar coccinellid larvae, which is consistent with decreased predation, and decrease the effectiveness of *Encarsia formosa*, the parasite of greenhouse whitefly. On this account, and to preclude selecting pest strains resistant to trichomes, a moderate trichome density is recommended rather than a very pubescent one [8.72].

Among various other examples of trichome-mediated chemical involvement in resistance is elicitation of mortality of *O. melanopus* larvae, through injection of cellulose and lignin by hairs protecting the surface; such hairs also kill by piercing the larval gut [8.75]. Hair-induced mortality in the larva of the Mexican bean beetle *Epilachna varivestis* was also elicited on soybean [8.36]. Glandular trichomes govern geranium resistance to the foxglove aphid *Acyrthosiphon solani* [8.73]. Trichomes of *Nicotiana* species exude allelochemicals such as nicotine, anabasine, and nornicotine that paralyze and kill aphids. Expression of nornicotine in hybrids from crosses of species that do or not produce it, indicated that the ability of insects to acylate this compound was inherited in a dominant pattern [8.27].

Surface waxes

Surface waxes and specifically a $C_{32}H_{66}$ alkane in leaves of the broad bean *Vicia faba* elicit probing by the pea aphid *Acyrthosiphon pisum*, but the alkane fraction from a nonhost deterred feeding [8.33]. Applying epicuticular wax from rice cultivar IR46, that is resistant to the BPH, to susceptible IR22 decreased probing and increased restlessness in this pest [8.77]. Seedling but not older sorghum plants are distasteful to *L. migratoria*, and deterrency is abolished by extracting the surface wax that contains alkanes and esters. Surface waxes may, however, also contain feeding stimulants [8.1].

Nutrients

Mean percentage N, P and K is significantly larger in susceptible germplasms of maize as compared with those resistant to the rice stem borer *Chilo zosellus*. Development of the grasshopper *Melanoplus mexicanus* is impaired if soil nitrogen levels cause foliage to contain less than 27 % protein. Other results tend to discount the importance of nutrients by showing, for example, that host and nonhost plants of the cotton boll weevil *Anthonomus grandis*, all contain the amino acids required by the pest, and that quantitative differences between them were insufficient to distinguish between hosts and nonhosts. In addition, it is claimed that phytophagous insects do not differ greatly in their nutritional requirements that are provided by almost every green plant [8.58, and references therein]. Nevertheless, the nutritionally-inadequate sap of the aphid-resistant raspberry cultivar Canby elicits retarded development, decreased size and fecundity, and increased mortality in the aphid *Amphorophora agathonica*. After feeding for 24 hr the aphids are restless and semistarved, apparently due to inadequate levels of sugars and nitrogen in the sap, and they depart the plant [8.31].

ANTIBIOSIS

Both biological and biochemical factors contribute to this type of plant resistance. Nutrients, physiological inhibitors, and toxicants influence antibiosis by affecting feeding and consequently, growth, reproduction and metamorphosis [8.3]. More recently, antibiosis has been categorized as consistent with one or more of the following: death of early instars; disturbances in physiology, morphogenesis and behaviour; and continuation of nutrient-poor food reserves [8.50].

Toxins

There are well documented examples of the importance of phenolics, their glycosidic derivatives and tannins (polymers and phenolic acids) in the constitutive resistance of plants to insects. Grape cultivars resistant to *Phylloxera vitifoliae* have higher tissue levels of phenolics than susceptible cultivars, and phenolics also affect resistance of sweet clover to insect attack [8.43; and references therein]. Other notable physiological inhibitors include, alkaloids (nicotine), terpenoids, juvenile hormone analogues and antijuvenile hormones.

The powerful compound 2,4-dihydroxy-7-methoxy-1,4-benzoxazin-3-one (DIMBOA) (Fig. 69) is a chemical factor in the resistance of maize *Zea mays* to first brood of the European corn borer (ECB) *Ostrinia nubilalis*. DIMBOA levels are high in most seedling stage maize plants, but decrease as plants mature. At the mid-whorl stage of development some lines retain high DIMBOA levels and are resistant to ECB, but DIMBOA levels are low in all lines at the time of second-brood infestation [8.34]. DIMBOA in glucoside form occurs in uninjured tissues but damage by insect feeding elicits enzymatic conversion to the aglycone that decomposes through intermediates to afford 6-methoxy-2-benzoxazolinone (MBOA), serving as a repellent or as a feeding deterrent. Incorporating MBOA in a meridic diet (1.5 mg/g) significantly extended time to pupation and adult emergence of the ECB. Further, the proportion of females in total emergence was decreased, as was their fecundity. Nevertheless, the pest may respond by jettisoning toxic deposits in the pupal case [8.10]. *N-O*-ME-DIMBOA (2-dihydroxy-4,7-dimethoxy-1,4-benzoxazin-3-one) occurs in whorl surface waxes of maize in concentrations larger than those of DIMBOA and MBOA. As concentrations of the compound are also larger in resistant than in susceptible lines, it was assigned a role in resistance to the southwestern corn borer *Diatraea grandiosella* [8.23].

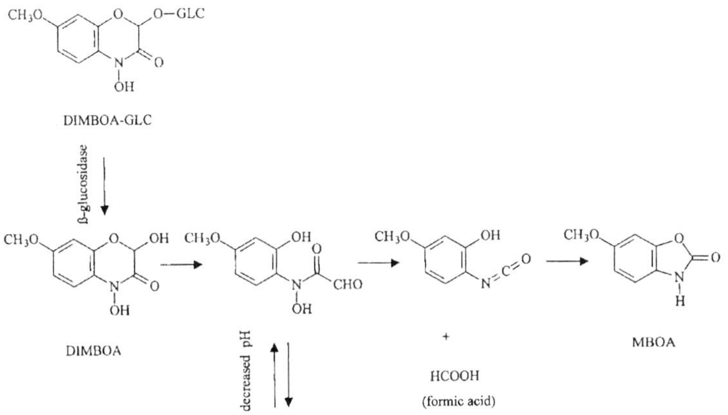

Figure 69. Production and breakdown of DIMBOA. Redrawn after Panda and Khush [8.50].

Increased levels of total glycoalkaloids (TGA) representing toxic, nitrogen-containing, steroidal glycosides in the foliage of 10 wild, tuber-bearing species of *Solanum* are significantly and positively correlated with resistance to nymphs of *E. fabae* [8.71] (Fig. 70). Values ranged from 13 to 688 mg/100 g fresh weight. The specific glycoalkaloid, tomatine, elicits stylet withdrawal, and concentrations as small as 0.05 % decrease the duration of salivation-ingestion [8.52]. Sixteen wild, tuber-bearing *Solanum* species, of potential value in potato breeding, evinced TGA concentrations of 0.2-39.0 mg/g dry weight, whereas the cultivated potato *S. tuberosum* cv. Katahdin was represented by the rather low level of 2.6 mg. Most species contained only α-solanine and α-chaconine, but glycoalkaloids not usually associated with commercial species were present in the wild ones. The very resistant *S. berthaultii* contained solasodine as the major constituent, in addition to compounds believed to be solamargine and solasonine [8.19]. Clearly, both total TGA content and the presence of specific glycoalkaloid constituents are significant for potato resistance.

Removal of glandular trichomes from an *S. tuberosum* and *S. berthaultii* F_3 hybrid increased aphid survival two-fold. In addition, adult longevity and fecundity were decreased by ca 63 % on the resistant hybrid, as compared with a susceptible cultivar. Further studies confirmed the heritability of this resistance by demonstrating that trichome exudate from the hybrid was as effective as that from *S. berthaultii* in immobilizing aphids [8.78]. *S. berthaultii* is also resistant to *L. decemlineata*, the major factor limiting potato production in the northeastern and mid-Atlantic states of the U.S.A. Specifically, there was a three-fold decrease in egg masses laid on *S. berthaultii* compared with

S. tuberosum cv. Katahdin, also egg numbers in individual egg masses were 38 % smaller [8.78].

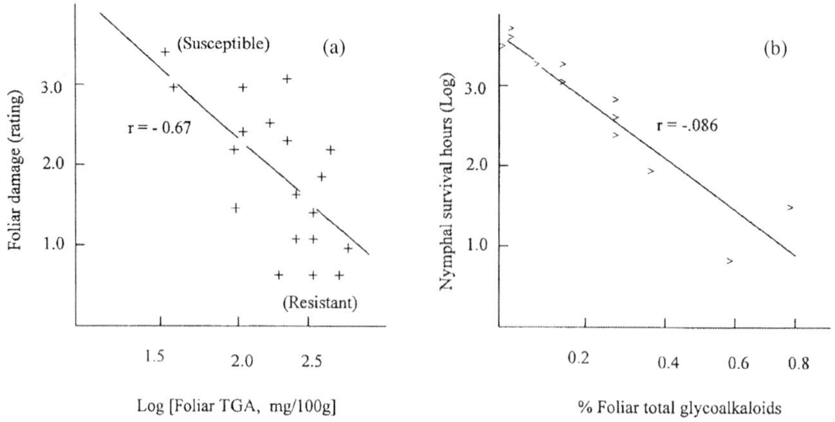

Figure 70. Association of: (a), nymphal survival of the potato leafhopper Empoasca fabae *with levels of total glycoalkaloids (TGA) in 12* Solanum *accessions [Redrawn after Raman* et al., *8.52]; and (b) foliar damage by Colorado potato beetle* Leptinotarsa decemlineata *and TGA levels in 20 clones of* Solanum chacoense. *Redrawn after Tingey [8.69].*

The Solanaceae provide a further example of convincingly demonstrated, chemically-based resistance, in that leaf-trichome concentrations of 2-tridecanone amounted to 45 µg and 0.1 µg in cultivated and wild tomato, resistant and susceptible respectively to the tobacco hornworm *Manduca sexta*. 2-Tridecanone, concentrated in the tips of glandular trichomes, is a feeding deterrent and is neurotoxic to *M. sexta*, tomato fruitworm *Helicoverpa zea*, *L. decemlineata*, and the aphid *A. gossypii*. Neonate larvae of *H. zea* firstly, exhibit paralysis and convulsions, then recover and begin feeding, only to die within days due to other factors associated with the leaf lamella [8.29] and not trichomes. Subsequent investigations implicated the 11-carbon methyl ketone, 2-undecanone which was established as necessary and sufficient in concentrations corresponding to those in planta to elicit pupal mortality and deformity, possibly through inhibiting metabolism of linolenic acid [8.16] Using resistant and susceptible lines of geranium *Pelargonium xhortorum*, and corresponding plants, washed with buffer to remove tall glandular trichome exudate, showed this was a critical factor for antibiosis expressed through decreased survival and fecundity [8.73]. For a detailed checklist of plant species exercizing resistance through glandular trichomes see 8.74.

The indole alkaloid gramine (N,N- dimethyl-3-aminomethylindole) is a toxin and a feeding deterrent associated with resistance of barley cultivars to the greenbug *Schizaphis graminum* and the aphid *Rhopalosiphum padi* [8.79]. Antibiosis of the grain aphid *Sitobium avenae* is associated with elevated levels of hydroxamic acids of which DIMBOA is the major constituent, and with phenols, but not with indole alkaloids [8.40] (Fig. 71). Antibiosis of the soybean looper *Pseudoplusia includens* through chronic inhibition of insect growth is mediated by coumestrol [8.61]. Resistance of wild Mexican beans *Phaseolus* spp. to the bruchid pest *Zabrotes subfasciatus* was attributed to the presence and antibiotic effect of a novel protein arcelin; resistance to another bruchid species *Acanthoscelides obtectus* was attributed to the presence of heteropolysaccharides [8.45]. A comparison of Asiatic cottons *Gossypium arboreum*, with commercial lines of *G. hirsutum* in terms of resistance to tobacco budworm *Heliothis virescens*, detected in the former cottons smaller concentrations of gossypol and larger concentrations of flavonoids. The most prevalent and toxic of these were gossypetin 8-0-glucoside and gossypetin 8-0-rhamnoside, neither of which occurred in *G. hirsutum* [8.24, and references therein]. Clearly, trichomes operate in at least two different categories antixenosis and antibiosis, and in the latter through at least two different modalities, toxins and growth inhibitors.

Rather fewer data are available for tree resistance. The concentration of specific terpenes in Douglas-fir *P. menziesii* foliage affects the success of the western spruce budworm *Choristoneura occidentalis* [8.11]. At a Montana site, nine variables explained 50 % of the variation in the budworm infestations and the following factors were inversely correlated with budworm population density: the acetate fraction of the terpenes; myrcene; an unidentified terpene; the timing of budburst; and bole radius. Positively correlated with budworm success were: evenness in terpene distribution in foliage; concentration of α-pinene; total nitrogen of foliage; and tree age. Quantitatively, the two most important chemical factors were the acetate fraction of terpenes and the distribution of terpenes in the foliage. Polyphenol concentrations were not relevant. Trees with an uneven terpene distribution were more resistant due likely to the distribution of the specific influential terpene factors, i.e., the acetate fraction of terpenes, myrcene and an unidentified terpene.

At Barley Canyon, New Mexico, introducing budworm to trees infested by few larvae showed that larger concentrations of: an unidentified terpene; total nitrogen; α-pinene; and of myrcene, significantly decreased dry weight of the female budworm. Similarly, larger concentrations of α-pinene, myrcene, bornyl acetate and its analogue citronellyl acetate, terpinolene, and of an unidentified terpene were associated with decreased dry weight production of male budworm/tree. Again, plant polyphenols were not a significant factor. Although nitrogen levels were implicated in resistance these did not seem large enough to elicit budworm toxicity. Such levels, therefore, were seen as merely reflecting enhanced tree productivity, that in turn increased overall chemical resistance to the budworm. Thus, increased concentration of terpenes provided the most important chemical variable associated with resistance. These data were inconsistent with the prediction that after qualitative defences were overcome, they would have less effect on larval growth than nutrient levels. It was quite clear that the designated terpenes were deleterious to budworm growth, i.e., were toxins and growth inhibitors [8.11].

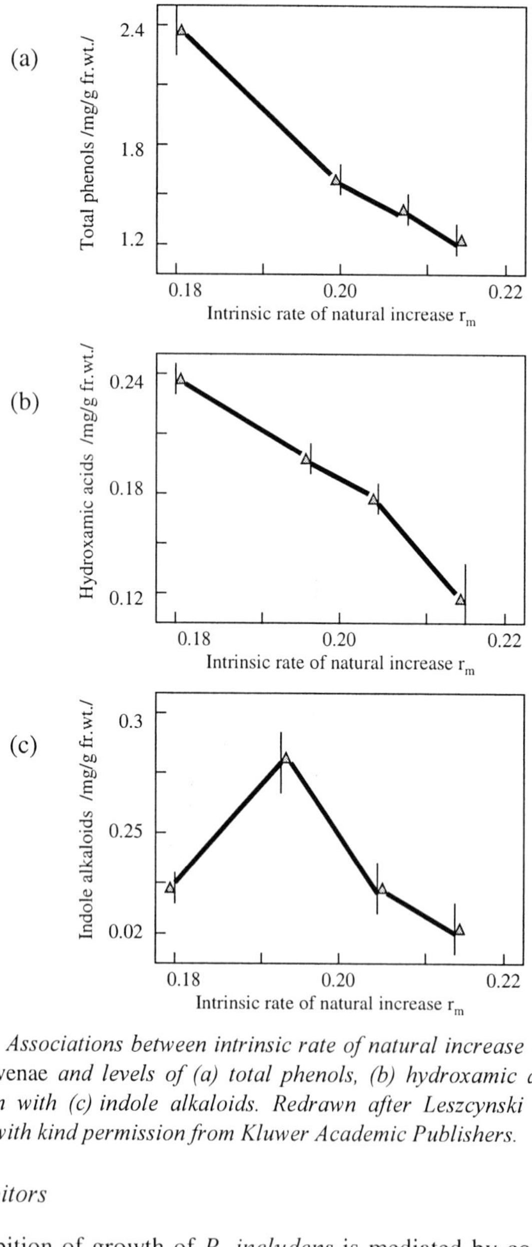

Figure 71. Associations between intrinsic rate of natural increase of the grain aphid Sitobion avenae *and levels of (a) total phenols, (b) hydroxamic acids. There is no association with (c) indole alkaloids. Redrawn after Leszcynski* et al. *[8.40] and reprinted with kind permission from Kluwer Academic Publishers.*

Growth inhibitors

Chronic inhibition of growth of *P. includens* is mediated by coumestrol. Gossypol, 0.1 % in a larval diet decreased by about 50 % the weight of 10-day old bollworm *Helicoverpa* sp., pink bollworm *Pectinophora gossypiella* and tobacco budworm *H. virescens*. Gossypol concentrations were negatively correlated with growth of the *H. virescens* larva. Similarly, the feeding rate of the cotton leafworm *Spodoptera littoralis*, was negatively correlated with gossypol concentrations [8.50, and references therein]. In regard to *H. virescens*, the effect of gossypol is enhanced by hemigossypolone, sesquiterpene quinones, and various heliocides. The monoterpene caryophyllene oxide at high concentrations synergizes the effect of gossypol. Specific biochemical mechanisms were: amylase and protease inhibition; inhibition of lipid peroxidation; increased microsomal N-demethylation activity; stimulation and inhibition of ATP-ase activity by small (10 μM) and by large gossypol concentrations respectively. These effects are obviously consistent with the observed effects on larval feeding and growth. Stimulating microsomal enzymes has, however, the rather undesirable side effect of enabling resistance to organochlorine and organophosphate insecticides, presumably through enhanced detoxification [8.66, and references therein]. The glycoalkaloid α-tomatine dose-responsively inhibits larval growth of *H. zea*, which is reversed in this species by equimolar doses of dietary cholesterol [8.8]. Clearly, individual compounds, such as gossypol and tomatine, may serve both antixenosis and antibiosis.

In addition, resistance of the Zapalote Chico cultivar of maize to *H. zea* has been attributed to the presence of maysin, a luteolin-C-glycoside flavone, operating through antibiosis in the silks. High performance liquid chromatography (HPLC) detected related flavones including apimaysin and 3'methoxymaysin. Bioassays of commercially available and related compounds indicated that the sugar moiety was not necessary for efficacy as luteolin was as potent as maysin, as was rutin (quercetin-3-rutinoside); galactoluteolin, apimaysin, 3'methoxymaysin, and chlorogenic acid were about 80, 50, 50 and 50% as efficacious respectively as maysin [8.76, and references therein].

The role of tannins has been explored in regard to soybean resistance to the agromyzid bean fly *Ophiomyia centrosematis*. Polyphenols including tannins, and polyphenols with tannins and lignins contribute to resistance of the undifferentiated and differentiated stem respectively; other characteristics associated with resistance included a narrow stem diameter and the presence of a purple pigment, the anthocyanidin malvidin. This was the first demonstration of an association between epidermal colour and the degree of plant resistance as judged by numbers of larvae and puparia [8.12]. Subsequently, antixenosis resistance in Cruciferae to the cabbage aphid *Brevicoryne brassicae*, was attributed to a red and glossy appearence [8.60]. Data from near-infrared (NIR) spectrophotometry indicated that high alcohol/aldehyde ratios, and shorter carbon chain lengths in the surface wax of sugarcane stalks was associated with resistance to the sugarcane borer *Eldana saccharina* through antibiosis, although a cause-and-effect relationship was not claimed [8.54].

Nutrients

Sugars and nitrogen are key dietary requirements for insects and deficiency levels are associated with antibiosis. Thus, the need of the ECB larva for glucose during the first three instars is such that a deficiency may elicit antibiosis. Both this species and *H. zea* distinguish between sugar concentrations, and plant resistance to the latter species is associated with sugar balance [8.35, and references therein].

Nitrogen levels in sap profoundly affect aphid feeding and survival. Resistance in green pea to *A. pisum*, is associated with statistically significant deficiencies in arginine, proline, serine, methionine and lysine, amino acids essential for growth of the aphid [8.2]. Population sizes of the grain aphid *Sitobion avenae* were directly proportional to the total content of free and essential amino acids, and larger cultivar resistance was associated with larger values for the ratio of free-phenol to free-amino acid content. The principal mechanism of resistance was antibiosis [8.13]. Protein content and protease activity of European wheat cultivars resistant to cereal aphids are substantially decreased compared to susceptible ones [8.40] In further regard to amino acids, decreased asparagine levels in the resistant rice cultivar Mudgo inhibited feeding by and fecundity of BPH [8.63].

Although it is abundantly clear that resistant cultivars are chemically well-defended against infestation, so-called susceptible cultivars are not without defensive resources. Larvae of *H. zea* fed on glanded (resistant) and glandless (susceptible) cotton, were compared with those raised on an artificial diet. Growth on the susceptible, glandless cotton was only 25 % of that on the diet. In addition to the likely involvement of structural and morphological factors in the glandless cotton, attention has been drawn to allelochemical-nutrient interactions. Defensive chemicals may decrease the bio-availability of nutrients, a relationship better understood for vertebrates than for invertebrates [8.53].

ANTIXENOSIS AND ANTIBIOSIS

In assigning plant resistance to one or other of the above, it must be recognized that several plant species are established as employing both. Thus maize hybrids resistant to *D. grandiosella*, and the ECB decreased feeding, weight gain, and survival, leading to the conclusion that both antixenosis and antibiosis were taking effect [8.14]. Significantly fewer larvae of the rice leaffolder *Cnaphalocrocis medinalis* settled on various species of resistant wild rice; larval weight was less on the resistant species and larval survival was also less when fed a diet enriched with powdered resistant plants [8.32]. The resistance of sorghum genotypes to the stem borer *Chilo partellus*, was attributed to ovipositional antixenosis, and anti-larval antibiosis [8.68]. Resistance of the summer forms of wheat, barley, and rye to the pale aphid *Metopolophium dirhodum* was attributed to both antixenosis and antibiosis [8.25]. Although resistance of cowpea *Vigna unguiculata* to *A. craccivora* is mainly through antixenosis, one cultivar also effected antibiosis through high nymphal mortality and low reproduction [8.17]. Two cultivars of rice resistant to *C. suppressalis* exhibited ovipositional antixenosis and greater toxicity in the vegetative stage (50 days) than in the reproductive stage (90 days) [8.15].

TOLERANCE

This expression of resistance, representing plant responses to insect attack, differs from the previous two that are based exclusively on response by the insect to the plant. Plant tolerance allowing genotype maintenance offers selective advantages. Practical gains include: the higher economic threshold for damage entails a lighter insecticidal load impacting less on the efficacy of natural enemies and on the environment; tolerance exerts no selective pressure on the pest and so does not promote development of resistant strains; tolerance may provide a temporarily stable situation in the event of pest strains developing such resistance to chemically-based antixenosis and antibiosis [8.50, and references therein].

INDUCED RESISTANCE

This is distinguished from constitutive resistance by relying on insect attack to induce the defensive response. Biological inducers are represented by plant pathogens and previous herbivory. Environmental factors include temperature, light, relative humidity, soil fertility, soil moisture, and air pollutants including ozone and sulphur dioxide. Among man-made inducers are insecticides, herbicides, fungicides, and growth regulators. A noteworthy feature of induced resistance is efficiency, which stems from confinement of the response to an actual attack, thus avoiding investment in defences that never may be required.

Tobacco cultivars sensitive to, and infected with, tobacco mosaic virus (TMV) are resistant to fungi and decrease by 11 % the reproductive rate of *M. persicae.* Growth rates of the fourth instar *M. sexta*, were decreased by 27 % on similarly-infested leaves having lesions, but by only 16 % on adjacent but symptomless leaves. When the striped cucumber beetle *Acalymma vittata* feeds on squash there is an accumulation of cucurbitacins that stimulate further attack by this pest. But the squash beetle *Epilachna tredecimnotata*, is deterred by such high concentrations. Feeding by the gypsy moth larva *Lymantria dispar*, on birch or oak leaves alters leaf physiology, thus decreasing growth and survival rates of other larvae feeding days later. Applying the growth regulator maleic hydrazide to broad bean plants decreases the survival and fecundity of bean aphids; the growth retardant chlormequat chloride (CCC) has a similar effect. Results from herbicides are less clear cut. 2,4-Dichlorophenoxyacetic acid (2,4-D) applied to barley decreases the fecundity of the grain aphids *R. padi* and *Macrosiphum avenae*. Applied to maize, however, 2,4-D had the converse effect on the corn leaf aphid *Rhopalosiphum maidis*, perhaps because food quality was improved [8.38, and references therein].

The six mechanisms proposed to explain induced resistance are: (1) phenological synchronization between plant and herbivores; (2) plant physiology; (3) nutrient concentration; (4) compensatory mechanisms; (5) plant allelochemicals; and (6) de novo synthesis of phytoalexins [8.38, and references therein].

Phenological Synchronization

Feeding by the spruce budworm *C. fumiferana* is decreased by induced delays in balsam fir budbreak which can be manipulated by applying abscisic acid with the result that larvae must feed on older, less edible leaves [8.38].

Plant Physiology

Plant bioregulators (PBRs) affect specific biochemical events with collateral morphological changes, whereas plant growth regulators (PGRs) may modify both plant biochemistry and growth. After timely applications of gibberellic acid to citrus fruits, both internal maturation and sugar concentration are unaffected but senescence of the peel is delayed. This maintains the concentration of peel terpenes conferring resistance against fruit fly larvae and post-harvest pests. PGRs and herbicides are also established as enhancing resistance to insects. Thus, the PGRs cycocel (CCC) and mepiquat chloride (PIX) increased pectin methoxy content and intercellular pectin levels, that decreased stylet penetration of sorghum by aphids. By increasing the relative condensed tannin concentration, terpenoid concentration, and astringency of cotton, PIX induced resistance to *H. zea*. Applying p-chloromercuriphenyl sulphonic acid to soybean induced the phytoalexin glyceollin serving as an antifeedant against *P. includens* [8.50]. Growth regulators such as CCC and gibberilin decrease levels of total nitrogen and sugar in apple with associated, and probably resultant, decreases in pest mite populations; these effects might also extend to insects [8.38].

Compensatory Mechanisms

Although compensation has been associated with tolerance [8.50], the fact that it is elicited by insect attack argues for its continuation under induced resistance. Herbivory has been reported as decreasing yield, and also as enhancing it through compensation. Increased yield reflects carbohydrate production which, with partitioning, is regulated by fructose-2,6-biphosphate, the expression of which, in turn, is affected by environmental factors. The presence of excess photosynthates allows leaf loss without yield loss. Conversely, source-limited plants would sustain a loss in yield. Similarly, net photosynthesis could variously respond to herbivory. It would decrease through decreased gross photosynthesis or increased respiration (net photosynthesis (P) = gross photosynthesis (Pg) - respiration (R)). Conversely, P would increase if herbivory increased the supply of cytokinins on account of decreased competition for plant hormones. Such an increase may increase CO_2 fixation. Additionally, decreased leaf area enables increased nitrogen availability and increased protein synthesis in the remaining leaves. Also yields may improve through removal of excess leaves and decreased respiratory demand [8.50].

Resource reallocation also features in compensatory reactions. Herbivory by *D. coarctata* on young wheat plants elicits tillering that more than compensates for lost shoots [8.56]. Mortality throughout larval life is relatable to the number of shoot/hatching larva, and is density-dependent. This mortality stabilizes population fluctuations [8.57]. After three defoliations of wheat grass by grasshoppers, tillering was enhanced beyond the level induced by hand defoliation; however, at 100 % defoliation, the converse was true. This indicated a feedback mechanism inhibiting tillering after very severe defoliation by grasshoppers [8.38]. Maize tolerant to the corn rootworm larva *Diabrotica* sp. has a well developed root system, enhanced by secondary roots developed by resource reallocation following larval attack [8.50].

Compensatory reactions are significantly affected by environmental factors such as water stress, temperature stress, and pollutants. The first two take effect by altering gas exchange and photosynthesis through stomatal closure. Some air pollutants such as acid rain serve as fertilizers that increase photosynthesis and enhance recovery from herbivory. Increased levels of CO_2 may enhance biomass build up, branching, and photosynthesis that all promote tolerance.

Allelochemical Concentrations

Feeding by the small bug *Lygus disponsi* on sugar beet leaves induces increased concentrations of plant phenolics and increased activities of polyphenol oxidase and peroxidase that potentially could decrease attack by other pests.

Synthesis of Phytoalexins

Low molecular weight antimicrobials occur in only small concentrations in unattacked plants. Following attack these concentrations may be speedily enhanced depending on the attack site, the extent of damage, and the inducers employed. Such phytoalexins especially in the Leguminosae, Compositae and Solanaceae affect insect herbivores, and plant-injury by herbivores may evoke their production. Experiments with the larva of *E. varivestis*, showed that although discs of soybean enriched in phytoalexins by ultraviolet (UV) light treatment were probed, they were not eaten; control discs were eaten avidly [8.20]. Consistent with this was the demonstration that vestitol, a phytoalexin extracted from roots of the forage legume *Lotus pedunculatus*, deterred feeding by larvae of the grass grub *Costelytra zealandica*. Other phytoalexins as pisatin, genistein and, coumestrol may diminish feeding in some herbivore species but not in others [8.38].

Similarly, attacks on Cruciferae by the flea beetle *Psylloides chrysocephala*, and by *D. floralis*, alter both the total concentration of glucosinolates and the relative proportions of aromatic and aliphatic chemicals. Specifically, attack by *D. floralis* on oilseed rape *Brassica napus* increased by two- to four-fold the levels of indole-base glucosinolates, and by four- to 17-fold the levels of 1-methoxy-3-indolyl methyl glucosinolate, the single compound most affected [8.7, and references therein]. The potential of induced proteinase inhibitors in pest control is described in Chapter 2.

The scope of induced resistance is a source of encouragement. Specifically, the fact that the reaction by the plant is attack-induced promotes both efficiency and manipulation. For example, it might be possible to: select cultivars with enhanced phytoalexin production; immunize plants to stimulate such production; apply phytoalexin-inducing chemicals as sprays timed before an insect attack; promote nutrient decreases and plant escape by use of herbicides and growth regulators. Exploitation of the induced response may offer a path around the dilemma of higher-yielding cultivars being frequently less resistant to insect attack [8.38, and references therein].

Overview of Resistance Mechanisms

Although the existing categories have served well the investigation of resistance, difficulty may occur in distinguishing between antixenosis and antibiosis. One criterion eliciting disagreement is when feeding is deterred to the extent that the insect dies; one perspective assigns this to antibiosis on the basis of death of the insect, but another disagrees on the basis that, given a choice, the insect would leave the resistant plant. Similarly, disagreement exists on whether inability of aphids to reach with their stylets the appropriate feeding location should be attributed to antixenosis or antibiosis. Accordingly, a decision-table has been drawn up comprising a sequence of steps represented by specific tests to enable an unambiguous conclusion [8.43]. The associated literature review in 1994 of 51 insect/crop interactions, of which several entailed more than one resistance mechanism, resolved as 45 expressions of antixenosis, 28 antibiosis and 14 tolerance. Tolerance drew special attention as, unlike antibiosis, it exerts no selection pressure on the pest population. The undesirable property of enabling a build up of the pest population might be countered by combining it with antibiosis in a single cultivar [8.43, and references therein]

INHERITANCE OF RESISTANCE

The stability of resistance in a plant to a particular insect species depends on: (1) the genotype of the plant, (2) the genotype of the insect; and (3) the genetic interaction between the plant and insect [8.18, and references therein]. Some resistance genes are dominant i.e., when the F_1 hybrids from resistant and susceptible parents are resistant. Others are recessive i.e., when corresponding F_1 hybrids are intermediate in resistance.

The inheritance of resistance in crop plants to insect attack has been widely investigated. For example, breeding line LJ 90234 of musk melon *Cucumis melo*, was resistant to natural infestations of western biotypes of *A. gossypii* in the field, and to controlled infestations in the glasshouse. However, the same line was very susceptible to western flower thrips *Frankliniella occidentalis*. Furthermore, musk melons resistant to southeastern biotypes of aphid were susceptible to the western biotype. A single dominant gene for resistance was identified and designated *Ag* [8.9].

Biotypes are populations of an insect species exhibiting varying abilities to exploit a crop plant. An insect pest with virulent genes exploits a host with one or more resistant genes; a pest unable to attack a host with a resistance gene has an avirulent gene or genes. More practically, in regard to possession of the former, some cultivars infested with the same insect biotype will evince resistance and others susceptibility, hence the alternative term biotype-specific resistance. In regard to possession of the latter, all resistant cultivars evince similar resistance to all biotypes, hence the corresponding alternative term, biotype-nonspecific resistance [8.18].

The gene-for-gene relationship suggests that for every major gene for resistance in the host species, there is a corresponding matching gene for virulence in the pest. The host plant shows a resistant reaction if it has a resistance gene, or if the insect has an avirulent allele at the corresponding gene locus. If the insect has a virulent gene at the corresponding locus, however, the plant is susceptible. For a specific insect to survive on a resistant plant it must undergo a genetic change that gives it the necessary virulence to overcome this resistance; for the host plant to survive this genetic change in the insect, it must change genetically to overcome insect virulence [8.18].

Two genetic systems exploited in developing crop cultivars resistant to insects are vertical resistance involving major genes or oligogenes, and horizontal resistance involving polygenes or minor genes or modifiers. The latter is considered more stable and permanent than vertical resistance because, with a single gene conditioning vertical resistance, only a single gene mutation in the insect is required to overcome it. Nevertheless, vertical resistance has been used successfully to control insects in many programs. For example, resistance to Hessian fly *Mayetiola destructor*, through the use of vertical genes, has almost eradicated this pest. Also, vertical resistance is easier to incorporate into new cultivars, and generally provides a high level of resistance. Horizontal resistance is difficult to incorporate into improved cultivars of a self-pollinated crop. Both types are important in crop improvement programmes and strategies for their use are being developed [8.18]. Of some 12 insect species expressing biotypes that have overcome plant resistance, seven are aphids [8.50] whose parthenogenetic reproduction facilitates such biotype development (this number is, of course, a fraction of the 500 or more insect species expressing insecticide-resistant biotypes).

The methodology to discern inheritance of resistance entails crossing resistant and susceptible parents, with some F_1 seeds serving to produce plants for further assessment and others serving to produce F_2. Plants of the next generation F_3 are expected to verify F_2 and to eliminate confounding effects of other pathogens. The same response is expected from all F_1 plants: a dominant gene prevails if these are resistant; a recessive one if susceptible; and an incompletely dominant one if the response is intermediate. In the F_2 a ratio of resistant to susceptible equivalent to: 3:1 indicates a single dominant gene for resistance; 1:3 indicates a single recessive gene; 15:1 is indicative of two dominant genes; 7:9 represents two recessive genes; and 13:3 a dominant and a recessive gene [8.50].

The genetic basis for virulence has been worked out for only three pests: *M. destructor*, on wheat; *S. graminum*, on wheat; BPH on rice [8.50]. Monogenic control prevails in the first two and polygenic control in BPH. For present purposes it is sufficient to examine a representative of monogenic and polygenic control.

Hessian Fly

When larvae of Hessian fly feed on susceptible wheat plants *Triticum aestivum* the leaves become stunted and turn dark green, and new leaves generally fail to form. Resistant seedlings show some leaf stunting at first, but they generally recover and remain light green like uninfested plants. Larvae that feed on resistant plants generally die; the few survivors remain small and do not stunt the seedlings. Thus, virulent larvae can be distinguished from avirulent ones by reaction of the plant to them, and by the reaction of the insect to plants of known genotype. Biotype distinction of the Hessian fly is based on the virulence/avirulence of larvae to different wheats having known genes for resistance. A biotype's phenotype (A, B, C, etc.) can be ascertained by scoring plant reaction to larvae of the same progeny, and larval ability to survive. Thus, wheat seedlings of known genotype are grown and infested with a pair of flies. The larvae are scored as dead or alive 15 days after egg laying has ceased, and when the lengths of the surviving larvae can be measured. The seedling reactions indicate progeny phenotypes.

When male Hessian flies form spermatozoa, the paternal haploid genome is eliminated and only the maternal chromosomes are transmitted. Consequently, F_2 and backcross progeny genotypes from crosses between adults of different biotypes vary according to the direction of the original cross between parents. Moreover, F_1 males breed as if homozygous because they transmit only maternal chromosomes; F_1 females breed as if heterozygous, showing normal transmission of both genomes.

The genetics of resistance has been explored by use of plants that are monosomic, disomic (normal chromosome complement), or trisomic (extra chromosome complement). In the first instance, crosses between different biotypes of Hessian fly were studied [8.21]. F_1, F_2 and backcross progeny from intercrosses of insect biotypes Great Plains (GP) and E, were tested on wheat cultivar Monon. This carries a single dominant gene H_3 conferring resistance to biotype GP but ineffective against biotype E. As all F_1 flies were avirulent on Monon and had the GP phenotype, the virulence is recessive. Also reactions of the segregating F_2 and backcross progeny showed that the difference between biotypes GP and E is caused by a single gene. In some crosses, no segregation occurred through elimination of the paternal genome. Thus, biotype E of Hessian fly has one recessive gene for virulence designated *m*, enabling it to overcome resistance governed by dominant gene H_3 in Monon. Five loci for virulence in Hessian fly were designated *t*, *s*, *m*, *k*, and *a*. Eight biotypes of the Hessian fly were identified (GP, A, B, C, D, E, F, and G) to which three more were subsequently added [8.50, and references therein].

Wild species represent a rich source of resistance genes but transfer into cultivated species is frequently compromised by abortion of embryos of interspecific hybrids. Nevertheless, transferring genes H_{13}, H_{22}, H_{23} and H_{24}, by backcrossing from *Triticum tauschii* into bread wheat *T. aestivum,* revealed two further independent and dominant genes, H_9 and H_{10}. Twenty six genes for resistance, all dominant except one, have now been identified in wheat [8.50].

The Brown Planthopper

Rice resistance to BPH is controlled by four genes, two (*Bph*-1 and *Bph*-3) inherited as dominant traits, and two (*bph*-2 and *bph*-4) as recessive. *Bph*-1 and *bph*-2 are closely linked and *Bph*-3 is nonallelic to and independent of *Bph*-1. In some rice cultivars, resistance is governed by the dominant *Bph*-3 and in others by recessive *bph*-4, both of which are located on chromosome 7. Additionally, unknown dominant and recessive genes affect resistance.

There are four BPH biotypes of which biotype 1 was widely-distributed in Southeast Asia before development of BPH-resistant cultivars. The Philippines was the source for biotype 2 capable of damaging the BPH-resistant cultivar IR 26 having the *Bph*-1 gene; laboratory selection of insects exposed to resistant cultivars with the *bph*-2 gene developed biotype 3; Southeast Asia is the source of biotype 4. Some cultivars resistant in India were susceptible in Southeast Asia and investigation of one ARC10550, detected a recessive gene *bph*-5, resistant to biotype 4 and independent of the other four genes. Subsequently, examination of seventeen cultivars, susceptible to biotype 1, but resistant to biotype 4, detected the dominant gene *Bph*-6 and the recessive gene *bph*-7. Cultivars from Thailand and Myanmar gave evidence of another recessive *bph*-8, and a dominant *Bph*-9. Hybridization of wild with cultivated rice has identified another dominant gene, *Bph*-10(t). Overall, cultivars resistant to biotypes 1 and 3 probably have *Bph*-1; those resistant to biotypes 1 and 2 have *bph*-2; and those resistant to the three biotypes have one of *Bph*-3, *bph*-4, *bph*-8 or *Bph*-9. Thus, new cultivars may be readily categorized by reference to resistance to biotypes, before launching a genetic study [8.50, and references therein].

Other Genetic Relationships

The genetic basis for trichome density and thus efficacy has been worked out for soybean species. The number of trichomes per leaf is controlled by a single gene and is fixed at an early stage of leaf development. Nevertheless, the final extent of leaf trichome density is determined by variations in leaf size. Essentially, the leaves of bean fly-susceptible soybean grow five to six-fold larger than bean fly-resistant soybean *G. soja*. Thus, although the actual number of trichomes is relatively constant, trichome density varies substantially. Furthermore, the larger leaf size offers additional ovipositional opportunities for the bean flies *Melanagromyza sojae* and *O. centrosematis*. In contrast, the action of polyphenols by its enzyme-dependent pathway is polygenic and is an example of horizontal resistance. On this account it is viewed as intrinsically more stable than single gene resistance [8.12, and references therein].

STRATEGIES FOR PRACTICAL USE

HPR offers six significant features consistent with Integrated Pest Management (IPM): specificity to the pest; cumulative effectiveness with succeeding generations; maintenance of effect for many generations; compatibility with the other recognized elements of IPM; lack of hazard to the environment; and ease of use as once developed, resistant cultivars may be reused without additional cost. Such advantages have been quantified. The use of resistant carrot cultivars allows the insecticidal dosage against *P. rosae* to be decreased by two-thirds with obvious advantages in terms of cost and decreased food contamination. Similarly, control of *D. floralis*, using resistant cultivars of swede with half the usual treatment of chlorfenvinphos, was as good if not better than that with full treatment on a susceptible cultivar [8.67]. Rice cultivars resistant to the BPH double the efficiency of hopper predators; such cultivars provide density-independent mortality of low-density populations, and density-dependent mortality of high density ones [8.8].

These advantages are not lost on governments of tropical countries and conspicuous successes have come from research institutes dedicated to the development of resistant cultivars. The International Rice Research Institute (IRRI) of the Philippines relies on resistant cultivars for its pest management system. Cultivars developed there and elsewhere (for example, Central Rice Research Institute, CRRI, Cuttack, India; Bangladesh Rice Research Institute, BRRI) are now grown on 50 million ha in Asia [8.5]. More than 100 insect resistant cultivars are grown in the U.S.A. and more than twice that number world-wide; most relate to the major food cereals, maize, sorghum and wheat. The potential for genetically engineering new, resistant cultivars is receiving increasing attention and is considered in Chapter 10 (Fig. 72).

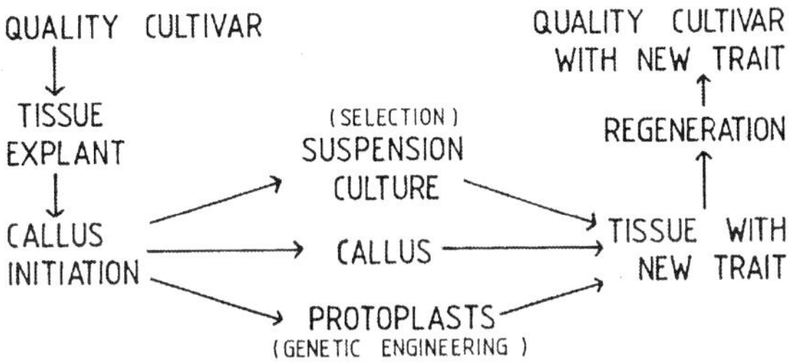

Figure 72. Schematic representation of stepwise development through genetic engineering of a new cultivar.

Constraints

Strategies for incorporating HPR into IPM programmes need to take account of HPR complexities with, for example, a single compound taking effect in several functional categories. In addition, the overall resistance profile of a particular cultivar may comprise a range of effects as in soybean resistance to *E. varivestis*. This entails antixenosis and antibiosis, each operating through several compounds, supplemented by induced resistance elicited by the isoflavonoid phytoalexin coumestrol. Not least is the effect of various environmental factors especially temperature that may: alter chemical concentrations in plants; modify pest physiology and behaviour; and affect plant responses to damage by feeding. Furthermore, levels of allelochemicals efficacious against a pest may impair development, fecundity and success of its parasitoid as evidenced by the relationship between cotton, and *Heliothis* spp. as pests, and the parasitoid wasp *Campoletis sonorensis*. Similarly, enhanced numbers of trichomes entrapping a pest may also impair mobility of its predator. 2-Tridecanone in trichomes is toxic to *H. zea* larvae but elicits pest resistance to the insecticide carbaryl presumably through enhanced activity of microsomal oxidases [8.29]. Allelochemical concentrations efficacious against one pest may attract another as indicated by gossypol in cotton, affecting *H. zea* and *A. grandis* respectively. Many allelochemicals, such as rutin and glucosinolates, are toxic and/or carcinogenic [8.50], although maximum tolerated doses for humans have not been worked out. Finally, a very detailed examination of soybean genotypes resistant to *Pseudoplusia includens* extended to four trophic levels. The predatory pentatomid *Podisus maculiventris* was as affected by antibiosis as the herbivore, through increased developmental time and decreased weight gain. Furthermore, overall reproductive capacities were decreased in the parasitoid *Telenomus podisi* emerging from eggs of *P. maculiventris* that were reared on *P. includens* larvae [8.48].

Principles

Nevertheless it is possible to approach some general principles. It seems more appropriate to produce crop cultivars taking effect through elevated levels of antifeedants than of toxins, as the former are more specific and thus less hazardous to other organisms. Computer modelling has indicated that the most fruitful category of resistance to deploy will depend on the invasion pattern of the pest. If invading in small numbers and taking several generations to cause significant damage, relatively modest antixenosis and antibiosis will elicit significant delays. Antixenosis will lessen initial and subsequent population increase; antibiosis will increase generation time, decrease survival and reproduction to the same ultimate effect. Although tolerance will not lessen population increase, it will elevate the population threshold at which insecticidal intervention becomes necessary [8.30]. Furthermore, antibiosis can offer complete control of a single major pest, such as Hessian fly, on a crop.

The computer model HELSIM was utilized to assess durability of various resistance patterns against *H. zea*. Apparently, antixenosis would not significantly decrease the pest population but antibiosis would elicit 50 % mortality of first and second-instar larvae. Increasing this mortality to 80 % would exert significant selection pressure on the pest leading to adaptation in only seven generations. Increasing antixenosis to 90 % had little effect, as 50 % of populations would adapt only after 23 generations. However, combining both categories allowed antibiosis to endure for some 32 generations and antixenosis for 100. A separate study estimated that in a pest with two genetic loci for overcoming resistance, this would develop quickly if only toxin-rich plants were encountered. In contrast, supplying either 10 or 30 % susceptible plants would defer development of a biotype for 150 and 500 generations respectively [8.30, and references therein]. This represents an affirmation of the practices of subsistence farmers in the use of cultivar mixtures.

Subsistence farmers produce 15-20 % of the planet's food and represent 60 % of its farmers. They rely heavily on mixed cropping and on landrace or varietal diversity especially in the tropics. Thus, for example, in Indonesia the Kantu people of Kalimantan will plant about 44 rice cultivars with an average of 17/household. North African farmers will grow mixtures of barley with tetraploid and hexaploid wheats in the same field. Some 2,000-3,000 cultivars of potatoes from eight *Solanum* spp. are cultivated in the Andes and individual farmers will use up to 30. Of the 5,000 cultivars of sweet potato known from the highlands of Papua, about 40 and 20 will be used per village and in a single garden respectively. Seven hundred cultivars of cassava are known by the Aguaruna Jívaru of Amazonian Peru and a sampling of 160 plants from one garden yielded 12 different cultivars. Up to 38 cultivars of yam may be sown in a single planting by African and Pacific farmers. Although strategies specifically against insects have been rarely studied, it is established that: mixtures of wheat cultivars offer potential for management of *S. graminum*; maize mixtures decreased infestation and damage by stem borers *Sesamia cretica*; and mixtures enhanced sorghum resistance to midge sorghum *Contarinia sorghicola*. Quite apart from the salutary lessons to be learnt from these farmers is their important contribution to protecting genetic diversity of world-significant crops [8.62]. Although there is no single panacea for insect pest control, an effective and environmentally-friendly option such as HPR should constitute a significant element in IPM programmes.

$$F = k\,(S - S_0) \tag{4}$$

or to take account of the Weber-Fechner law (response is proportional to the logarithm of the stimulus amplitude) as:

$$F = k\,.\,\log\,(S - S_0) \tag{5}$$

where F is the frequency of action potentials, k is a constant for the family of neurons, S is the actual stimulus intensity, and S_0 is the lowest stimulus that first produces a response, the threshold stimulus. As there must be an upper limit to the frequency of action potentials, the relationship is written more realistically as a power function:

$$F = k\,.\,\log\,(S - S_0)^n \tag{6}$$

where n is constant within each family of receptor cells. Given the physical upper limit constraining the generation of action potentials, n values are usually less than unity thus producing an upper asymptote. Conventionally, this Stevens relationship is converted to a double-logarithmic form:

$$\log F = \log k + n\,(\log S - S_0) \tag{7}$$

which gives a straight line plot of slope n that may quantify stimulus/response relationships: the more the slope approaches unity the more sensitive the response is to changes in stimulus concentration [5.35] (Fig. 41).

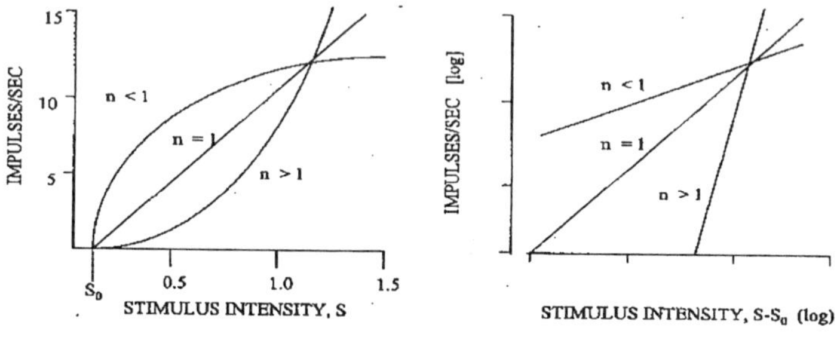

Figure 41. Intensity function characteristic of a chemoreceptive neuron. Left, untransformed values; right, log transformed values, S_o, stimulus threshold. Redrawn after Schmidt [5.35].

Central Processing

Two principal mechanisms, labelled lines and cross-fibre patterning, have been proposed to explain how a burst of action potentials is interpreted in the central nervous system (CNS). The former suggests that as specific stimuli are detected by

ERRATUM

Michael F. Ryan
Insect Chemoreception: Fundamental and Applied
ISBN 1-4020-0270-X

Please turn over for the corrected page 142.

REFERENCES

8.1 Atkin, D.S.J. and Hamilton, R.J. (1982) The effects of plant waxes on insects. *J. Nat. Prod.* **45**, 694-696.

8.2 Auclair, J.L., Maltais, J.B. and Carter, J.J. (1957) Factors in resistance of peas to the pea aphid, *Acyrthosiphon pisum* (Harris) (Homoptera: Aphididae). II Amino acids. *Can. Entomol.* **10**, 457-464.

8.3 Beck, S.D. (1965) Resistance of plants to insects. *Ann. Rev. Entomol.* **10**, 207-232.

8.4 Berenbaum, M.R. and Zangerl, A.R. (1992) Genetics of secondary metabolism and herbivore resistance in plants. In Rosenthal, G.A. and Berenbaum, M.R. (eds.) *Herbivores: Their Interactions with Secondary Plant Metabolites*, Vol II. *Ecological and Evolutionary Processes*. Academic Press, New York, pp. 415-438.

8.5 Bergman, J.M. and Tingey, W.M. (1979) Aspects of interaction between plant genotypes and biological control. *Bull. Entomol. Soc. Am.* **25**, 275-279.

8.6 Bernays, E. and Chapman, R.F. (1975) The importance of chemical inhibition of feeding in host-plant selection by *Chorthippus parallelus* (Zetterstedt) *Acrida* **4**, 83-93.

8.7 Birch, A.N.E., Städler, E., Hopkins, R.J., Simmonds, M.S.J., Baur, R., Griffiths, D.W., Ramp, T., Hurter, J. and McKinlay, R.G. (1993) Mechanisms of resistance to the cabbage and turnip root flies: collaborative field, behavioural and electrophysiological studies. *Bull. OILB/SROP* **16**, 1-5.

8.8 Bloem, K.A., Kelley, K.C. and Duffey, S.S. (1989) Differential effect of tomatine and its alleviation by cholesterol on larval growth and efficiency of food utilization in *Heliothis zea* and *Spodoptera exigua*. *J. Chem. Ecol.* **15**, 387-398.

8.9 Bohn, G.W., Kishaba, A.N., Principe, J.A. and Toba, H.H. (1973) Tolerance to melon aphid in *Cucumis melo* L. *J. Am. Soc. Hortic. Sci.* **98**, 37-40.

8.10 Campos, F., Atkinson, J., Arnason, J.T., Philogène, B.J.R., Morand, P., Werstiuk, N.H. and Timmins, G. (1988) Toxicity and toxicokinetics of 6-methoxybenzoxazolinone (MBOA) in the European corn borer, *Ostrinia nubilalis* (Hübner). *J. Chem. Ecol.* **14**, 989-1002.

8.11 Cates, R.G., Redak, R.A. and Henderson, C.B. (1983) Patterns in defensive natural product chemistry: Douglas fir and western spruce budworm interactions. In Hedin, P.A. (ed.) *Plant Resistance to Insects.* ACS Symposium Series 208, American Chemical Society, Washington, DC, pp. 3-19.

8.12 Chiang, H.-S. and Norris, D.M. (1984) "Purple Stem", a new indicator of soybean stem resistance to bean flies (Diptera: Agromyzidae). *J. Econ. Entomol.* **77**, 121-125.

8.13 Ciepiela, A. (1989) Biochemical basis of winter wheat resistance to the grain aphid, *Sitobion avenae*. *Entomol. Exp. Appl.* **51**, 269-275.

8.14 Davis, F.M., Ng, S.S. and Williams, W.P. (1989) Mechanisms of resistance in corn to leaf feeding by southwestern corn borer and European corn borer (Lepidoptera: Pyralidae). *J. Econ. Entomol.* **82**, 919-922.

8.15 Dhaliwal, G.S., Pathak, M.D. and Veja, C.R. (1988) Effect of plant age on resistance in rice variety to *Chilo suppressalis* (Walker) – allelochemical interactions. *J. Insect Sci.* **1**, 142-148.

8.16 Farrar Jr., R.R. and Kennedy, G.G. (1988) 2-Undecanone, a pupal mortality factor in *Heliothis zea*: sensitive larval stage and *in planta* activity in *Lycopersicon hirsutum* f. *glabratum*. *Entomol. Exp. Appl.* **47**, 205-210.

8.17 Firempong, S. (1988) Components of resistance to *Aphis craccivora* on some cowpea cultivars. *Entomol. Exp. Appl.* **48**, 241-246.
8.18 Gallun, R.L. and Khush, G.S. (1980) Genetic factors affecting expression and stability of resistance. In Maxwell, F.G. and Jennings, P.R. (eds.) *Breeding Plants Resistant to Insects.* Wiley and Sons, New York, pp. 63-86.
8.19 Gregory, P., Sinden, S.L., Osman, S.F., Tingey, W.M. and Chessin, D.A. (1981) Glycoalkaloids of wild tuber-bearing *Solanum* species. *J. Agric. Food Chem.* **29**, 1212-1215.
8.20 Hart, S.V., Kogan, M. and Paxton, J.D. (1983). Effect of soybean phytoalexins on the herbivorous insects Mexican bean beetle and soybean looper. *J. Chem. Ecol.* **9**, 657-672.
8.21 Hatchett, J.H. and Gallun, R.L. (1970) Genetics of the ability of the Hessian fly, *Mayetiola destructor*, to survive on wheats having different genes for resistance. *Ann. Entomol. Soc. Am.* **63**, 1400-1407.
8.22 Hedin, P.A. (1976) Seasonal variation in the emission of volatiles by cotton plants growing in the field. *Environ. Entomol.* **5**, 1234-1238.
8.23 Hedin, P.A., Davis, F.M. and Williams, W.P. (1993) 2-Hydroxy-4, 7-dimethoxy-1,4-benzoxazin-3-one (*N-O*-ME-DIMBOA), a possible toxic factor in corn to the southwestern corn borer. *J. Chem. Ecol.* **19**, 531-542.
8.24 Hedin, P.A., Jenkins, J.N. and Parrott, W.L. (1992) Evaluation of flavonoids in *Gossypium arboreum* (L.) cottons as potential source of resistance to tobacco budworm. *J. Chem. Ecol.* **18**, 105-114.
8.25 Hinz, B. (1988) Resistance of the summer forms of triticale, wheat and rye to the pale aphid, *Metopolophium dirhodum* (Walk.) (In German). *Nach. Pflanzenschutz* **42**, 132.
8.26 Hoxie, R.P., Wellso, S.G. and Webster, J.A. (1975) Cereal leaf beetle response to wheat trichome length and density. *Environ. Entomol.* **4**, 365-370.
8.27 Huesing, J., Jones, D., Deverna, J., Myers, J., Collins, G., Severson, R. and Sisson, V. (1989) Biochemical investigations of antibiosis material in leaf exudate of wild *Nicotiana* species and interspecific hybrids. *J. Chem. Ecol.* **15**, 1203-1217.
8.28 Kelsey, R.G., Reynolds, G.W. and Rodriguez, E. (1984) The chemistry of biologically active constituents secreted and stored in plant glandular trichomes. In Rodriguez, E., Healey, P.L. and Mehta, I. (eds.) *Biology and Chemistry of Plant Trichomes.* Plenum Press, New York, pp. 187-241.
8.29 Kennedy, G.G. and Dimock, M.B. (1983) 2-Tridecanone: a natural toxicant in a wild tomato responsible for insect resistance. In Miyamoto, J. (ed.). *IUPAC Pesticide Chemistry (Human Welfare and the Environment)*. Pergamon Press, New York, pp. 123-128.
8.30 Kennedy, G.G., Gould, F., de Ponti, O.M.B. and Stinner, R.E. (1987) Ecological, agricultural, genetic and commercial considerations in the deployment of insect resistant germplasms. *Environ. Entomol.* **16**, 327-338.
8.31 Kennedy, G.G. and Schaefers, G.A. (1975) Role of nutrition in the immunity of red raspberry to *Amphorophora agathonica* Hottes. *Environ. Entomol.* **4**, 115-119.
8.32 Khan, Z.R., Rueda, B.P. and Caballero, P. (1989) Behavioral and physiological responses of rice leaffolder *Cnaphalocrocis medinalis* to selected wild rices. *Entomol. Exp. Appl.* **52**, 7-13.
8.33 Klingauf, von F., Nöcker-Wenzel, K. and Klein, W. (1971) Einfluβ einiger Wachskomponenten von *Vicia faba* L. auf das Wirtswahlverhalten von

Acyrthosiphon pisum (Harris) (Homoptera: Aphididae). *Z. Pflanzenkrankh. Pflanzenschutz* **78**, 641-648.
8.34 Klun, J.A., Tripton, C.L. and Brindley, T.A. (1967) 2,4-Dihydroxy--7-methoxy-1,4-benzoxazin-3-one (DIMBOA), an active agent in the resistance of maize to the European corn borer. *J. Econ. Entomol.* **60**, 1529-1533.
8.35 Knapp, J.L., Hedin, P.A. and Douglas, W.A. (1966) A chemical analysis of corn silk from single crosses of dent corn rated as resistant, intermediate, and susceptible to the corn earworm. *J. Econ. Entomol.* **59**, 1062-1064.
8.36 Kogan, M. (1986) Natural chemicals in plant resistance to insects. *Iowa State J. Res.* **60**, 501-527.
8.37 Kogan, M. and Ortman, E.E. (1978) Antixenosis – a new term proposed to replace Painters 'nonpreference' modality of resistance. *Bull. Entomol. Soc. Am.* **24**, 175-176.
8.38 Kogan, M. and Paxton, J. (1983). Natural inducers of plant resistance to insects. In Hedin, P.A. (ed.) *Plant Resistance to Insects.* ACS Symposium Series 208, American Chemical Society, Washington, DC, pp. 153-171.
8.39 Lee, Y.I. (1983) The potato leafhopper, *Empoasca fabae*, soybean pubescence, and hopperburn resistance. Ph D Thesis, University of Illinois.
8.40 Leszcynski, B., Wright, L.C. and Bakowski, T. (1989) Effect of secondary plant substances on winter wheat resistance to grain aphid. *Entomol. Exp. Appl.* **52**, 135-139.
8.41 Maki, A. and Ryan, M. F. (1989) Root-mediated effects in carrot resistance to the carrot fly, *Psila rosae*. *J. Chem. Ecol.* **15**, 1867-1882.
8.42 Maki, A., Kitajima, J., Abe, F., Stewart, G. and Ryan, M. F. (1989) Isolation, identification, and bioassay of chemicals affecting nonpreference carrot-root resistance to carrot-fly larva. *J. Chem. Ecol.* **15**, 1883-1897.
8.43 Manglitz, G.R. and Danielson, S.D. (1992) A re-appraisal of Painters mechanisms of plant resistance to insects, with recent illustrations. *Agric. Zool. Rev.* **6**, 259-276.
8.44 Maxwell, F.G. and Jennings, P.R. (1980) *Breeding Plants Resistant to Insects.* Wiley and Sons, New York.
8.45 Minney, B.H.P., Gatehouse, A.M.R., Dobie, P., Dendy, J., Cardona, C. and Gatehouse, J.A. (1990) Biochemical bases of seed resistance to *Zabrotes subfasciatus* (bean weevil) in *Phaseolus vulgaris*: a mechanism for arcelin toxicity. *J. Insect Physiol.* **36**, 757-767.
8.46 Mooney, H.A. Gulmon, S.L. and Johnson, N.D. (1983) Physiological constraints on plant chemical defences. In Hedin, P.A. (ed.) *Plant Resistance to Insects.* ACS Symposium Series 208, American Chemical Society, Washington, DC, pp. 21-36.
8.47 Neal, J.J., Steffens, J.C. and Tingey, W.M. (1989) Glandular trichomes of *Solanum berthaultii* and resistance to the Colorado potato beetle. *Entomol Exp. Appl.* **51**, 133-140.
8.48 Orr, D.B. and Boethel, D.J. (1986) Influence of plant antibiosis through four trophic levels. *Oecologia* **70**, 242-249.
8.49 Painter, R.H. (1958) Resistance of plants to insects. *Ann. Rev. Entomol.* **3**, 267-290.
8.50 Panda, N. and Khush, G.S. (1995) *Host Plant Resistance to Insects.* CAB International, Wallingford, U.K.
8.51 Pillemer, E.A. and Tingey, W.M. (1978) Hooked trichomes and resistance of *Phaseolus vulgaris* to *Empoasca fabae* (Harris). *Entomol. Exp. Appl.* **24**, 83-94.

8.52 Raman, K.V., Tingey, W.M. and Gregory, P. (1979) Potato glycoalkaloids: effect on survival and feeding behavior of the potato leafhopper. *J. Econ. Entomol.* **72**, 337-341.
8.53 Reese, J.C. (1983) Nutrient-allelochemical interactions in host plant resistance. In Hedin, P.A. (ed.) *Plant Resistance to Insects.* ACS Symposium Series 208, American Chemical Society, Washington, DC, pp. 231-243.
8.54 Rutherford, R.S. and Van Staden, J. (1996) Towards a rapid near-infrared technique for prediction of resistance to sugarcane borer *Eldana saccharina* Walker (Lepidoptera: Pyralidae) using stalk surface wax. *J. Chem. Ecol.* **22**, 681-694.
8.55 Ryan, J.D., Gregory, P. and Tingey, W.M. (1982) Phenolic oxidase activities in glandular trichomes of *Solanum berthaultii. Phytochemistry* **21**, 1885-1887.
8.56 Ryan, M.F. (1973) The natural mortality of wheat-bulb fly larvae. *J. Appl. Ecol.* **10**, 875-879.
8.57 Ryan, M.F. (1975) Fluctuations and regulation of wheat-bulb fly populations at Rothamsted. *Ann. Appl. Biol.* **81**, 83-86.
8.58 Ryan, M.F., Guerin, P.M. and Behan, M. (1977) Possible roles for naturally occurring chemicals in the biological control of carrot fly. In Duggan, J.J. (ed.) *Proc. Symp. Biol. Control.* Royal Irish Academy, Dublin, pp. 130-143.
8.59 Saxena, R.C. (1986) Biochemical bases of insect resistance in rice varieties. In Green, M.B. and Hedin, P.A. (eds.) *Natural Resistance of Plants to Pests: Roles of Allelochemicals.* ACS Symposium Series 296, American Chemical Society, Washington, DC, pp. 142-159.
8.60 Singh, R. and Ellis, P.R. (1993) Sources, mechanisms and bases of resistance of Cruciferae to the cabbage aphid, *Brevicoryne brassicae. Bull. OILB/SROP* **16**, 21-35.
8.61 Smith, C.M. (1989) *Plant Resistance to Insects: A Fundamental Approach.* Wiley and Sons, New York.
8.62 Smithson, J.B. and Lenné, J.M. (1996) Varietal mixtures: a viable strategy for sustainable productivity in subsistence agriculture. *Ann. App. Biol.* **128**, 127-158.
8.63 Sogawa, K. and Pathak, M.D. (1970) Mechanisms of brown planthopper resistance in Mudgo variety of rice (Hemiptera: Delphacidae). *Appl. Entomol. Zool.* **5**, 145-148.
8.64 Städler, E. (1986) Oviposition and feeding stimuli in leaf surface waxes. In Juniper, B.E. and Southwood, T.R.E. (eds.) *Insects and the Plant Surface.* Arnold, London pp. 105-121.
8.65 Stipanovic, R.D. (1983) Function and chemistry of plant trichomes in insect resistance: protective chemicals in plant epidermal glands and appendages. In Hedin, P.A. (ed.) *Plant Resistance to Insects.* ACS Symposium Series 208, American Chemical Society, Washington, DC, pp. 69-100.
8.66 Stipanovic, R.D., Williams, H.J. and Smith, L.A. (1986) Cotton terpenoid inhibition of *Heliothis virescens* development. In Green, M.B. and Hedin, P.A. (eds.) *Natural Resistance of Plants to Pests: Roles of Allelochemicals.* ACS Symposium Series 208, American Chemical Society, Washington, DC, pp. 79-94.
8.67 Taksdal, G. (1993) Resistance in swedes to the turnip root fly and its relation to integrated pest management. Bull. OILB/SROP **16**, 13-20.
8.68 Taneja, S.L. and Woodhead, S. (1989) Mechanisms of stem borer resistance in sorghum. Int. Workshop Sorghum Stem Borers. Patancheru, India. pp.137-143.

8.69 Tingey, W.H. (1984) Glycoalkaloids as pest resistance factors. *Am. Potato J.* **61**, 157-167.

8.70 Tingey, W.M., Leigh, T.F. and Hyer, A.H. (1975) Glandless cotton: susceptibility of *Lygus hesperus* Knight. *Crop Science* **15**, 251-253.

8.71 Tingey, W.M. and Sinden, S.L. (1982) Glandular pubescence, glycoalkaloid composition, and resistance to the green peach aphid, potato leafhopper, and potato fleabeetle in *Solanum berthaultii. Am. Potato J.* **59**, 95-106.

8.72 Van Lenteren, J.C. (1991) Biological control in a tritrophic system approach. In Peters, D.C. and Webster, J.A. (eds.) *Aphid-plant Interactions: Populations to Molecules*. Oklahoma State University Press, Stillwater, Oklahoma, pp. 3-28.

8.73 Walters, D.S., Craig, R. and Mumma, R.O. (1989) Glandular trichome exudate is the critical factor in geranium resistance to foxglove aphid. *Entomol. Exp. Appl.* **53**, 105-109.

8.74 Webster, J.A. (1975) Association of plant hairs and insect resistance. An annotated bibliography. *USDA-ARS Misc. Publ.* **1297**, 1-18.

8.75 Wellso, S.G. (1973) Cereal leaf beetle: larval feeding, orientation, development, and survival on four small-grain cultivars in the laboratory. *Ann. Entomol. Soc. Am.* **66**, 1201-1208.

8.76 Wiseman, B.R. and Snook, M.E. (1996) Flavone content of silks from commercial corn hybrids and growth responses of corn earworm (*Helicoverpa zea*) larvae fed silk diets. *J. Agric. Entomol.* **13**, 231-241.

8.77 Woodhead, S. and Padgham, D. (1988) The effect of plant surface characteristics on resistance of rice to the brown planthopper, *Nilaparvata lugens. Entomol. Exp. Appl.* **47**, 15-22.

8.78 Wright, R.J., Dimock, M.B., Tingey, W.M. and Plaisted, R.L. (1985) Colorado potato beetle (Coleoptera: Chrysomelidae): expression of resistance in *Solanum berthaultii* and interspecific potato hybrids. *J. Econ. Entomol.* **78**, 576-582.

8.79 Zúñiga, G.E. and Corcuera, L.J. (1986) Effect of gramine in the resistance of barley seedlings to the aphid *Rhopalosiphum padi. Entomol. Exp. Appl.* **40**, 259-262.

CHAPTER 9

PHEROMONES IN PLANT PROTECTION

Enthusiasm for the use of pheromones in plant protection was ignited by both dismay at the undesirable environmental effects of the misuse of pesticides, and by the recognition of pheromones as specific and very potent modifiers of insect behaviour. There is, however, a considerable gap between the ideal and the reality of pest control by behaviour modification. Significant constraints include maintenance of sufficiently high pheromone concentrations for continuous effectiveness, and insect adaptation. Furthermore, the case for pheromone usage is rather weakened by a relative scarcity of data on yield enhancement and on cost/benefit relationships, contrasting unfavourably with the ready availability of such data for insecticides. Although scarcity of data may be attributed to the fact that pheromone usage is still in a development phase, it seems unlikely that pheromones will replace insecticides as a control method, but a judicious combination of these could decrease insecticidal loadings.

TECHNICAL CONSIDERATIONS

A key issue is controlled-release on account of the rather high volatility of pheromones which, if uncontrolled, would rapidly dissipate the compound. There are four principal devices enabling controlled-release: the microcapsule; the plastic laminate flake; the hollow fibre; and the twist-tie rope [extensively reviewed in 9.15].

Microcapsules

Essentially, microencapsulation modifies a volatile, hydrophobic fluid into a stable, water-soluble dispersion appropriate for spraying. Although it was anticipated that a continuous distribution or 'fog' of pheromone-loaded microcapsules would be required for maximum effectiveness, some studies indicate more aggregated distributions would be more effective [9.14, and references therein].

Microcapsules comprise small (1-1000 μm) droplets or particles of pheromone enclosed in a polymeric cyst frequently, but not always, of polyamide or polyurea: other wall materials are gelatin, gum arabic, cellulose esters, and styrene-maleic anhydride copolymers. Usually the pheromone is dispersed in a continuous phase with the wall-forming polymer that is then precipitated by coacervation. Formation of coacervate droplets, an intermediate phase in the formation of a polymer into a solid precipitate is achieved, primarily by physicochemical methods and also by mechanical means. The former may entail changes in pH, in solvent, or in temperature. In the latter, adding a hot solution of a polymer, such as polyvinyl chloride, to an emulsion of pheromone in water will, on cooling, elicit formation of a polymer film around the pheromone. In interfacial polycondensation, when the pheromone and a monomer are emulsified in an aqueous solution of the other monomer, there is condensation at the pheromone/water interface. In situ interfacial polycondensation interacts in water the pheromone and a polyisocyanate that on addition to alkali, hydrolyzes to the carbamic acid. Decomposition of this liberates carbon dioxide with formation of the corresponding amine that, on reaction with the remaining isocyanate, forms a polyurea that is pressure resistant. Potentially toxic excess isocyanate groups are related with ammonia. Nevertheless, pheromones, such as unprotected alcohols that react with a capsule wall monomer, are not appropriate for this method of dispersion.

Microencapsulation by mechanical methods may utilize spray drying in which a pheromone-rich solution of polymer is subjected to the passage of hot, inert gas. Solvent removal deposits the polymer on the pheromone droplets. Spray congealing disperses a polymer melt containing core material into cooler liquid or air. Alternatively when pheromone, dissolved in oil and emulsified in water-soluble acrylic resin is sprayed and dried, the result is formation of capsules each containing thousands of droplets of the pheromone/oil mix.

In multiple-orifice methods, passing pheromone through a film of polymer creates polymer-coated pheromone droplets, that are made permanent by drying or cooling. A form of this process forces pheromone from a rotating disc through multiple orifices in a centrifuge bowl fed with a stream of polymer.

Electrostatic forces could be harnessed by giving opposing electric charges to pheromone and polymer respectively, before allowing them to combine. However, the equipment requirements and conductivity specifications may be difficult to satisfy.

Gelatin microcapsules were employed in the earliest approach to pheromone microencapulation using disparlure from the gypsy moth *Lymantria dispar*. However, this completely saturated molecule, with an epoxide as the only functional group, is atypical of pheromones in general, that usually are unsaturated alcohols, aldehydes and esters. Exposure to environmental conditions especially sunlight increase pheromone loss, apparently due to deterioration of the gelatine wall that may be extremely thin (1 μm). Coating with surfactant as a sticker decreased this loss.

In contrast, capsules formed by interfacial polycondensation were less prone to such deterioration, although it still occurred. Indeed, the thinness of typical capsule walls is such that release by degradation may be as significant as release by diffusion. In fact, the diffusion rate of a non-polar structure, such as most pheromones, through a highly polar substance such as gelatin, polyurea, or polyamide, would always be slight. The relationship governing such diffusion may be derived from Fick's first law as:

$$J = \frac{4\,KD\,\Delta C r_o r_i}{(r_o - r_i)} \qquad (14)$$

where: J is the rate of release or flux in g/cm^2/s; K is a distribution constant analogous to the liquid-liquid partition coefficient; D is the diffusion coefficient of permeant in the membrane (cm^2/s); ΔC the concentration difference between permeant in the capsule and that adjacent to the capsule surface; and for a small microcapsule, r_o and r_i are the external and internal radii respectively (Fig. 73).

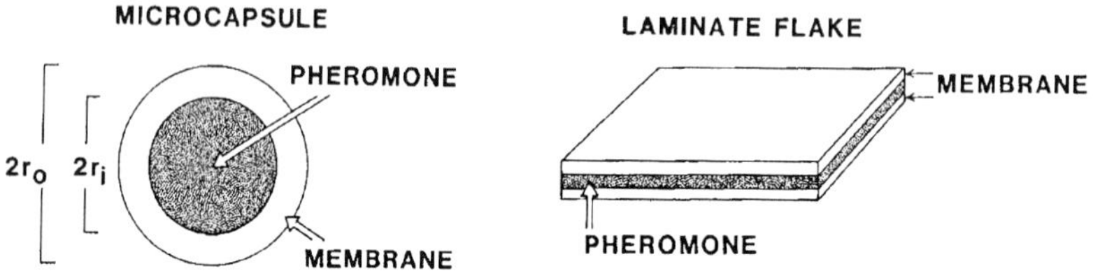

Figure 73. Schematic representations of a microcapsule (r_i and r_0 = inner and outer radii respectively), and of a laminate flake.

Accordingly, if C is kept constant, release should be zero-order (at a constant rate). In practice, as the pheromone is usually mixed with stabilizers or other additives, there is a decrease in the internal concentration, and release is decreasing and first-order (concentration-dependent). Furthermore, pheromone that migrates during storage is released first causing a longer initial release. [9.14].

Laminate flakes

As dispensers of pharmaceuticals, insecticides, fragrances, and pheromones, laminate flakes consist of two layers of vinyl between which sits a porous layer loaded with permeant. The dispenser is manufactured by coating a sheet of polymeric film with pheromone enriched polymer and by completing the sandwich with another sheet of polymeric film. The configuration bonds when placed in an atmosphere of appropriate temperature and pressure, after which laminated sheet is sliced into various shapes, such as strips, flakes, or confetii. For dispersion from aircraft, these are coated with a inert adhesive to stick them to foliage. Flakes may be loaded with only pheromone, or with pheromone and a pyrethroid insecticide. Pheromone emission is regulated by factors such as pheromone concentration and layer thickness as the two most important, and also by layer stiffness, edge-volume ratio, presence of codiffusants, molecular weight of pheromone, and prevailing wind and temperature conditions [9.24, and references therein].

Given the geometry of a laminate flake (Fig. 73), and the distribution coefficient between the reservoir layer and the barrier membrane as unity or larger, the reservoir-barrier system forms a single homogenous film. Concentration of permeant in the reservoir will not remain constant but will decrease over time, thus providing an unsteady state, where the release rate is first-order (varies with time). Fifty percent of 4.3 mg disparlure dissipated from 5 mm x 25 mm flakes in 12 weeks (i.e. half-life), and 1.2 and 0.22 mg exhibited half-lives of 8.0 and 6.5 weeks respectively.

Pheromone release is inversely proportional to outer layer thickness, conventionally expressed in 'mil' (1 mil = 0.001 in or 0.0254 mm). Thus, in the greenhouse, disparlure half-lives in laminates made with 2 mil and 5 mil vinyl, were 29 and 54 days respectively.

Pheromone release is also inversely proportional to outer layer stiffness. Essentially, the pheromone molecule must reorient regions of the polymer chain of the barrier layer in order to diffuse. The stiffer (more crystalline) polymers are less easy to rearrange, thus decreasing pheromone diffusion. Speed of diffusion is, in descending order, through flexible polyvinyl chloride, rigid polyvinyl chloride, acrylic, polypropylene and nylon, and non-permeable polyester. Pheromone half-lives are usually shorter where prevailing temperatures are higher.

The presence of additives affects permeation rates through modification of the polymer matrix of the outer layer. Thus, phenol or ethylene glycol phenyl ether enhance emission rates and decrease pheromone half-life. Such diffusion is inversely proportional to the square root of the molecular weight of diffusant (Graham's law). As pheromone diffusion occurs through reorientation of the polymer chain by the pheromone, then a low molecular weight pheromone with less segments to reorient will diffuse faster than a high molecular weight compound.

In regard to significant environmental factors, temperature and wind speed predominate although humidity may also be relevant. An increase in air temperature from 21.1 to 37.6 °C served to increase pheromone vapour pressure by up to 200%, and also altered the physical characteristics of the barrier film. As wind removes disparlure from the flake surface, it is replaced by pheromone migration from the reservoir until the concentration there is significantly depleted.

Flake configuration is relevant as the ratio of flake perimeter to surface area, the edge effect, increases pheromone release from smaller, as compared with larger flakes that are otherwise similar. Coating flakes with sticker will attach them to foliage but will also alter pheromone release rates by leaching, although this may serve to synchronize pheromone release with a well-defined mating period. Pheromone release rate from flakes may be represented by:

$$\frac{dM}{dT} = M_{\infty}\left(\frac{2DC}{C_0 l^2 t}\right)^{1/2} \qquad (15)$$

where D = diffusion coefficient; t = time; C_o = total concentration of pheromone in the matrix; C_s = solubility of pheromone in matrix; M_∞ = total mass of pheromone; and l = thickness of the reservoir wall [9.25].

Hollow Fibres

These usually comprise short pieces of thermoplastic, non-reactive, impermeable tubing sealed at one end, and loaded with pheromone. This is released at the liquid-air interface by convection, following diffusion through the air column to the open end of the fibre. As the first discharge of pheromone from these fibres is followed by release at a more or less constant rate (pseudo zero-order kinetics), they were considered an improvement on microcapsules and flakes that release according to first-order kinetics (concentration-dependent) [9.27, and references therein]. Release, excluding leakage through the wall, occurs in three stages entailing: evaporation from the liquid-vapour interface; diffusion to the end of the capillary through the vapour-air column; and finally convection from the capillary end; of these, diffusion seems to be the rate-limiting step. Provided that the lumen wall is smooth and not scratched, fibre diameter is uniform along its length, and there is little or no absorption into the fibre wall, diffusion may be represented by transport equations describing the rate of meniscus regression (dl/dt) as:

$$\left[-\frac{McD}{2\rho}\ln\left(1-\frac{P_{vap}}{P}\right)\right]^{1/2} t^{-1/2} \qquad (16)$$

where M = molecular weight of liquid charge; c = molar density of vapour-air column; D = diffusion coefficient; ρ = density of the liquid; P_{vap} = vapour pressure of the liquid; and P = atmospheric pressure; and t = time. P_{vap} will either exceed or be less than P where the meniscus is convex and concave respectively. In practice, it is usually concave on the vapour side thus decreasing the value of vapour pressure. Mass trapping and population monitoring programmes require: the authentic pheromone blend; controlled release of pheromone in the appropriate blend and over a satisfactory time interval; additives such as stabilizers, and insecticides to kill entrapped insects; and an insect trap of optimal efficiency. Secondly, area-wide control requires: the authentic pheromone or a major component; controlled-release; additives as above; a system for dispersal and for sticking fibres to the crop plant. A further practical consideration is that the fibre material should be approved by the FDA for food contact in the U.S.A. Other countries may have corresponding requirements.

An ideal fibre material such as polyethylene terephthalate should exhibit: (1) mutual insolubility of pheromone and fibre material, and non-permeability of fibre to the pheromone and to additives such as antioxidants; (2) ready suitability of the fibre material to yield lengths with a consistently-sized lumen, unscratched internal walls, and clear cut endings; (3) suitability for contact with human foodstuffs; and (4) photo-, bio-, or environmental-degradability.

The use of hollow fibres failed to become widespread primarily due to inconsistent costing. True costs included those associated with polymer and pheromone procurement and production, internal volume of fibres, and overheads. Pricing policy, however, was inconsistent as it was based on what a grower would otherwise pay for conventional insecticide control of a particular crop [9.27, and references therein].

Twist-tie Ropes

These are plastic fibres about 15 cm long sealed at both ends and equipped with a hollow channel loaded with pheromone, and with a wire spine for tying. This arrangement differs from those of the other three systems by accommodating a large dose (30-300 mg) of pheromone that is point-released over a long time (30-200 days). Nevertheless, the design principles and constraints of twist tie ropes are similar to those of hollow fibres [9.27, and references therein].

Twist-tie ropes or ties proved to be the method of choice for applying a blend of (*Z*)-9-dodecenyl acetate and (*Z*)-11-tetradecenyl acetate, as the sex pheromone of the grape berry moth *Endopiza (= Paralobesia) viteana.* This is a pest of grapes from which wine, jelly and jam are produced in upstate New York. A difficulty associated with ties is the need to fix them by hand, but this is not a significant problem in viniculture or with other crops that are intrinsically labour intensive. At the New York State Experiment Station in Geneva, ties were initially applied at a density of 9000/ha but a density of 500/ha was sufficient to substantially decrease successful matings and subsequent grape damage. However, the rather intimate contact between pheromone-loaded ties and the crop plant facilitates contamination. This hindered the granting of an Experimental Use Permit within the U.S.A. for plants the products of which are destined for human consumption. Accordingly, this approach may best serve labour intensive crops, in societies where such labour is abundant and cheap, and on crops such as cotton that are not for human consumption [9.11, and references therein]. For an account of use of this method against infestations of the pink bollworm *Pectinophora gossypiella*, on cotton, see *Mating Disruption* below.

Traps

Design

The insect trap, frequently a key ingredient in the use of pheromones in crop protection, should be cheap and easy to examine. Trap designs range from the very simple, such as suspended plates, to more complex winged designs, and to those equipped with one-way valves (Fig. 74). Sticky traps are coated with adhesive the efficiency of which declines with exposure, essentially due to occlusion by entrapped insects. Trap efficiency may be enhanced by use of colour, especially yellow, depending on the insect species.

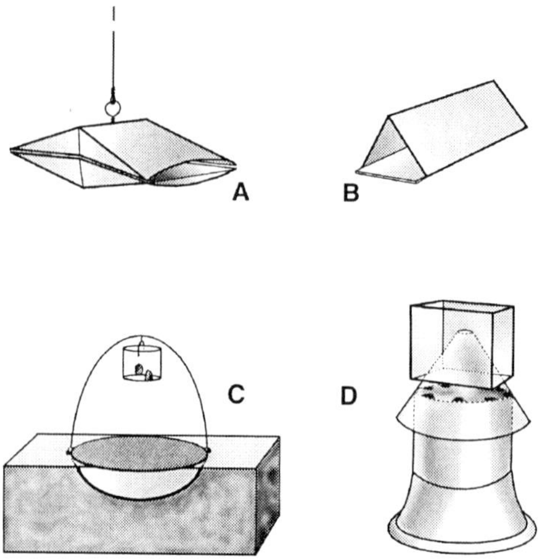

Figure 74. Schematic representations of some pheromone traps: (A) wing design appropriate for many lepidopteran species; (B), delta design as employed to trap the pea moth Cydia nigricana; *(C), puddle trap for bark beetles (redrawn after Byers [9.4]) (D), Hardee trap for cotton boll weevil* Anthonomus grandis grandis *(A, B, and D redrawn after Wall [9.28]).*

Pheromone traps attract insects from downwind but there have been relatively few detailed studies on the effect of trap design on insect behaviour. These showed inter alia that wind direction is a key factor by influencing the structure of the pheromone plume arising from the trap. This in turn affects flight, pheromone perception, and landing behaviour of the target insect. Consistency between traps is achievable by use of self-orienting designs, or by the cheaper procedure of using several traps differently oriented at each location.

A combination of factorial field tests and wind tunnel observations clarified the factors affecting catches of the light-brown apple moth *Epiphyas postvittana*, by delta sticky traps [9.12]. Essentially, both trap width and trap length were significant as more males were caught in traps with the largest surface area. Furthermore, largest numbers were caught in the two longest traps with a significant relationship between catch and base length. This was independent of changes in glue efficiency over time. Wind-tunnel observations established that although some 40% of males exited downwind from the regular sized (18 cm long) trap, within 1 min of entering, only 7% did so from the 36 cm long trap. This difference in retention was proportional to the difference in field catches by regular and longer traps. Apparently, the longer walk toward the pheromone source within the longer trap enhanced entrapment. As expected, catches were always higher in traps in which the sticky bases were changed regularly. The provision of vertical entrance barriers enhanced catches in the regular traps both by impeding males from exiting, and by forcing them to land more upwind necessitating a larger walk back to the downwind exit; catches in long traps were unaffected by the presence of barriers. Thus trap catches reflect not merely numbers of insects entering the trap, but also within-trap features that affect distance to be traversed in glue to the exit, and ease of exit [9.12] (Fig. 74).

An experimental effort to improve trap design focussed on the slope of the ramp, 60°, 30°, or 0°, leading to a cockroach trap. Rather minor modification significantly affected trap catches as, irrespective of whether the design was a delta shape or had a lower flat roof, decreasing the slope from 60° to 30° served to increase the catch by more than 50%; but decreasing the slope to 0° decreased the catch. Insects hesitated before both slopes, but were deterred by the steeper one and turned back; they did not hesitate before the flat ramp, did not turn back, but escaped more readily [9.22].

Location
Three factors affect optimal trap location: height; position in relationship to vegetation; and trap density. In regard to height a compromise may be necessary between the height that achieves maximum catches and that readily manageable by the operator; also a growing crop may necessitate adjustment of trap height. In conjunction with the factors listed in the previous section, the investigation of *E. postvittana* also examined, in a wind-tunnel, the effect of height of the pheromone source above the surface on landings. Increasing source height, by as little as 1.5 cm, above a horizontal surface was associated with proportionally increased landings by males on the source. The interaction of the pheromone plume with the landing surface merited further investigation [9.12].

Trap density is a compromise between the number necessary for meaningful sampling of the population and that decreasing individual trap catches. Trap interactions vary according to trap type and to the insect species being trapped, and will obviously reflect the range of action for a particular trap. This parameter has been variously termed: active space; range of attraction; range of stimulation; range of sampling; and, more simply, effective sample radius that, derived from relative trapping by baited and unbaited traps, is expressed by the effective radius of the former. Mark-recapture methods provided a tool by which the maximum range of sampling (R_s) and corresponding sampling area, where sampling efficiency is unity, was derived for Uni-traps and the fall web worm moth *Hyphantria cunea*.

Essentially three Uni-traps baited with pheromone were placed on bamboo poles at a height of 1.8-2.0 m, 15 m apart, and perpendicular to the prevailing wind. Marked insects were released 30, 50, 80, and 250 m downwind on a line with the central trap of the three. The extra distance travelled to reach the outer two traps from any release point was derived by triangulation but, as there was no interaction between trap position and distance, data were pooled for distance. Recaptures declined linearly with distance indicating by regression a maximum R_s of 344 m (190-710 m, confidence interval). The sampling area, derived by integrating recapture probability from the source over each annulus to the maximum R_s, was estimated as 6.9 ha, corresponding to a circle of radius 148 m. However, the actual area sampled will be significantly less than this, given a stable prevailing wind direction and that only downwind insects are attracted. The R_s value for *H. cunea* is, by exceeding 100 m, broadly consistent with those for several other Lepidoptera, although values less than this are known for other lepidopteran species [9.30].

Bark beetles seem to have shorter ranges than moths and this has been investigated in field experiments and by computer simulation [9.4]. Field data were derived from use of a very simple puddle trap involving a looped wire framework, surmounted by a pheromone dispenser, and anchored by an attached polyethylene rubbish bag filled with water. This is embedded in the forest floor (Fig. 74). Traps arranged in two sites each in a 7 x 7 grid, using spacings of 1.5, 3.0, 6.0, or 12.0 m, were baited with constituents of the aggregation pheromone of either the smaller European spruce engraver *Pityogenes chalcographus*, or of the spruce bark beetle *Ips typographus*.

Beetles either struck the dispenser cup and fell into the water, or landed directly in it. Trap catches were appreciably higher in this puddle-trap than in sticky traps, pipe traps, or window traps. Proportionally more insects were caught by the outer ring of 24 traps than by the inner ring of 16, which in turn caught more than the central ring of nine. Evidently, beetles entering the ring were more likely to be taken by the first traps encountered, the decreased number escaping would then be exposed to the next ring, and similarly for the next. The probability of escaping reflects the effective size of traps, which in turn depended on the amount of pheromone released.

Computer simulation varied the effective catch radius from small, where approximately the same catch would be taken by all traps whether outer or inner, to very large, where outer traps caught infinitely more. This gave a quadratic term, the reciprocal of which gave the relationship between trap radius and outer/inner catch ratio. An acknowledged anomaly was that although, in principle, the effective catch radius should be the same irrespective of grid spacings, this was not the outcome. Nevertheless and as an example, at 6-m spacing the effective catch radii were 0.47- and 2.04 m for *P. chalcographus* and *I. typographus* respectively. Traps placed closer together than these respective distances should incur inter-trap competition [9.4].

BIOLOGICAL CONSIDERATIONS

Exploiting fully the potential of pheromones for pest control obviously requires insight into the factors that influence pheromone emission and perception. These include: insect age and mated status; inputs from plants and pathogens; as well as physical factors such as temperature, light intensity, photoperiodicity-calling, wind speed, and other atmospheric factors (extensively reviewed in 9.19; the emphasis below is on the male response, as this is most frequently the target for manipulation).

Insect Age and Mated Status

Females of various species initiate calling at various ages and for time periods that are not fixed for a particular species. Thus, older females may gain a competitive advantage by commencing calling earlier in the scotophase than younger ones. Calling may be advanced or delayed by the presence of conspecific female pheromone according to the species (see below). Older females tend to call for shorter time intervals and to release pheromone at significantly lower rates. In complementarity, male response increases with increasing age.

Mating serves to decrease both pheromone synthesis and calling frequency, in species the female of which mates only once; in contrast, calling recommences after some days, in species the females of which engage in multiple matings. Multiple matings are characteristic of male Lepidoptera and lack of previous mating activity may be inferred in some species from a dark coloured fluid in the primary simplex of the virgin male. Previous mating does not seem to diminish male responsiveness to the female pheromone.

Host Plants and Pathogens

The quality of female diet may influence pheromone production rates, specially in oligophagous species. Also the presence of appropriate plant volatiles may be a necessary requirement for calling. The presence of sunflower pollen elicits calling significantly sooner after emergence by the sunflower moth *Homoeosoma electellum*, the neonate larva of which requires pollen as an essential source of food. Male responsiveness to female pheromone may be enhanced by the presence of volatiles from the host plant, especially when this is mature, and such receptivity may be age-dependent.

Failure to complete the sequence of behavioural steps leading to mating by western spruce budworm *Choristoneura occidentalis* males, infected with spores of the pathogen *Nosema fumiferanae*, increased with increasing spore load. Electroantennogram (EAG) responses were unaffected, so the peripheral nervous system was not implicated. Although flight duration in a wind tunnel also seemed unaffected, flight speed, which may be of greater significance in male competition for females, was not assessed. Overall, the effects of such infections on pheromone emission and on male responses seemed to merit further investigation.

Temperature

In general, decreases in temperature during pupal development delay initiation of calling. Regimes of decreased temperature after adult emergence elicit earlier calling in nocturnal species, but in diurnal species calling is deferred until there is an increase in ambient temperature, or in body temperature as influenced by basking. Duration of calling varies in response to temperature according to species.

Low temperatures decreased cabbage looper *Trichoplusia ni* male responsiveness to female pheromone, and also decreased upwind flight of the male true armyworm *Pseudaletia unipuncta*; this is consistent with captures of the latter males in light- but not in pheromone traps during late summer and early autumn. Correspondingly, increased temperatures are associated with a proportional increase of recaptures of marked *L. dispar* males. Lower temperatures promote trap captures of crepuscular and nocturnal species. Many male Lepidoptera exhibit diel periodicity in flight activity, but it is not clear if there is always a corresponding periodicity in receptivity to female pheromone.

Daylength and Light Intensity

A shorter daylength may delay maturation of lepidopteran males as increased light intensity decreases the activity of males of several species, in response to unaltered pheromone concentrations. Such effects may be temperature-sensitive, as rice stem borer *Chilo suppressalis* males at 30° C are entirely inhibited by 20 lux, but at 15° C some are responsive in 700 lux.

Wind Speed

Although of obvious significance for male flight, relatively few studies have explored the precise effect of wind speed. Absence of wind seemed to prevent *T. ni* males locating a pheromone source, and wind speeds of 2-3 m/s decreased captures.

Significance for the Male Response

Diurnal changes in the above conditions may, in addition to modifying the configuration of the pheromone plume, subsequently alter male behaviour as to diminish trap catches. In response to temperatures declining to near the flight threshold (10ºC), the nantucket pine moth *Rhyacionia frustana* focusses its searching behaviour at a higher level in the tree canopy. Such responses would affect the effectiveness of traps established at particular heights. It is long established that competition between females and traps is influenced by protandry (earlier male emergence), and by changes in female population density over time. The significance of the foregoing factors may be assessed within the framework of competition window-width between traps and females, as represented by the duration of female calling. Given the time and plasticity of female calling in response to these factors, then competition width is equally variable. Pheromone traps may be viewed as conspecifics and female responses range from earlier and longer calling and thus more competition in *C. occidentalis*, to decreased calling and less competition in the small tea tortrix *Adoxophyes* sp. and the oriental tea tortrix *Homona magnanima*. Male receptivity although variable seems less susceptible to environmental conditions, with the exception of temperature and wind speed [9.19, and references therein].

USES OF PHEROMONES

The uses for pheromones in crop protection are usually categorized under the heading of: surveying and monitoring; mass trapping; and mating disruption.

Surveying and Monitoring

As pheromone-loaded traps entrap insects at small population densities they may be used as an early warning system. Furthermore, a series of strategically-placed traps may delineate the geographical distribution of the pest. Difficulties arise in attempting to use trap catches to quantify the pest population. Confounding factors include: attraction of insects from adjacent but distinct populations; spuriously large catches resulting from increased mobility rather than increased population density; competition between traps for male insects; variations resulting from trap-design, trap-efficiency, and trap-location within the crop; and use of pheromones of inadequate purity. Nevertheless, a threshold value of 5 moths/trap/week was successfully used for the codling moth *Cydia* (=*Laspeyresia*) *pomonella* in the U.S.S.R., as the number of pesticide spray applications required for pest control was decreased by 50-75% [9.5]. Perhaps one of the most convincing systems of monitoring linked to spraying was developed in the U.K. in relationship to the pea moth *C. nigricana.*

The Pea Moth

Peas, ranked fifth in value as a crop in the U.K. after cereals, potatoes, rape and sugar beet, are grown as: combining peas for human consumption and or seed and are harvested dry; vining peas harvested green for freezing or canning; and protein peas for animal feed. Damage by *C. nigricana* larvae, tunnelling into pods and attacking seeds, may be tolerated in combining crops, as the damaged peas are removed after harvest, but the grower is cash-penalized. However, little damage is tolerated in vining peas as they are not segregated after harvest. In contrast, damage to crops grown for animal feed is widely tolerated as yield is relatively unaffected. The pea moth is univoltine and although pupation occurs from late April to early May, adult emergence from late May is very variable. The only life-stage susceptible to insecticidal spraying is the newly hatched, first-instar larva, prior to invading the pod. The variable emergence by the adult necessitates a local monitoring system that is readily useable by individual growers.

The sex pheromone of the female *C. nigricana* is (*E*,*E*)-8, 10-dodecadien-1-yl acetate (*E*, *E*8, 10-12 : Ac) and its slow release is regulated using 3 mg on a rubber stopper, with an antioxidant added as stabilizer against the effect of sunlight; under these conditions attraction continues for three months. Delta-type traps, equipped with a removable base, were essential for this species as the wings are held in a steep roof-like pattern, thus presenting only leading edges of the wing for adhesion. Optional trap height, 2/3 that of the growing crop, is maintained by use of a sliding clip on a metal pole. One trap/crop is usually sufficient for areas up to 50 ha, above which two are required. For vining peas, one trap/36 ha is recommended. Spraying is most effective if applied about ten days following trap-catches of at least ten moths per two days, in either of a trap pair in the same field [9.17]. However, this did not take account of subsequent and variable larval development as affected by temperature. Such development is critical for the timing of spraying, as larvae are vulnerable to insecticides only after hatching and before pod invasion.

The hatching of *C. nigricana* eggs was related to temperature in a nomogram indicating that: daily percentage development (t) = 0.609 (maximum temperature + minimum temperature (all °C) - 11.72). From this a disc calculator was derived, suitable for use by growers after the threshold trap catch was recorded (Fig.75). This displayed in a window the daily percentage of egg development as driven by temperature thus giving, by summation over a period of days, the time of larval emergence. Comparing the effects of sprays applied at different dates indicated that the calculator erred on the side of caution by eliciting an early rather than a late spraying. A second spray one week later gave effective protection against the moth i.e. damage levels in the range of 0.05 to 0.5% damaged peas. Furthermore, sustained trap catches of ten were identified as indicative of the threshold; catches of ten followed by less gave unreliable results.

The relationship between trap catch and infestation levels was acknowledged as tenuous not least because: only males are trapped; trap efficiency is low as only one in four moths at a trap enter it at the first approach; fecundity is not estimated; and mortality of eggs and larvae is unknown in the 12-week interval before damage assessment. Nevertheless, the prediction system works efficiently for combining peas, using the following model derived from the temperature-development relationship, expressed as an accumulated development curve (Fig. 75):

$$T = \frac{100}{\tan\,\theta} = 100\,\frac{(t_2 - t_1)}{(d_2 - d_1)} \qquad (17)$$

where: T = number of days from threshold to spray date; t_1 = threshold date; t_2 = date on which prediction is made; d_1 = accumulated development at t_1; d_2 = accumulated development at t_2; θ = slope of egg development curve.

When a grower, detecting a threshold by use of the disc calculator, telephones the local office of the National Agricultural Advisory Service he receives a recommended spray date derived by computer and based on the date of his threshold; alternatively, he is given the appropriate computer page on a viewdata system. This prediction is based on both recorded and forecast temperatures and is updated within several days. Most growers now run their own traps and follow minimal, but timely, spray schedules [9.18].

For vining peas, the following system was introduced in 1987 on a trial basis. Single traps loaded with 200 μg of female pheromone plus antioxidant are located 100 m into the crop and 600 m from the nearest trap. As there is a linear relationship between the percentage of peas damaged and cumulative catch at full flower, a regression gives confidence limit contours for future catches and associated damage. This indicates if there is a need to spray [9.29].

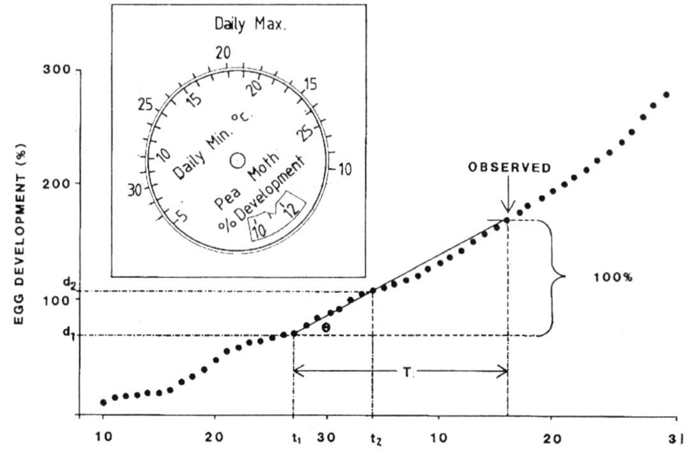

Figure 75. Accumulated development curve for C. nigricana, *based on maximum and minimum temperatures. Redrawn after Macaulay* et al. *[9.18]. Inset, schematic representation of disc calculator (see text for details).*

The foregoing data confirm the utility of pheromone-loaded traps in surveying and monitoring, especially in conjunction with use of insecticides, by minimizing the number of sprayings or by indicating the optimal time for spraying, or both. For large scale growers, the financial savings from fewer sprayings might not be a major consideration but they stand to gain from the deferment of pest resistance. There is an obvious environmental benefit from decreasing the insecticide load deposited on soil and on crops destined for human consumption. For small scale growers and in underdeveloped economies, where insecticides represent an expensive investment if not a luxury, the financial savings from fewer sprayings might be considerable.

Mass Trapping

This approach is cleaner than surveying and monitoring as the pheromone should attract only the target species, thus precluding the broadcast use of insecticide with associated effects on innocuous and beneficial fauna and on soil contamination. Mass trapping may be considered in the context of Dutch elm disease (DED) that, caused by the fungus *Ceratocystis ulmi* and vectored by bark beetles, has virtually eliminated the elm tree from the Northeastern U.S.A. [9.16]. Five categories of semiochemical interactions are recognized as governing relationships between trees and bark beetles [9.3] (Fig. 76) (see also Chapter 4). From these arise six possible control strategies: (1) prevent pheromone production by removing susceptible trees, as few trees are mass attacked; (2) prevent the perception of semiochemicals by permeating the atmosphere with them; (3) exploit repellent allomones; (4) attract entomophages; (5) exploit kairomones and aggregation pheromones by mass trapping; (6) exploit antiaggregation pheromones [9.3]. Of these, (5) and (6) seem to have the greatest potential.

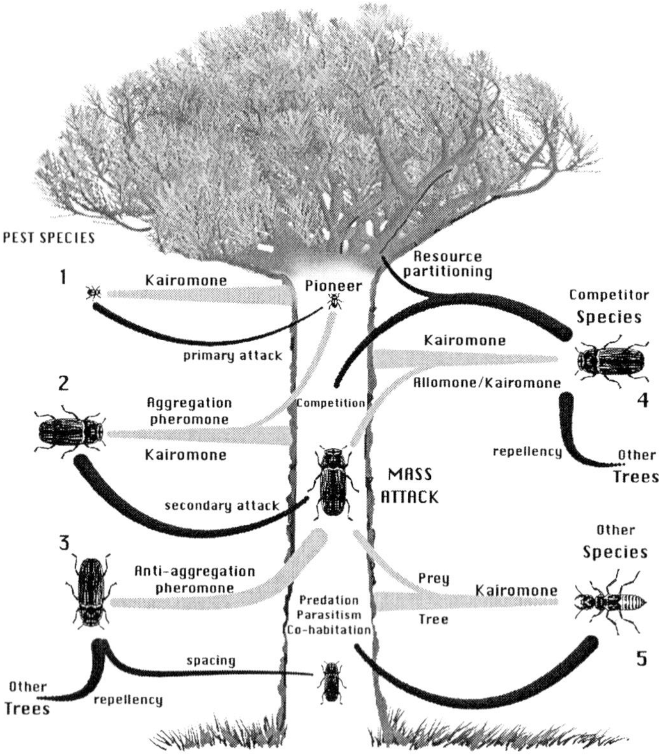

Figure 76. Schematic representation of five categories of semiochemical interactions between trees, bark beetles and other insect species; light tone represents chemical signals and dark tone represents insect responses. Redrawn after Borden [9.3].

The efficiency of mass trapping is affected by trap spacing, dose/trap, and vaporization rate. The efficacy of mass trapping has been dogged by questions regarding adequate controls when very large populations are addressed, despite the fact that baited trap catches may be numbered in millions or billions. The method seems to give the most clear cut effects when used against populations that are small and isolated. The first successful operation was directed against the western pine beetle *Dendroctonus brevicomis* infesting 283 trees in 65 km^2 of Southern California. In excess of half a million beetles were captured in some 341 attractant-baited traps during the flight period of one generation. Tree mortality decreased by 66% and did not spread from the treatment site for a further two years. Similarly, mass trapping against the smaller European elm bark beetle *Scolytus multistriatus*, the principal vector of DED, in 12 isolated groves of the American elm groves abruptly decreased the infection. Sawmills provide another environment sufficiently restricted to favour this method. Accordingly, population decreases of 65% and 44-77% were recorded for the ambrosia beetle *Gnathotrichus sulcatus* and the striped ambrosia beetle *Trypodendron lineatum* respectively [9.3, and references therein].

In contrast, the capture of almost one million beetles in 421 traps within 1 km^2 of Detroit city was associated with increased incidence of DED. Contributing factors were given as: low rates of beetle capture (20%); immigration by beetles; competition from attractants in naturally damaged trees; and increased attacks on baited trees [9.8]. In possibly the largest mass trapping campaign ever undertaken, the use of 605,000, 650,000 and 220,000 baited traps captured respectively 2.9, 5.1 and 1.0 billion *I. typographus* in Norway and Sweden. Although, trees killed by beetles declined from three million, two years before the campaign, to 100,000 one year following it, the effect was nevertheless attributed to: wet, cool weather; improved tree resistance consequent on recovery of water balance; removal of infested trees; and mass trapping [9.1, and references therein].

A variant of mass trapping is use of baited or trap trees which takes advantage of a practice followed by European foresters for at least three centuries: exposing felled or girdled trees, that beetles prefer to standing timber, and then destroying them by burning, debarking, or applying insecticide. Trees baited with semiochemicals served to concentrate infestations of the Douglas-fir beetle *Dendroctonus pseudotsugae*, before logging. Similarly, lodge pole pine trees baited at four trees/ha ensured that emerging mountain pine beetles *D. ponderosae* were within 35 m of a bait. This inhibited beetle emigration, promoted immigration, and the infestation was concentrated. Debarking at the mill killed the beetles [9.3].

An alternative to applying insecticide to the baited tree is using an arboricide to kill terminally diseased or unwanted trees, and thus to create trap-trees. An example is killing the American elms *Ulmus americana* with cacodylic acid, and then baiting them with synthetic aggregation pheromone of *S. multistriatus*, the main DED vector. This elicits attack by female beetles whose pheromones then elicit a mass attack. The trap-tree is as attractive as an attacked dying elm tree and competes effectively for the flying beetle population, an advantage over conventional mass traps. An additional positive feature is the conversion of diseased elms from a liability to an asset. Beetles reproduce within the bark but the brood fails to develop, thus decreasing both the population density of the next generation and the incidence of DED. This method controlled DED in two locations in central New York state. Trap-trees sustained 4,333 and 9,467 attacks/tree in each site respectively, representative of a beetle population twice that size (females begin the attack, sex ratio is 1:1). Larval populations decreased by 97 and 88% respectively, and in the first location DED decreased significantly from 7.7 to 4.3%, whilst increasing from

5.0 to 6.3% in a corresponding control area; however, the estimation of DED in the second site was confounded by incidence of elm phloem necrosis [9.21]. This success contrasts with a mass trapping campaign using multilure, a racemic mixture of the aggregation pheromone, on sticky traps that captured millions of beetles but gave contradictory results in terms of DED control. Essentially, protection was achieved in only areas with little brood wood (freshly-infested wood) to act as competing sources. Furthermore, multilure does not attract the native elm bark beetle *Hylurgopinus rufipes* that also vectors DED [9.21]. Nevertheless, multilure serves to monitor beetle activity and to pinpoint brood sources.

It is salutary that the use of cacodylic acid-treated trap-trees proved ineffective in a different geographical location, northern England and Scotland. The situation there is more complex with at least one additional vector, the smaller elm bark beetle *Scolytus scolytus*, and possibly a third species *S. laevis*, together with at least seven species/cultivars of elm. And climate is problematical. In the U.S.A. the time of beetle flight is easier to predict which is significant, as treating diseased trees at this time gives best results. In northern England, Scotland, flight days (when maximum temperature equals or exceeds 22°C) are relatively few (typically 20 between May and September), vary considerably from year to year, and are difficult to predict. Thus, some weeks may elapse between treatment with cacodylic acid and beetle attack. Furthermore, cacodylic acid promotes growth of the bark saprophytic fungus *Phomopsis oblonga* that, colonizing the inner bark of dying trees, competes directly with bark beetles for this resource. As time elapses between tree treatment and beetle flight, this fungus exploits the habitat before beetles can arrive. By preventing mass attack this undermines the trap-tree rationale [9.20].

Exploitation of antiaggregation pheromones is exemplified by the use of 3-methyl-2-cylohexen-1-one (MCH), the antiaggregation pheromone of *D. ponderosae*, that effectively abolishes attraction after beetle numbers established in a tree are sufficient to overcome its intrinsic resistance. The adaptive significance would seem to be prevention of excessive colonization and competition for space between broods. A controlled-release formulation of MCH in rod-shaped, granules of polyamide, dispersed in May by an aerial spreader suspended from a helicopter, at a rate of 4.5 kg/ha, gave a ground-measured rate of 2.0 to 2.7 kg/ha; this decreased infestation by 96% in the following month. In addition, attacks by the spruce beetle *D. rufipennis*, were decreased by 55% on fewer, intermingled trees of Engelmann spruce [9.13].

Mating Disruption

The concept of pheromone usage to disrupt sexual communication and mating involves neuronal modulation, either through adaptation of peripheral pheromone receptors, or habituation at the level of the central nervous system. The net effect is to raise the response threshold of the male insect, such that pheromone concentrations naturally emitted by the calling female are too small to elicit a behavioural response. Effectiveness is enhanced if the artificially high pheromone concentration elicits arrestment of upwind flight. Practical difficulties include maintaining sufficiently high concentrations in the field, where pheromone is dissipated by natural diffusion and dispersive air currents. Also high insect populations will promote chance encounters that facilitate mating [9.2].

The second mechanism is disorientation or confusion of the male by the competing pheromone trail such that the calling female is not located. The artificial pheromone source dislocates typical insect flight response i.e. optomotor anemotaxis linked with programmed counter turning resulting in zigzag upwind flight (see Chapter 3). Specific effects could involve: modification of the filamentous pattern of the pheromone plume emanating from the female; swamping the plume in a background of synthetic pheromone; providing sufficient point sources of pheromone to out-compete calling females; and dislocation in time by eliciting male flight before the female commences calling. As previously, dense populations of insects may lessen the efficiency of each of these effects [9.7, and references therein].

Supplementary strategies may involve use of: unbalanced blends of the pheromone bouquet to diminish perception of the natural, high-potency blend; use of parapheromones that mimic pheromone action presumably by binding to the appropriate site on the neuronal membrane of appropriate sensilla (see Chapter 6). Interacting mechanisms may be significant as exposure to a high pheromone dose will, in addition to decreasing female detection, promote a male response to only a correspondingly high dose. Whatever strategy is chosen, the ecology of the insect is relevant.The male of *P. gossypiella*, by spending significant amounts of time below the soil surface, is less likely to encounter the confusing signal. That species is probably the major pest of cotton, as larvae feeding on bolls and flower buds are inaccessible to insecticides. Mating disruption of this species provides a case history in widely different geographical locations, worthy of detailed scrutiny [9.6, 9.7].

USA

The true pheromone, (*Z,Z*)- and (*Z,E*)-7,11-hexadecadienyl acetate (1:1), has proved more efficacious than the parapheromone, 7,11-hexadecadienyl acetate, and in 1978 its dispersal in hollow fibres was registered for use by the EPA of the U.S.A.. Subsequently, a laminate flake formulation was given an experimental use permit (EUP) and a combined total of 50,000 ha, sprayed with insecticide, were treated in 1981. In 1982, use of the pheromone was made mandatory in the Imperial Valley of California resulting in fewer insecticidal sprayings, larger yields, and 5% crop damage as compared with 30% from insecticide-treated fields. When the compulsory approach was dropped in 1984 the expectation that growers would voluntarily continue with the pheromone programme was not fulfilled; most reverted to conventional insecticidal programmes.

The use of twist-tie formulations at the rate of 1,000/ha, equivalent to 78 g active ingredient/ha with an expected half-life of 58 days, decreased insecticide sprayings by 40% and significantly decreased larval populations. In a further success, management from 1990-93 of 11,000 ha of cotton in Arizona using twist-ties or fibres decreased boll damage by larvae, from 23% in one year prior to 0% in some 25,000 bolls in the final year of the programme [9.6].

Egypt

In 1981, field trials established that microencapsulated pheromone gave control equivalent to conventional insecticidal use, in terms of crop damage and yield of seed cotton. The following year a direct comparison of three formulations, microcapsules dropped from helicopters, hollow fibres, and laminate flakes attached by hand, indicated that all three elicited yields comparable to those derived from cotton treated by only insecticides.

A large-scale field trial in 1983 treated 250 ha of cotton with helicopter-released microcapsules on five occasions (10g/ha each) every 2/3 weeks; a control was provided by 160 ha of cotton in the same locality using conventional insecticidal control sprayed every two weeks. Infestation levels were assessed in both areas as were numbers of predatory insects. Essentially, the pheromone-treated region never exceeded the economic threshold of 10% damage but this was exceeded on three occasions where insecticides were used. Furthermore, predatory insects were much more abundant in the pheromone-treated area. In 1984, pheromone formulations sold to the Egyptian government commanded prices similar to those for insecticides. In 1986, pheromone treatment hitherto confined to central Egypt was extended to the Nile Delta. However, *P. gossypiella* infestations were much larger that year both in pheromone- and in insecticide-treated areas, and when damage exceeded threshold in the pheromone-treated areas, these were over-sprayed with insecticide. Subsequently, yields from areas treated only with insecticide were significantly better than from those over sprayed, perhaps because the insecticidal overspray had adversely affected predators (Fig. 77).

The disappointing result of 1986 cooled official fervour for mating disruption and a much smaller area was treated in 1987 when again pheromone-treated areas were oversprayed in two distinct regions. The pest population was larger than in 1986 in one, but was lower in the other. The larger population elicited larger infestations in pheromone-treated as compared with the insecticide-treated areas; the converse was true for the smaller infestations. These data, taken in conjunction with the previous years results, suggests a pattern of decreased efficacy for mating disruption against denser populations, an unacceptable liability for a proposed control method. Given that the official response is to overspray with insecticide, with likely negative effects on predators exacerbating the situation, the future of this approach against *P. gossypiella* seemed uncertain. However, by 1993, pheromone-treated areas had gradually increased to 50,000 ha [9.6, and references therein] (Fig. 78).

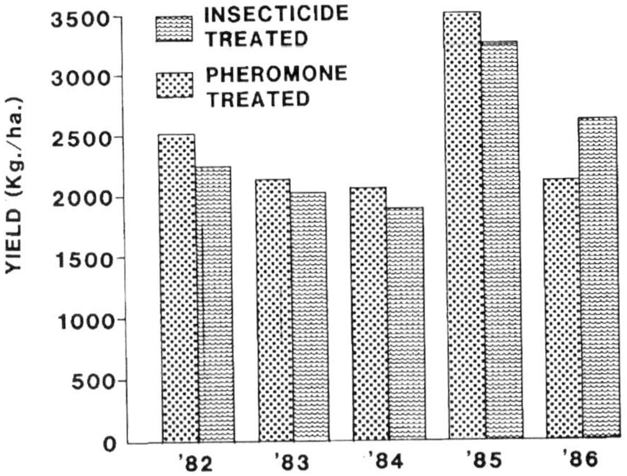

Figure 77. Yield of cotton seed from Egyptian sites insecticide-treated, and pheromone-treated followed by an insecticide overspray. Redrawn after Campion et al. *[9.6].*

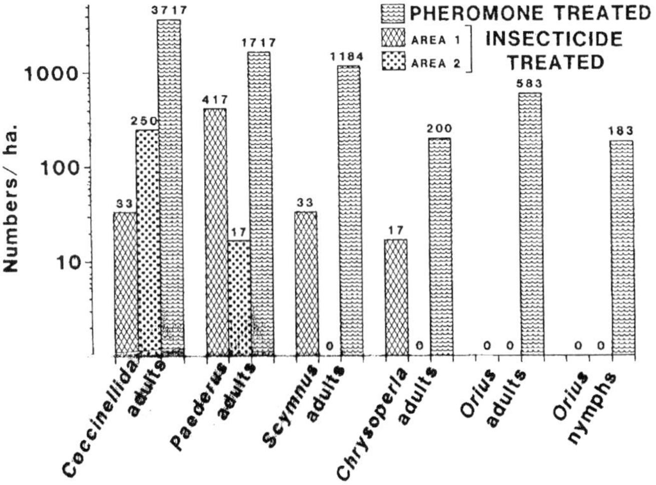

Figure 78. Schematic representation of the effect on predatory insects of insecticide applied in two cotton areas as compared with pheromone applied in a third area in Egypt (numbers/ha derived from suction sampling). Compiled from Campion et al. *[9.6].*

As indicated earlier, the application of pheromones to pest control is still in a development phase. The list of other species for which unambiguous data are available on satisfactory field control by mating disruption includes: oriental fruit moth *Grapholita molesta*; tomato pinworm *Keiferia lycopersicella*; *E. postvittana*; currant clearwing moth *Synanthedon tipuliformis*; European grape moth *Eupoecilia ambiguella*; *E. viteana*; *C. pomonella*; and a complex of leafroller moths on apple [9.7].

No doubt this list will be extended as investigations advance from definition of a particular blend to replicated field trials, emphasizing statistical analysis of both elicited changes in population density and associated crop yield. The efficacy of this approach against any putative target species will always be influenced by factors such as: efficiency of the male in mate location; population density of the pest; and rate and duration of calling by the female. Finally, mating disruption of one particular pest of a crop is not a panacea as other pest species, freed from competition, may intensify their effect.

Other Applications

Interesting possibilities have been opened by the discovery that plants produce and release chemicals, identical to pheromones, that advantageously manipulate insect behaviour. For example, leaves of the wild Bolivian potato *Solanum berthaultii* release (*E*)-β-farnesene, the aphid alarm pheromone, that repels the peach-potato aphid *Myzus persicae* at a distance of 1-3 mm from the leaves. *S. berthaultii* is resistant to many potato pests including flea-beetles, leafhoppers, beetles, thrips, mites, and aphids. Such resistance is associated with type A and type B trichomes on the leaves and steams (see Chapter 2). Type A hairs occur rarely in cultivated potato and type B are entirely absent.

Headspace analysis of *S. berthaultii* leaves afforded 50 mg (*E*)-β-farnese/20 ml air, from 1 g of leaves, or sufficient to elicit the alarm response [9.10]. An obvious desirable step is introduction of the genes for type B hairs into cultivated potato that is achievable by conventional methods as *S. berthaultii* readily hybridizes with it. Furthermore, alate aphids, associated with most virus transmission, are especially sensitive to *(E)*-β-farnesene. This work was extended to the generation of propheromones, pheromone derivatives that are less readily oxidized and less volatile or more stable then the pheromone, and capable of releasing it under field conditions. 1,4-Cycloaddition reactions of farnesene with a dienophile such as SO_2 produced an unstable and phytotoxic sulfolene; such reactions with RO.CO.C $\equiv$ CCO.OR produced relatively involatile didecyloxycarbonyl compounds and higher homologs. Twelve derivatives were assayed in terms of: the alarm response elicited; the extent to which aphids settled on treated leaves; and the acquisition by treated plants of aphid-transmitted viruses. A relatively high molecular weight, 1,4-cycloaddition product from *(E)*-β-farnesene and didecyl acetylenedicarboxylate: significantly decreased ($P<0.001$) the number of aphids settling; significantly decreased virus acquisition from insecticide-resistant aphids ($P<0.001$); persisted for at least one week; and exhibited no phytotoxicity. Controlling the diverse range of insect pests of the cultivated potato through breeding-in genes for *(E)*-β-farnesene production, and through direct application propheromones represent interesting possibilities [9.9, and references therein].

Regulatory Constraints

Although many entomologists and agronomists, mindful of the undesirable environmental and other consequences of insecticide usage, view the use of pheromones as environmentally and ethically superior, regulatory agencies are obliged to treat pheromones objectively i.e. as pesticides and putative toxins. This does not, preclude recognition by regulatory agencies of the facts that unlike conventional pesticides pheromones are: rarely applied directly to food crops; dispensed in only minutes quantities; and the chemical structure of most sex and aggregation pheromones are closely related to these of many other compounds (e.g. monoterpenes), naturally occurring in edible green leaves, not subject to restrictions. Such recognition is associated with a 1992 proposal from the EPA of the U.S.A. entitled "Incentives for Development and Registration of Reduced Risk Pesticides". However, objections to this document include: unfairness of advance categorization of any new pesticide as intrinsically more environmentally correct than another; active promotion of such an approach by a Federal Agency could elicit intervention by other regulatory agencies and objections from interested parties i.e. the pesticidal industry; formally defining some conventional insecticides as high risk in contradistinction to pheromones as low risk would fuel public pressure for banning the former, despite an obvious and continuing need for them, not least as standby oversprays when pheromones fail. As of 1992, only 14 pheromones have been registered by the EPA [9.23].

Furthermore, applying the rigours (toxicology and residue studies) of registration to both the individual components ($\bar{n}$ =3) and the blend of a range of pheromones will entail very significant expense. This will be virtually impossible to recover given that each preparation will likely be specific to only a particular insect pest, and will be required in only trivial amounts per grower. This contrasts unfavourably with the capacity of a standard conventional insecticide, with perhaps one associated analogue, to kill a broad range of pests by use of larger and profit-generating volumes [9.26]. It is perhaps melancholy to conclude that in circumstances prevailing at time of writing, replacement of insecticides by pheromones in the highly regulated U.S.A. and European Union is an increasingly uncertain aspiration.

REFERENCES

9.1 Bakke, A. and Lie, R. (1989) Mass trapping. In Jutsum, A.R. and Gordon, R.F.S. (eds.) *Insect Pheromones in Plant Protection*, Wiley and Sons, New York, pp. 67-87.

9.2 Bedard, W.D. and Wood, D.L. (1981) Suppression of *Dendroctonus brevicomis* by using a mass-trapping tactic. In Mitchell, E.R. (ed.) *Management of Insect Pests with Semiochemicals*. Plenum Press, New York, pp. 103-114.

9.3 Borden, J.H. (1989) Semiochemicals and bark beetle populations: Exploitation of natural phenomena by pest management strategists. *Holarctic Ecology* **12**, 501-510.

9.4 Byers, J.A. (1993) Orientation of bark beetles *Pityogenes chalcographus* and *Ips typographus* to pheromone-baited puddle traps placed in grids: a new trap for control of scolytids. *J. Chem. Ecol.* **19**, 2297-2316.

9.5 Campion, D.G. (1985) Survey of pheromone uses in pest control. In Hummel, H.E. and Miller, T.A. (eds.) *Techniques in Pheromone Research.* Springer, New York, pp. 405-469.

9.6 Campion, D.G., Critchley, B.R. and McVeigh, L.J. (1989) Mating disruption. In Jutsum, A.R. and Gordon, R.F.S. (eds.) *Insect Pheromones in Plant Protection.* Wiley and Sons, New York, pp. 89-119.

9.7 Cardé, R.T. and Minks, A.K. (1995) Control of moth pests by mating disruption: Successes and constraints. *Ann. Rev. Entomol.* **40**, 559-585.

9.8 Cuthbert, R.A., Peacock, J.W. and Cannon, W.N. (1977) An estimate of the effectiveness of pheromone-baited traps for the suppression of *Scolytus multistriatus* (Coleoptera: Scolytidae). *J. Chem. Ecol.* **3**, 527-537.

9.9 Dawson, G.W., Gibson, R.W., Griffiths, D.C., Pickett, J.A., Rice, A.D. and Woodcock, C.M. (1982) Aphid alarm pheromone derivatives affecting settling and transmission of plant viruses. *J. Chem. Ecol.* **8**, 1377-1388.

9.10 Dawson, G.W., Griffiths, D.C., Pickett, J.A., Smith, M. and Woodcock, C. M. (1982) Improved preparation of *(E)-β*-farnesene and its activity with economically important aphids. *J. Chem. Ecol.* **8**, 1111-1117.

9.11 Dennehy, T.J., Clark, L.G. and Kamas, J.S. (1991) Pheromonal control of the grape berry moth: an effective alternative to conventional insecticides. *N.Y. Food Life Sci. Bull.* **135**, 1-6.

9.12 Foster, S.P. and Muggleston, S.J. (1993) Effect of design of a sex-pheromone-baited delta trap on behaviour and catch of male *Epiphyas postvittana* (Walker). *J. Chem. Ecol.* **19**, 2617-2633.

9.13 Furniss, M.M., Markin, G.P. and Hager, Y.J. (1982) Aerial application of Douglas - fir beetle antiaggregative pheromone: equipment and evaluation. *USDA For. Ser. Gen. Tech. Rep.* INT-137.

9.14 Hall, D.R. and Marrs, G.J. (1989) Microcapsules. In Jutsum, A.R. and Gordon, R.F.S. (eds.) *Insect Pheromones in Plant Protection.* Wiley and Sons, New York, pp. 199-248.

9.15 Jutsum, A.R. and Gordon, R.F.S. (eds.) (1989) *Insect Pheromones in Plant Protection.* Wiley and Sons, New York.

9.16 Lanier, G.N. (1979) Protection of elm groves by surrounding them with multilure-baited stick traps. *Bull. Entomol. Soc. Am.* **25**, 109-111.

9.17 Lewis, T. and Macaulay, E.D.M. (1976) Design and elevation of sex-attractant traps for pea moth, *Cydia nigricana* (Steph.) and the effect of plume shape on catches. *Ecol. Entomol.* **1**, 175-187.

9.18 Macaulay, E.D.M., Etheridge, P., Garthwaite, D.G., Greenway, A.R., Wall, C. and Goodchild, R.E. (1985) Prediction of optimum spraying dates against pea moth, *Cydia nigricana* (F.), using pheromone traps and temperature measurements. *Crop Protection* **4**, 85-98.
9.19 McNeil, J.N. (1991) Behavioral ecology of pheromone-mediated communication in moths and its importance in the use of pheromone traps. *Ann. Rev. Entomol.* **36**, 407-430.
9.20 O'Callaghan, D.P. and Fairhurst, C.P. (1983) Evaluation of the trap tree technique for the control of Dutch elm disease in northwest England. In Burdekin, D.A. (ed.) *Research on Dutch Elm Disease in Europe*. For. Comm. (U.K.) Bull. no. 60, pp. 23-38.
9.21 O'Callaghan, D.P., Gallagher, E.M. and Lanier, G.N. (1980) Field evaluation of pheromone-baited traps to control elm bark beetles, vectors of Dutch elm disease. *Environ. Entomol.* **9**, 181-185.
9.22 Phillips, A.D.G. and Wyatt, T.D. (1992) Beyond origami: using behavioural observations as a strategy to improve trap design. *Entomol. Exp. Appl.* **62**, 67-74.
9.23 Plimmer, J.R. and Parry, R.M. (1994) Registration of biopesticides. In Hedin, P.A., Menn, J.J. and Hollingworth, R.M. (eds.) *Natural and Engineered Pest Management Agents*. ACS Symposium Series 551, American Chemical Society, Washington, DC, pp. 509-515.
9.24 Quisumbing, A.R. and Kydonieus, A.F. (1989) Plastic laminate dispensers. In Jutsum, A.R. and Gordon, R.F.S. (eds.) *Insect Pheromones in Plant Protection*. Wiley and Sons, New York, pp. 149-171.
9.25 Sher, H.B. (1977) Microencapsulated pesticides. In Scher, H.B. (ed.) *Controlled Release Pesticides*. ACS Symposium Series 53, American Chemical Society, Washington, DC, pp. 127-144.
9.26 Spittler, T.D. (1994) Effect of regulation of pheromones as chemical pesticides on their viability in insect control. In Hedin, P.A., Menn, J.J. and Hollingworth, R.M. (eds.) *Natural and Engineered Pest Management Agents*. ACS Symposium Series 551, American Chemical Society, Washington, DC, pp. 509-515.
9.27 Swenson, D.W. and Weatherston, I. (1989) Hollow-fibre controlled-release systems. In Jutsum, A.R. and Gordon, R.F.S. (eds.) *Insect Pheromones in Plant Protection*. Wiley and Sons, New York, pp. 173-197.
9.28 Wall, C. (1989) Monitoring and spray timing. In Jutsum, A.R. and Gordon, R.F.S. (eds.) *Insect Pheromones in Plant Protection*. Wiley and Sons, New York, pp. 39-66.
9.29 Wall, C., Garthwaite, D.G., Greenway, A.R. and Biddle, A.J. (1986) Prospects for pheromone monitoring of the pea moth, *Cydia nigricana* (F.), in vining peas. *Aspects of Applied Biology* **12**, 117-125.
9.30 Zhang, Q-H. and Schlyter, F. (1996) High recaptures and long sampling range of pheromone traps for fall web worm moth *Hyphantria cunea* (Lepidoptera: Arctiidae) males. *J. Chem. Ecol.* **22**, 1783-1796.

CHAPTER 10

GENETIC ENGINEERING

If earlier Chapters indicate the difficult prospects for allelochemicals, especially pheromones, as candidate replacements for pesticides, rather more positive prospects are suggested for genetic engineering in two forms. Firstly, inserting into crop plants the genes coding for insecticidal toxins such as that from the entomocidal, spore-forming bacterium *Bacillus thuringiensis*. Secondly, disrupting coding sequences or promoters that affect genes governing such vital processes as pheromone production and perception, and kairomone perception.

Microbial organisms offer several positive features in regard to insect pest control. Highly specific they: are harmless to other insects including predators, parasitoids and innocuous species; leave no toxic residue on crops or in soil; are the products of evolution and of coevolution with insects and so should be less likely than synthetic chemicals to elicit resistance; can reproduce and amplify through the pest population [10.21, and references therein]. For these features to find satisfactory expression in the field the organism should: compete successfully with other organisms in the same ecological niche; flourish under prevailing weather conditions including extremes of temperature, humidity and sunlight; exhibit high rates of reproduction, transmission, infectivity, and lethality; and be easy and cheap to produce. As relatively few microorganisms meet these requirements, attention has turned to engineering their toxins into crop plants. Sources include viruses, fungi and bacteria.

VIRUSES

Their small and well-characterized genomes are extremely conducive to genetic engineering and attention has focussed on the nuclear polyhedrosis viruses (NPVs). Affecting Lepidoptera in particular, the ingested virus releases virions that invade midgut epithelial cells and enter the nucleus where replication of viral DNA takes place. The progeny exit into the haemolymph and generate secondary infections in haemocytes, fat body, and hypodermis eliciting inclusion bodies and insect death 2-5 days after infection. A well defined form is the *Autographa californica* NPV or AcNPV that comprises some 128 kb pairs of a double-stranded, covalently sealed, circular DNA genome in a cylindrical nucleocapsid. Polyhedrin constitutes some 95% dry weight of this occluded form that in turn comprises up to 30% dry weight of the infected insect. The polyhedrin gene has been cloned and deletion variants have been constructed [10.21, and references therein].

FUNGI

Of the 400 species of entomocidal fungi most take effect through spores that attach to the insect cuticle. Hyphae extend to and penetrate the cuticle to proliferate within the haemocoel. Insect death is attributed to toxic metabolites or organ disruption. Some such as *Nomuraea rileyi* are associated with Lepidoptera. However, the rather complex nature of fungal toxins does not readily lend itself to genetic manipulation.

BACTERIA

Of the more than 100 known species of entomocidal bacteria, five from the family Bacilliaceae are of particular interest, with attention now focussed on *B. thuringiensis* and its various subspecies. The principal *Bt* toxin is an insoluble crystal or δ-endotoxin that can amount to 20-30% of cell dry weight. An additional, water soluble, β-exotoxin is mildly toxic when ingested by birds. The endotoxin, with a half life of only eight days under nonweathering conditions, is approved by the EPA (U.S.A.) for use against some 50 lepidopteran pests in the field. The toxin from var. *kurstaki* is specifically toxic to Lepidoptera whilst that from var. *israelensis* is toxic to Diptera but not to Lepidoptera. The toxin affects the surface of midgut epithelial cells, swelling the apices and inhibiting feeding, followed within a few days by cell rupture and insect death.

Gene Transfer Systems for Crops

Agrobacterium tumefaciens is a bacterium that infects numerous dicotyledonous plants, transferring a portion of its genetic information into plant cells during infection. The transferred DNA (T-DNA) is stably integrated into the host genome. Expression of genes in the T-DNA elicits the synthesis of plant hormones, which cause uncontrolled growth of the plant tissue at the site of infection or tumour formation. This T-DNA is located on a plasmid called tumour-inducing (Ti) plasmid inside the *Agrobacterium*. After "disarming" the T-DNA from its cancerigenous genes, gene-cloning sites are constructed within the disarmed T-DNA and this new Ti plasmid transfers foreign genes to the DNA of target plants. Although a well-established method to create transgenic dicotyledonous plants, this transfer system does not succeed with monocotyledonous plants, among which cereals constitute a major food source for humans. Alternative gene transfer systems include: direct gene transfer in which protoplasts are incubated in DNA solutions and which succeeds with cereals; biolistics or use of the particle gun which delivers DNA on microprojectiles directly into cells and which also succeeds with cereals; pollen transformation, with a variety of different approaches using pollen and pollination as the biological vector system into the zygote; the pollen tube pathway using the cut-off pollen tube as a guiding tube for DNA into the zygote; incubation of seeds or tissues in DNA or agroinfection; DNA viruses; electroporation or use of electric currents; liposome fusion; macroinjection into inflorescences, embryos sacs, ovules or meristems; laser treatments for opening holes in cell walls or membranes; and electrophoresis to force DNA molecules through cell walls [10.17, and references therein].

BT TOXINS

Although *Bt* was originally considered as efficacious against only a few species of Lepidoptera, various strains of the bacterium collectively produce a range of δ-endotoxins that affect Coleoptera, Diptera and Nematoda [10.22, and references therein]. Essentially, CryI proteins affect Lepidoptera, CryII Lepidoptera and Diptera, CryIII Coleoptera, and CryIV Diptera: another affects both Lepidoptera and Coleoptera, another a Hymenoptera, and yet another affects Nematoda, but their respective names have yet to be agreed.

The toxins of each class are grouped according to sequence homology with CryI proteins represented by various 130-140 kDa entities that are cleaved proteolytically by the insect host to yield active proteins of 60-70 kDa. The CryI proteins comprise such groups as IA(a) 133 kDa, IA(b) 131, IA(c) 133, IB 138, IC 135, IC(b) 134, ID 133, IE 133 and so on. Toxin specificity reflects its solubilization, activation, and binding to particular proteins on the brush border of the outer membrane of gut epithelial cells, with the latter step viewed as particularly significant. Nevertheless solubilization and proteolytic cleavage are not trivial. Thus, the Colorado potato beetle *Leptinotarsa decemlineata* is not affected by the full length CryIB protein, as the more neutral pH of the gut does not enable solubilization. Ingestion of the IC1 protein by the yellow fever mosquito *Aedes aegypti* (Diptera) yields a 53 kDa active moiety whereas ingestion by the cabbage white butterfly *Pieris brassicae* (Lepidoptera) provides a 55 kDa activated toxin: such differences are attributable to differential processing of the protoxin by gut proteases. However, the presence of multiple binding sites within various domains of the toxin could explain much of the observed specificity of effect. Thus, CryIIA, toxic to the tobacco hornworm *Manduca sexta* and *A. aegypti*, exhibits 87% homology with CryIIB which is active against only *M. sexta*; the specificity toward *A. aegypti* is associated with 76 amino acids between residues 307 and 382 of CryIIA. As these two proteins differ in this region by only 18 amino acids, small changes in appropriate domains may govern specificity of effect.

An active area of research is isolation and cloning of a gut-binding protein for the δ-endotoxin. Such a step will enable the attribution of activity against different insects either to the toxin binding to multiple sites in the gut protein, or to the presence of a particular binding site in the different species. The former seems more likely given that the same toxin does not affect closely related insect species. Further evidence for the significance of multiple binding sites derives from the fact that the development of resistance to one toxin is associated with increased susceptibility to another through an increased number of binding sites.

An exciting possibility is engineering δ-endotoxin molecules by introducing new domains that are associated with activity against additional target species. In a further step, δ-endotoxins may be linked to protein from *A. californica* multiple nuclear polyhedrosis virus (AcNPV) to derive proteins with new properties, although the stability of the chimera was problematical [10.22, and references therein].

Bt toxins, crystallized by the bacterium in liquid medium in large fermentation vessels, have been deployed as wetted powders in foliar applications since the 1940's, but their use is not trouble-free. The toxin does not persist in the open air and so repeated applications are required. And it does not reach burrowing insects, or euedaphic pests under the soil surface; maintenance in water to control aquatic pests is also problematical. Given that each toxin is encoded by a single gene, and the facility with which genes may be introduced into plants, it is not surprising that *Bt* toxins became the focus for the first genetically engineered, insect-resistant plants. On cotton alone decreased yield and cost of control of insects specially Lepidoptera amounted to ca US$ 650 million a year [10.28], so a lepidopteran pest was an obvious target at that time. The power and constraints of this approach may be illustrated by reference to specific case histories.

Insect-Tolerant Transgenic Tomato

The toxin protein comprises subunits of 130 kDa and the encoding genes have been isolated and sequenced, thus, setting the stage for their introduction into crop plants. Pioneering research at Monsanto Corporation introduced the *Bt* gene into tomato plants with evidence for conferred insect resistance as follows [10.10].

A preliminary stage was the isolation of a toxin gene from *Bt* var. *kurstaki* HD-1 (*Btk* gene) and its introduction into plasmid pMAP4. *Btk* comprises 3734 base pairs and is highly homologous with toxin genes identified from other strains of *Bt* The N-terminal half, but not the C-terminal half, is required for toxicity and a truncated protein comprising the first 565 residues was ineffective, but one representing the first 646 residues was lethal.

DNA manipulations

Essentially, chimeric truncated *Btk* genes were constructed by altering deletion variants in various vectors though the addition of a linker, containing a BglII site. These genes were then inserted into the BglII site of pMON316, a plant expression cassette vector consisting of the 35S promoter of cauliflower mosaic virus (CaMV), and the 3' end of the nopaline synthase (NOS) gene from the Ti plasmid of *A. tumefaciens*, with a BglII site containing the truncated toxin gene between these i.e. adjacent to the CaMV promoter (Fig. 79). These vectors, designated pMON9711 and pMON9713 were introduced and integrated into the disarmed Ti.

These enriched vectors were used to transform tomato explants from which transgenic tomato plants were subsequently recovered. Primary transgenic plants regenerated from tissue culture were designated Ro which were selfed, and first generation progeny plants were designated R1. These were assayed for inheritance of the transforming DNA by scoring for nopaline synthase and for kanamycin resistance. R1 plants with the transforming DNA were selfed and the second generation plants were designated R2. Putative homozygous R1 plants were recognized as those not segregating any nopaline-negative, kanamycin-sensitive progeny in the R2 generation.

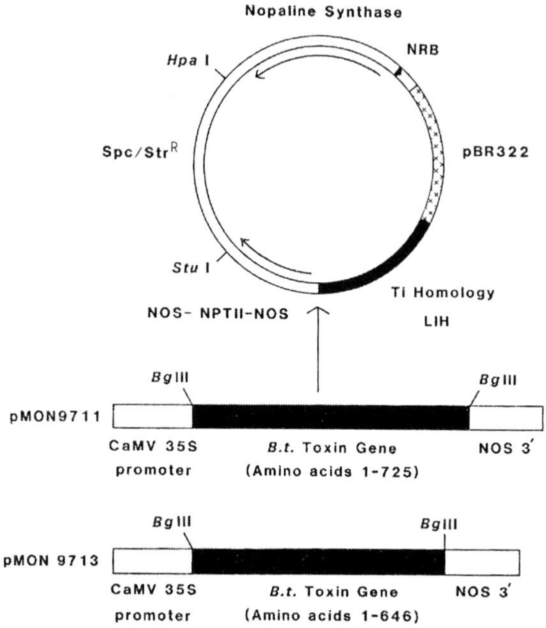

Figure 79. Significant features of engineered vectors pMON9713 and pMON9711 for transformation of tomato plants by inclusion of Btk *genes. For both vectors (l to r): CaMV 35S is the cauliflower mosaic virus promoter; Bg III is a linker site from a restriction enzyme;* Bt *is a gene encoding toxins of two different sizes; and NOS 3' is the nopaline synthase gene providing polyadenylation signals. Circular diagram of functional regions: NOS-NPTII-NOS is a chimeric gene for kanamycin resistance; Spc/Str^R is a bacterial gene conferring resistance to spectinomycin/streptomycin; NRB is the right border of the T-DNA from a nopaline Ti plasmid; pBR322 is a replication site from that plasmid; LIH is a region of the octopine Ti plasmid with homology for integration with disarmed Ti plasmids. Redrawn after Fischhoff* et al. *[10.10].*

Chimeric *Btk* genes containing the CaMV 35S promoter, and either the full length *Btk* protein sequence or that for active truncated variants, were expressed in plants, although expression of the *Btk* gene was rather low; the significance of this will be considered later.

Isolated leaves or whole plants used in feeding assays indicated clear-cut effects. After four days feeding on a control there was no mortality and larval size of *M. sexta*, increased 5-fold; feeding on pMON9711, transgenic plants elicited 100%

mortality within 48 hr, with little evidence of feeding on leaf. Such early inhibition of feeding is consistent with *Bt* poisoning.

Furthermore, in seven days, 10 *M. sexta* larvae completely defoliated a nontransformed tomato plant extending their feeding to the stems as well as the leaves without insect mortality. In contrast, larvae feeding on the pMON9711 transgenic plant were all killed within 72 hr without damage to the leaves. When 17 kanamycin-resistant, transgenic pMON9711 plants were assayed, 10 were lethal (90-100% mortality), two were less toxic (20-50% mortality) and five were inactive. Four of five pMON9713 transformed plants were highly toxic. However, such transformations were less effective against the corn earworm *Helicoverpa zea*, and the tobacco budworm *Heliothis virescens*. Nevertheless, their larvae were severely stunted, as control-fed larvae doubled in size compared with no weight gain for those fed on *Btk*-transfected plants; there was no mortality or feeding inhibition on plants transfected by vectors without the *Btk* gene.

The *Btk* gene segregated with nopaline production and kanamycin resistance as expected and as a single dominant Mendelian marker; the progeny proved at least as lethal as the primary transformants. Thus, transgenic plants containing either of the truncated chimeric *Btk* genes expressed the toxin at levels sufficient to kill or significantly inhibit feeding by lepidopterous larvae. However, as *Btk* genes were less abundant than expected, the chimeric transplant may be unstable in plant cells.

Btk activity also occurs in tomato fruit as *H. virescens* larvae consumed it at only half of the control rate. However, complete control would require adequate *Btk* expression in fruit as larvae are likely to attack this in addition to foliage, especially if it were less protected than foliage. In fact, greater protection of fruit is required as the presence of even a single larva in fruit is consistent with economic damage, compared with several larvae required to damage foliage. Extrapolation from the amount of leaf material eaten in quantified assays indicated feeding inhibition was elicited by 50 ng *Btk*/g leaf tissue (fresh weight) [10.10].

Extending these observations to field trials indicated that *Btk*-expressing tomato gave excellent control of a sensitive species *M. sexta*. However, more economically important pests, *H. zea*, and the tomato pinworm *Keiferia lycopersicella*, were substantially less affected, the former proving 100-fold less sensitive than *M. sexta*. Such data clearly indicated the need for higher levels of *Btk* expression for routine control of these significant pests [10.6].

Insect Resistant Btk-*enriched Cotton*

The rather low level of *Btk* expression (equivalent to only 0.001% of total soluble protein) was directly addressed, in this study of a more valuable crop, by use of a combination of factors [10.28]. Firstly, expression was enhanced 5- to 10-fold by use of an improved promoter in *Agrobacterium* Ti transformation vectors, and by incorporating the Ca MV 35S promoter with a duplicated enhanced region.

Secondly, expression was independently enhanced 100-fold by use of a variant of the structural gene highly modified to take account of putative regulatory sequences, mRNA secondary structure, and codon frequency. The result was a hybrid gene, CryIA(c), containing the first 453 amino acids (CryIA(b)) of the gene used in transgenic tomato, linked to the modified and more active form of the region from amino acids 454-615. The transgenic cotton thus derived was bioassayed with the cabbage looper *Trichoplusia ni* which is *Btk* sensitive, and the beet armyworm *Spodoptera exigua* which is 100-fold less sensitive, and *H. zea*.

The leaves of cotton plants enriched by the wild-type gene were susceptible to attack by *T. ni* and *S. exigua*, but those containing the highly modified version were resistant, killing the larvae of both species. Satisfactory protection would, however, require toxin expression in buds, squares and bolls. This was assayed with high infestation levels of *H. zea*, i.e. 100-fold larger than the threshold for insecticidal treatment and eliciting up to 100% damage of control plants. Expression of the highly modified gene elicited up to 75% of bolls surviving and *Btk* levels equivalent to 0.1% of total soluble protein. The continuation of these efficacies into the second generation indicated the commercial potential of this approach to such a valuable crop [10.28].

Enhanced resistance to insects has been associated with *Bt* expression in other significant crops including tobacco and maize, the immature embryos of which were enriched by a synthetic *Bt* gene, CryIA (b), using microprojectile bombardment [10.22, and references therein].

Transgenic Insect Resistant Tobacco(I)

The second engineered category of toxin derives from joint investigations at the University of Durham and the Plant Breeding Institute, Cambridge, U.K. [10.16] claiming that the identification of useful genes is the real problem as methods for introducing them are routine. They viewed *Btk* transfections as limited in potential by the high specificity of the toxin (affecting only Lepidoptera) and the fact that transformed plants were still not proven as efficacious in the field. Accordingly, they chose a naturally occurring mechanism, proven to resist insects in the field, i.e. the trypsin inhibitors or CpTIs from cowpea *Vigna unguiculata*. These are small polypeptides of some 80 amino acids belonging to the Bowman-Birk type of doubled-headed serine protease inhibitors, and are the products of a small gene family. Their metabolic target is the catalytic site of an enzyme so it was claimed that insect resistance based on site mutation should be small. The level of CpTI within seeds of cowpea is correlated with field resistance to their major pest the cowpea weevil *Callosobruchus maculatus*; incorporated into diets they are anti-metabolic agents against a wide range of insects including the genera, *Heliothis*, *Spodoptera*, *Diabrotica* and *Tribolium*. Although usually cooked before consumption, cowpea seeds can be eaten raw by humans, and meal from raw seeds was not toxic to rats.

The CpTI gene used was derived from plasmid pUSSRc3/2, part of a complementary DNA library prepared from cowpea cotyledon polyadenylated RNA (Fig. 80). The following constituents were transferred to the Sma I site of *A. tumefaciens* Ti plasmid binary vector, pRok2: an AluI-ScaI restriction fragment, 550 bp long, containing the 240 bp coding sequence for the mature inhibitor; a long leader sequence with four in-phase methionine codons; and a 96 bp 3' non-translated sequence. This differs from the pRokI vectors primarily in having a multipurpose cloning site between the strong, constitutive CaMV 35S gene promoter, and the nopaline synthase gene-transcription termination sequence.

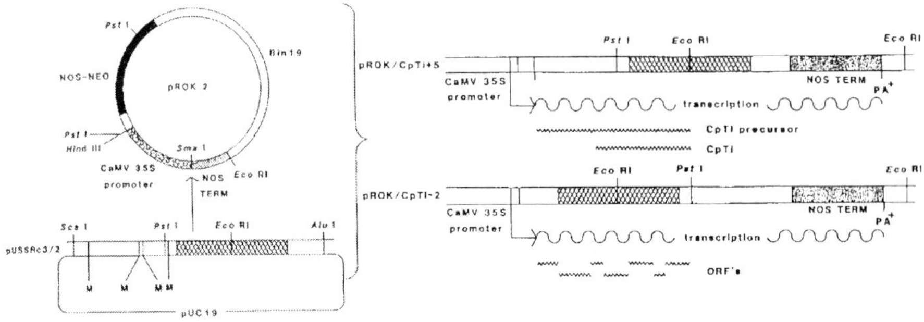

Figure 80. Vector construction for engineering CpTI genes into tobacco with essential details as for Fig. 79.

Clones recognized as containing an insert in the correct orientation relative to the CaMV promoter were designated pRok/CpTI + 5 and they produced CpTI; others in the 'reverse' orientation pRok/CpTI – 2, served to produce control transformants that were mobilized into *A. tumefaciens* and used to transform leaf discs of *Nicotiana tabacum*. Transformants were selected by their resistance to kanamycin, and transformed plants were regenerated from shootlets by transfer to a root-inducing, kanamycin-containing, agar medium.

The plants studied contained 3-7 unrearranged copies of the construct that were simply and stably inherited. Expression levels in pRok/CpTI + 5 transformants ranged from below detection limits to ~1% of total soluble protein that is consistent with use of the CaMV 35S gene promoter: CpTI was not expressed in pRok/CpTI - 2 transformants. Western blotting of soluble leaf proteins from the plant evincing highest CpTI expression (+5/5), detected a polypeptide corresponding to one of the isoinhibitors from cowpea seed. Protein species of 13 kDa and higher, representing either trypsin inhibitor precursors or their aggregates, occurred in both +5/5 and cowpea seed extracts but not in -2/8. The plant-produced CpTI was confirmed as functional by an in vitro assay.

In the insect bioassay, young transformed plants were infested with newly-emerged larvae of *H. virescens*, a significant pest of tobacco. Plant damage was very variable which is consistent with variable efficacy of transferred genes.

Nevertheless, insect survival on and damage to 20% of the CpTI +5 transformants was clearly decreased compared to controls; also the transformants had highest levels of CpTI.

Resistant transformants were replicated as stem cuttings to provide sets of 12 genetically identical clonal plants for feeding trials. These confirmed the CpTI-producing plants as more resistant to insect attack; control plants were devastated by this level of infestation and were reduced to a stalk in 7 days. Although larvae began to feed and did limited damage to the CpTI-producing plants, +5/5 and +5/21, they either died, or failed to develop [10.16].

Accordingly, the incorporation of this gene was proposed for other important crops such as cotton, maize and rice, recognizing that the ultimate test would be the demonstration of resistance to insects in the field. It must also be acknowledged that CpTI is quite ineffective against aphids, the most important insect pests of cereals in Western Europe.

Transgenic Insect Resistant Tobacco(II)

The third case history also refers to tobacco and to serine proteinase inhibitors, but to different molecules and to a different target pest. Specifically, the inhibitors derived from tomatoes and potatoes are known as: inhibitor I (monomer 8.1 kDa) with a single reactive site that inhibits chymotrypsin, and trypsin only weakly; inhibitor II (monomer 12.3 kDa) with two reactive sites for chymotrypsin and trypsin respectively. Both inhibitors are expressed in leaves subjected to insect or mechanical damage, and the larva of *M. sexta* was the target pest [10.19].

Essentially, the genes for inhibitor I and II were separately incorporated with appropriate terminators, the CaMV 35S promoter, and a chimeric nopaline synthase-neomycin phosphotransferase fusion, into plasmids and verified in potato leaves according to the principles and procedures described above. Larvae fed on leaves expressing inhibitor I at $\leq$ 130 μg/g leaf tissue grew as well as those fed control leaves. However, larval growth was severely inhibited by leaves expressing $\geq$ 100 μg inhibitor II, concentrations within the range (50-300 μg/g leaf) for leaves naturally wounded. The relative failure of expressed inhibitor I that only weakly inhibits trypsin compared with the potency of expressed inhibitor II, indicates that trypsin inhibition may have elicited the observed effect [10.19]. This is consistent with the data from the previous case history [10.16]. It was subsequently shown that growth inhibition could not be simply ascribed to decreased proteolysis, as gut proteinase activities were the same or enhanced in the presence of inhibitor. Rather, the effect was attributed to proteinase hyperproduction, elicited by a feedback mechanism responding to originally decreased activity, that depleted essential amino acids [10.29].

Transgenic Insect Resistant Rice

Rice, one of the world's most important cereal crops, sustains insect damage amounting to several billion US dollars p.a. Three widely-used rice cultivars have been engineered to express the gene (*pin* 2) coding for the potato proteinase inhibitor II (PINII), a serine proteinase that is wound-inducible [10.7]. *Pin* 2 expression is regulated by its own *pin* 2 promoter and a 3' terminator sequence. The plasmid pTW was constructed by inserting the first intron of the rice actin I gene (*act* 1) between the *pin* 2 promoter and *pin* 2 gene. To the latter was linked the usual CaMV 35S promoter linked in turn to *bar*, the gene for bacterial phosphinothricin acetyl transgenase serving as a selectable marker for rice transformation. Vector introduction was by bombardment of rice cell cultures, with tungsten particles coated with plasmid, from which plants were regenerated. DNA hybridization gels indicated which plants contained the introduced gene, and assays for trypsin activity confirmed the presence of wound-induced proteinase inhibitor II. Homozygous transgenic cell lines were obtained and fifth-generation plants from these exhibited significantly enhanced resistance in the greenhouse to the pink stem borer *Sesamia inferens*. Furthermore, larvae on control plants achieved a 3- to 4-fold weight gain, in contrast to essentially none in those on transgenic plants [10.7].

It should be noted, however, that in common with the proteinase inhibitor from cowpea in a previous case history, this inhibitor could ultimately elicit in insect field populations a modification of the proteinase profile of the gut, sufficient to overcome the inhibitor [10.8].

OTHER TOXINS

Lectins

Accounting for 1-8% of total protein in vacuoles of cotyledons, lectins bind with high affinity to carbohydrates including glycoproteins, glycolipids, and polysaccharides. Well-known examples include phytohemoaglutinin (PHA), soybean agglutinin, concanavalin A, pea lectin, and favin. The PHA family comprises four homologous polypeptides (PHA-E, PHA-L, α-amylase inhibitor (αAI), and arcelin). Pure PHA does not seem to be toxic to insects but a minimum of 10% (w/w) arcelin decreased emergence of the bean weevil *Zabrotes subfasciatus* from 71% to 18%. αAI is a potent inhibitor of insect amylases, as 0.21% and 0.8% (w/w) elicited delayed development and 100% mortality respectively of larvae of *C. maculatus* [10.3].

Seeds of many graminaceous species contain lectins that bind N-acetylglucosamine, its oligomers and its polymer chitin. The hexaploid wheat lectin WGA inhibits development of *C. maculatus*, the European corn borer *Ostrinia nubilalis*, and the spotted cucumber beetle *Diabrotica undecimpunctata*. Such lectins bind to glycosytated proteins in brush border membranes of the insect midgut, and this could inhibit absorption of nutrients [10.3, and references therein].

For effectiveness, lectins must be incorporated in μg amounts in the insect diet, which is at least 10-fold larger than quantity required for the *Bt* toxin. This severely constrains the utility of this family of toxins. Transgenic maize separately armed with three conspicuous lectins, wheat agglutinin, jacalin, or rice lectin, elicited little mortality and only moderate inhibition of larval growth. Unlike *Bt* endotoxin that lyses cells at a rate of a few molecules/cell, lectins and proteinase inhibitors are unlikely to bind to a single target molecule that is vital for cell integrity. The long term exposure to large amounts of toxin thereby necessitated seems beyond the scope of presently achievable expression rates [10.8, and references therein].

Vegetative insecticidal proteins (Vips)
These occur in the culture supernatant fluid of *Bacillus* species in log-phase or vegetative growth before sporulation and as early as 15 hr after culture initiation, compared with 36 hr for Cry type proteins. Vip 1 and Vip 2 isolated from *B. cereus* and Vip 3 from *B. thuringiensis*, comprise a toxin group structurally separate from the endotoxins. Vip 3A(a) and its homologue Vip 3A(b) have been assayed against the black cutworm *Agrotis ipsilon* that damages more than 50 crops, including cereal grains, and is relatively insensitive to *Bt* endotoxins. The active component is a secreted protein with a high molecular mass (ca 88 kDa) having no homology to known proteins. Recombinant *Escherichia coli* enriched with Vip 3A genes exhibited insecticidal activity when fed in ng amounts to larvae of the fall armyworm *S. frugiperda*, *S. exigua*, *H. virescens*, and *H. zea*. These species collectively elicit crop losses valued in excess of US$1 billion p.a. The mode of action resembles that of endotoxins by eliciting lysis of midgut columnar epithelium and gut paralysis, but seems slower [10.9].

Cholesterol oxidase

Random screening of *Streptomyces* cultures detected a cholesterol oxidase (CO) (52.5 kDa) with an LC_{50} value of 20.9 μg/ml against larvae of the cotton boll weevil *Anthonomus grandis*, the single most destructive pest of U.S.A. cotton. Cholesterol oxidase (EC 1.1.3.6) is a bifunctional enzyme that: catalyzes cholesterol oxidation to 5-cholesten-3-one and reduces oxygen to hydrogen peroxidase; and serves as a ketosteroid monooxygenase that is flavoprotein-linked and catalyzes hydroxylation of cholesterol to 4-cholesten-6-ol-3-one [10.2, and references therein]. It seems to modify cholesterol or a related sterol present in insect midgut membranes to elicit the observed effect of midgut cell lysis.

Dietary concentrations corresponding to those effective against neonates also control second instar larvae of *A. grandis*. Furthermore, when fed to adult females prior to oviposition, CO decreased numbers of eggs laid and subsequent larval survival by 83% and 97% respectively. Treated females exhibited underdeveloped ovaries with few developing oocytes. Such compromised oogenesis could result from decreased nutrient assimilation, and the observed atrophy of the fat body in affected females supported this hypothesis [10.13]. In a further significant step, the COAgene (*choA*) was expressed in tobacco cells using a plasmid containing the 35S promoter introduced by particle gun bombardment. All transformed cell lines consistently exhibited 1.5- to 8.0-fold higher enzyme activity than the untransformed cell line [10.8, and references therein].

The foregoing selection of toxins, actually or potentially transformable into various plant species, presents an early outline of developments in this exciting field. Quite obviously many more studies will be required to satisfactorily assess the long-term efficacy of this approach. Nevertheless, it is possible at this preliminary stage to glimpse some significant features especially as revealed by use of *Bt*, the most widely used toxin.

FIELD EFFICACY OF *BT*-ENGINEERED SPECIES

After the pioneering work of Fischhoff at Monsanto, *Bt* was subsequently engineered into all major crop species to provide resistance in: potatoes against the *L. decemlineata*; cotton against primarily the pink bollworm *Pectinophora gossypiella*; and maize against *O. nubilalis* that causes at least US$10 billion annually in crop losses [10.8].

On 1 March 1995, the EPA of the U.S.A. gave premarket registration to Ciba seeds, Mycogen Plant Sciences, and Monsanto for *Bt*-engineered, insect-resistant potato, maize, and cotton. This permitted in principle the planting in 1996 of 60,000 acres (24,240 ha) of engineered seed in the U.S.A. In recognizing the danger of resistance developing to the *Bt* toxin, the decision called for a resistance management plan. This entailed the planting of "refugia", or adjacent sites with non-transformed seed to maintain non-*Bt* resistant insects that could cross-breed with toxin-resistant ones. This would minimize development of a strain homozygous for toxin resistance. In addition, there was a commitment from the companies to monitor the crops for development of *Bt* resistance. The registration decision was seen as reflecting a desire within the EPA to minimize use of chemical pesticides [10.18].

In 1996 two million acres (808,000 ha) of *Bt* cotton were planted, equivalent to about 13% of the total US crop, and the most economically important of all *Bt* crops planted that year. It was also an important undertaking for individual growers who paid US$32/acre as a licence fee for use of the *Bt* seed. At least 20,000 acres (8,080 ha) of cotton in eastern Texas succumbed to bollworm attack, raising serious doubts about the effectiveness of the methods and fears of early evolution of *Bt*-resistant strains. Critics also claimed the cotton expressed insufficient toxin to kill bollworms. An additional concern was that failures such as this could accelerate the evolution of toxin-resistant strains that would render *Bt* sprays ineffective. In reply, Monsanto drew attention to unusually hot weather, and increasing planting of maize as a breeding opportunity for bollworms. Fears of *Bt*-resistance were discounting by emphasizing that the bollworm feeds on a range of other crops [10.20, 10.26]. Irrespective of whichever explanation prevails, the experience showed that the case for genetically engineered insect-resistant crops was then essentially unproven. This does not preclude ultimate success for this approach but the strategy of relying on toxicity alone must be questioned, as insects have evolved resistance to every toxin directed at them by man.

Resistance

Although not necessarily indicative of putative resistance to *Bt*-engineered plants, insect resistance to sprayed δ-endotoxins is relevant. This was first described from laboratory cultures of sprayed field populations. Such cultures represent an imperfect model of field conditions as rare alleles may not be included, and inbreeding in the cultures favours partial resistance alleles becoming homozygous, and also polygenic resistance. Nevertheless, the cultures are useful in indicating the biochemical basis for *Bt* resistance which in the Indian-meal moth *Plodia interpunctella* entails binding proteins with dissociation constants some 50-fold less for the CryIA(b) toxin than in susceptible strains. Although there was no alteration in binding affinity for the CryIC toxin, the concentration of binding proteins was increased 3-fold in the resistant strain. Accordingly, this early example of *Bt*-resistance entailed a modification of the target site that is one of the two commonest resistance mechanisms to conventional insecticides; the other is enhanced metabolism of the toxicant [10.22].

The first report of resistance in a field population related to the diamondback moth *Plutella xylostella*, and indicated the elicitation of resistance by repeated application of the same δ-endotoxin in somewhat isolated areas in Hawaii. A subsequent study of the same species on brassicas in Florida suggested variable expression of the *kurstaki* toxin in different years, and increasing resistance associated with intensive use. This study concluded that resistance to *Bt* sprays would necessitate the use of supplementary measures such as crop rotation, resistant cultivars, biological and cultural controls [10.33].

Bt *Gene Flow*

The general issue of quantifying *Bt* gene flow has been approached by reference to transgenic oilseed rape (OSR) *Brassica napus* and consequential pollination of male sterile plants in a realistically-sized agricultural area (70 km^2) in Scotland; concomitant observations were made on pollen captures in motorized traps. Bait colonies comprised 10 cytoplasmic male sterile plants located 0 to 4000 m from a recognized pollen source. After harvesting and germinating mature seed, DNA was extracted and profiled.

Pollen counts declined sharply over distance. Specifically, at 170 m counts were some 25% of these at fields edge, and at distances greater than 100 m 82% of pollen occurred as single grains (Fig. 81). At distances greater than 150 m, however, densities were not significantly correlated with distance. Pollination also declined over distance, from approximately 90% 1 m from a field's edge to 5% at the furthest distance, in a relationship that was statistically significant ($P<0.0001$). An interesting anomaly was that one set of bait plants was fertilized by pollen from a source 4 km away, but little pollen (<1 grain m^{-3}/day) was detected at that site. Such long range pollination was attributed to insects, especially bees. In terms of pollination by non GM and GM plants, values of the former were relatively constant within a 1000 m range, but values for the latter declined from 75 to 10 % at distances of 100 and 930 m respectively. These data, from a realistic set up employing an established OSR crop as the pollen source, leave little doubt of a facile flow of transgenes to near plants and a non-trivial flow to distant ones [10.36]. The issue then arises as to the likely significance of this.

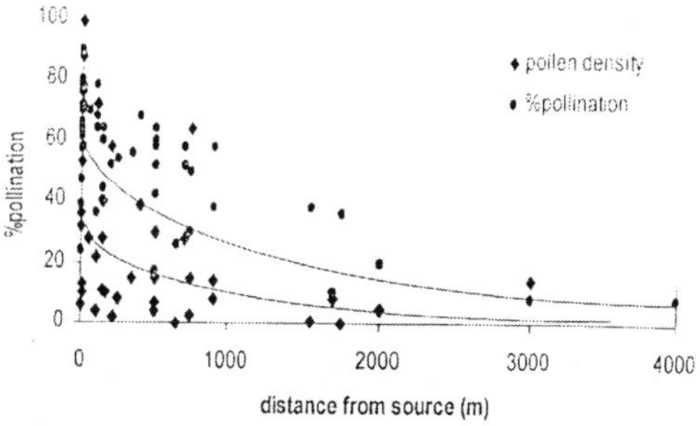

Figure 81. Decay curves for pollination (%) and density of airborne pollen of oilseed rape Brassica napus *in relationship to distance from nearest pollen source. Redrawn after Thompson* et al. *[10.36].*

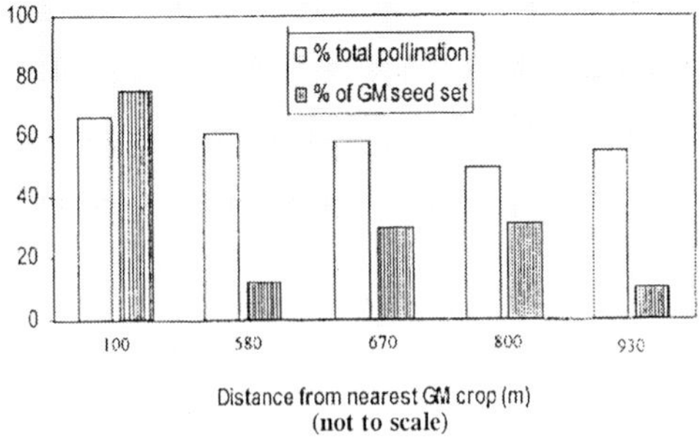

Figure 82. Total pollination of oilseed rape Brassica napus *(%) and pollination by genetically modified (GM) rape in relationship to distance from nearest GM rape crop. Redrawn after Thompson* et al. *[10.36].*

Consequences of Pollen Dispersal

A further issue is the deposition of pollen from *Bt*-transformed plants on those of a different species. Pollen from, for example, *Bt*-transformed maize could be ingested by non-target insects feeding on the second plant species. This was explored in the laboratory using the milkweed *Asclepias curasavica* and the larva of the monarch butterfly *Danaus plexippus* [10.24].

Specifically, pollen from both transformed and untransformed maize was gently applied by spatula to milkweed leaves, such that the pollen density matched that of field-collected leaves. Essentially, pollen from *Bt*-transformed and untransformed maize elicited 44% and 0% mortality respectively (P=0.008); leaves with no pollen applied also elicited no mortality. Consumption by monarch larvae, of leaves dusted with *Bt*-transformed pollen was 51% of that of leaves treated with untransformed pollen (P<0.001) that, in turn, was 70% of control (no pollen) leaves (P=0.001). Apparently, the non-host pollen deterred feeding by this oligophagous species, but *Bt*-enriched; non-host pollen further suppressed it. It was acknowledged that it would be inappropriate to extrapolate from such preliminary laboratory data to a field situation [10.24].

Initially, these data gave cause for concern for the future survival of *D. plexippus*, as in the U.S.A. the milkweed grows on the perimeter of maize fields that shed pollen between June and mid-August when the larvae feed. Furthermore, 50% of the population of this insect occurs in the mid-west of the United States, colloquially known as the 'corn belt'. The risk was assessed by a 2-yr research collaboration between several U.S.A. states and Canada.

Essentially, monarch population densities were as high in agricultural habitats, including corn fields, as in non-agricultural habitats. Within corn fields, 95% of milkweed leaves had less than 600 corn grains/cm^2, 54-86% of which were removed by one shower of rain. Furthermore there is limited overlap between pollen shed and larval feeding. Growth was inhibited and larvae were killed by the high levels of CryIA toxins in event 176 pollen, but this event will be out of use from 2003. Events *Bt* 11 and Mon 810, that express toxin levels almost 100-fold less than event 176, did not adversely affect larval survival. Pollen contaminants, especially from anthers fractured during processing, are toxic to larvae, thus rendering erroneous the attribution of all the observed toxicity to pollen. Overall, these data indicate little danger to the monarch from the *Bt* corn pollen [10.32, and references therein]. Corresponding results were reported from separate studies on the effects of event 810 as compared with event 176, on black swallowtail caterpillars [10.37, and references therein].

Tritrophic Effects

Although there is no conclusive field evidence of negative effects of transgenic insect-resistant plants at the tritrophic levels, i.e. on predators and parasitoids of the target pest, laboratory studies give varying results. In regard to predators, feeding *Bt* pollen to the ladybird *Coleomegilla maculata*, the anthocorid bug *Orius insidiosus* and the lacewing *Crysopa carnea* elicited no acute negative effects on growth, development, and survival of immature stages. However, peach-potato aphid *Myzus persicae* reared on potatoes expressing the lectin GNA and fed to the two spotted ladybird *Adalia bipunctata*, decreased fecundity of the latter by 38%, longevity by 50%, and egg viability by 3- to 4-fold. It was unclear whether these effects were attributable to GNA toxicity or to decreased quality of the prey [10.31, and references therein].

In regard to parasitoids, parasitism by the braconid wasp *Cotesia plutella* on *P. xylostella* was the same whether on *Bt* or on wildtype (WT) oilseed rape. In an assessment of effects on parasitoids of non-target pests such as aphids that are not controlled by *Bt*, the efficacy of the parasitoid *Diaeretiella rapae* was compared on *Bt* and WT oilseed rape in laboratory cages. There were no significant differences between numbers of live aphids and of parasitoid mummies on each plant type. The interesting possibility that *Bt* plants might release a modified profile of volatiles was investigated in windtunnel experiments. Adults of the parasitoid *C. plutella* did not distinguish between leaves of *Bt* and WT oil seed rape [10.31].

Future Monitoring of Transgenic Flow

The phenotypic and biochemical methods described above are likely to be superseded in time by use of genes for the non-toxic, green fluorescent protein (GFP) from jellyfish. When excited with ultra-violet (UV) light (360 – 400 nm) or blue (440 – 480 nm) light this 27 kDa monomer fluoresces green without requiring an enzyme, substrate, or co-factor. As it can be visualized non-destructively using a hand-held UV source and as non-transformed plants visualize red due to the reaction of chlorophyll, it will serve as an ideal whole-plant marker for transformed plants. Furthermore, it is stably inherited by progeny. Finally, linking the *Bt* and GFP genes on the same plasmid, prior to introduction from non-transformed plants, has served to discriminate transformed from non-transformed plants with no evidence of cost to the former [10.35, and references therein].

Side Effects of Marker Genes

Rapid discrimination of transformed plants usually entails the use of marker genes of bacterial origin conferring resistance to antibiotics (*nPtII*) or to herbicides (*bar*), or encoding readily detectable enzymatic activity (*gus*). However, marker genes in potato have elicited pleiotropic effects such as alterations in tuber number, tuber weight, and plant height. Feeding by *L. decemlineata*, has been systematically investigated on potato transformed by *A. tumefaciens* armed with plasmids bearing *gus* and *nptII* controlled by the TR promoter, or *nptII* controlled by the NOS promoter, and CryIIIA controlled by the 35S promoter with a double enhanced sequence. Specifically 2 cm d leaf discs cut from 5 month old plants, transgenic and controls, and exposed to individual second instar larvae for 20 hr were weighed to ascertain food uptake. There were no differences between control and CryIIIA-transformed plants. However, plants expressing the marker *nPtII-gus* construct enhanced foliage consumption by 50%. It was not clear if the response was assignable to a direct influence of the construct, or to an indirect effect of it on metabolic activity of the transformed potato. A further possibility is that as insertion of T-DNA in the plant genome, is essentially random, this could occur in a coding sequence that regulates physiology of the plant. Whichever mechanism is ultimately indicated, these data raise doubt regarding the long term consequences of using apparently safe but untested marker gene constructs [10.23].

NOVEL POSSIBILITIES

Plantibodies

Given the uncertainties associated with toxin-engineered plants, it is surely appropriate to consider the option of engineering other chemicals, of which those modifying pest behaviour constitute a primary alternative. One approach would employ plant-produced antibodies (plantibodies) to bind, for example, the key chemical host-location cues and to deny pest access to the kairomone, and thus to the plant. This remarkable technique of generating plantibodies exposes a random-primed cDNA expression library (commercially available) to the selected host location cue. These libraries are so extensive (containing up to n x 10^{12} entities) that they will contain coding sequences for any behaviour-modifying or other chemical. Plasmids are then constructed and are recognized in the conventional manner, i.e. by selection of antibiotic-resistant forms. Monoclonal antibodies (Mabs) raised against these enriched plasmids are inserted and incorporated in seeds of the appropriate host plants. In growing and mature plants, these Mabs could recognize and bind to the specific host location cues, thus excluding them from the plant bouquet. A conspicuous advantage of this method is that chemicals are being engineered 'out' not 'in'. No one familiar with insect evolutionary versatility will expect this effect to last indefinitely. Obvious insect behavioural shifts include recognizing and adapting over time to other constituents of the plant bouquet. However, the speed and elegance of this technology is such that once the identity of the new cues is ascertained, a corresponding set of antibodies could be constructed. This paradigm could, in principle, also apply to the sequestration of precursors for sex- and aggregation pheromones.

Engineered Arthropods

The ready manipulation of the *Drosophila* genome has promoted exploration of other species in terms of modifying pest DNA with appropriate recombinant DNA in order to dislocate reproduction. Homologous recombination inserts DNA at a locus evincing homology with the insect. On the other hand, transposons (transposable elements, jumping genes) transpose and integrate foreign DNA by becoming stable genome components, usually without detriment to the organism. Comprising more than 20% of the genome of higher eukaryotes and 10% of the human genome, transposons are associated with a wide range of DNA modifications including insertions, deletions and rearrangements of chromosomes by jumping from one chromosome to another.

Possible effects include behaviour modification and autocidal responses. In the former, a knowledge of the factors governing the phases of immigration from take off to landing could indicate the action of genes switched on by temperature. These could be modified to no longer respond to the temperature threshold for migration. Autocidal effects are feasible in association with genes regulating diapause. These could be linked to a promoter engineered to trigger diapause in response to environmental stimuli such as application of salt ions, or natural phenomena such as temperature or UV rays in sunlight.

Until very recently the genome of only *D. melanogaster* (ca 15K genes) was amenable to routine genetic transformation. Transformation of *D. melanogaster* was enabled by use of a transposon, the P element. Structurally, each end of active P elements is contiguous with a series of nucleotides in reverse orientation or inverted terminal repeats. Functionally, the P element encodes a transposase that facilitates incorporation into the recipient genome. However this transposase has to be removed in order to preclude further jumps so it is usually supplied to facilitate the initial integration, either on a helper plasmid or with the transposon [10.34]. Despite many attempts it gradually became clear that this P element was not appropriate to insert genes into the genome of insects other than *D. melanogaster*. However, the stable transformation of the Mediterranean fruit fly *Ceratitis capitata* [10.25] and *A. aegypti* [10.4] seemed to presage the transformation of many additional species. The search for non-P transposons and for novel ones yielded *Hermes*, *piggyBac* (from insect baculovirus), and *mariner* which occurs in some 400 species of arthropods.

A further requirement difficult to satisfy was a satisfactory marker to detect transformants as *neo* (neomycin resistance), *lacZ* (β-galactosidase), *OPD* (organophosphate resistance), and *Rdl* (dieldrin resistance) all proved unsatisfactory. Ultimately, the *white-eye* mutation of *Aedes* proved appropriate [10.4, and references therein]. This particular marker gene enabled rapid and unambiguous identification of transformants by complementing a mutant gene with an eye-colour phenotype. The deep purple of the mosquito eye is attributable to ommochrome pigments that are synthesized from tryptophan in a sequence of well-defined and enzymatically catalyzed steps. A *white-eye* mutant of *A. aegypti* is elicited by mutations in the kynurenine hydroxylase that catalyzes the conversion of kynurenine to 3-hydroxy-kynurenine; this is designated *kynurenine hydroxylase-white* (kh^w). It may be homologous to the wild-type *cinnabar* (*cn*) gene reported from *D. melanogaster*. Experimentally injecting embryos of white-eye *A. aegypti* with a wild-type copy of the *cn* gene from *D. melanogaster* partially restored eye colour in adult mosquitoes. Accordingly, the first reported transformation of *A. aegypti* employed the class-II transposon *Hermes* from the *hAT* family of elements occurring in the housefly, in association with the *cn* gene.

Transformation efficiency was 8% in terms of fertile founders (G_0) or 0.3% in terms of injected embryos giving stable transmission over at least 10 generations. As eye colour changes segregated in a Mendelian pattern, they were associated with chromosome insertions. Southern blots indicated the insertion of only a single copy of the transgene. This significant step opens the way to mosquito control by genetic manipulation. The wider implications are that if the *D. melanogaster cn* gene could similarly serve as a transformation marker for other Diptera, or for orders expressing kynurenine hydroxylase mutants, then a corresponding genetic strategy could be employed against them [10.4]. Specifically, this could in time include the creation of transformants deficient in genes for pheromone synthesis and perception, and for kairomone perception [for a recent overview see 10.14].

THE SOCIAL DIMENSION

Although the present text is a scientific one, it would be quite unrealistic to discuss genetic engineering without regard to the social dimension, given the depth of scepticism and opposition. There should be little public opposition to the control of conspicuously unpleasant arthropod pest, such as flies, mosquitoes, and biting midges, by genetic engineering rather than by pesticides. (The association in the public mind between cancer and pesticide usage has been considered in Chapter 1). However, it is by no means clear that the European public will accept a widening occurrence and increasing doses of bacterial or other toxins in their diet, even if they replace pesticides. At time of writing the outcome is in the balance, but some omens are not propitious. The Green Movement, now a partner in the Government of Germany, is hostile. Other environmentalist and conservationist groups have raised objections with emphasis on possible adverse ecological effects [10.30]. Principal among these are: development of insect resistance which is, of course, inevitable; rearrangement of the introduced toxin genome to produce a form virulent against humans which seems unlikely; allergenic effects developing over time, either directly from the plant or from transgenic pollen introduced via honey, which seems unlikely as *Bt* does not survive peptic and tryptic digestion, which would be necessary to elicit an allergenic response, and *Bt* exhibits no homology with known allergens; toxic effects on innocuous Lepidoptera considered above, and predatory Coleoptera, that are difficult to rule out; interactions between viruses naturally infecting the modified plant and fragments of the introduced genome especially coat protein RNA, to produce a new virus type which is a matter of debate [10.12].

Europe

Against this tide of concern the response of the engaged scientific community has been rather less than impressive. Downplaying consumers' concerns on the grounds that they lack scientific proof or credentials is hardly a winning strategy, given the experience with bovine spongiform encephalopathy (BSE). A British government denied the association between prion-infected beef and the disease, claiming that there was no scientific evidence that BSE could be communicated from cattle to humans. But there was, and they possessed it; this was subsequently acknowledged, after a Parliamentary enquiry, by an official scientific spokesman of a later government. These events shocked public opinion in the U.K. and elsewhere in Europe creating an atmosphere of acute distrust. Public trust is associated with perceptions of accuracy, knowledge, and concern with public welfare, with the single overriding factor represented by whether information supplied subsequently proves right or wrong, and whether the source is unbiased [10.11].

These concerns led to the formulation of a policy framework or matrix for hazard identification and risk assessment comprising: characteristics of the donor gene, acceptor gene, and of genetically-modified (GM) plants; coupled with the conditions of the receiving environment and the interaction of modified plants with it. Accordingly in October 1991, the European Union by directive 90/220/EEC laid down procedures governing the field-release of transgenic organisms. In October '98 the British government agreed with life-science corporations a moratorium on commercialization of genetically-engineered crops. By this measure, a three-year delay will ensue before introducing insect-resistant crops, during which time effects on biodiversity and interactions with pesticides will be independently monitored. More radical is the creation of an 'environmental stakeholders forum' to access the views of environmentalist and conservation groups.

Subsequently, France suspended sales of GM crops. Then in an unexpected development in June '99, the environment ministers of the European Union meeting in Luxembourg introduced a moratorium on the approval of all new GM food products until procedures were in place to reassure consumers. Effectively, this will delay all new commercial production of GM foods in the EU until 2002 at the earliest. In the interval, risk assessment and monitoring of GM crops were to be strengthened, and labelling made more stringent. On 1 July '99, the extensive British retailer Marks and Spencer announced that, forthwith, all its food products would be free of GM ingredients, a step reported as that firm's biggest single food project; subsequently, the large scale grocers Sainsbury's followed suit. These gestures were clearly in response to consumer mistrust and fears, that may have little connection in reality with GM crops.

North America

Public opinion, in this continent, distinguishes between BSE disease on the one hand, and GM crops on the other, and is sanguine but not complacent about the difference. A conspicuous but not the only reason is surely the energy, application, and professionalism in disseminating science-based information, exhibited by professional bodies including, for example, the American Medical Association. A review by its Council on Scientific Affairs encouraged safety-related research on GM foods in terms of: unintended side effects; altered nutrient and toxicant levels; alternatives to antibiotic resistance markers; and putative allergenicity. In terms of environmental impact it endorsed research on: effects compared with standard methods on nontarget organisms; assessment of gene flow especially into weeds; and monitoring and management of pest resistance; and other possible consequences not indicated by field trials. Recognizing the many benefits of this technology, it did not support a moratorium on planting [*http: // www . ama – assn . org / ama / pub / article / 2036 – 3604 . html*]. This exemplary study is but one of a steady stream of sober, science-based reports fed to and reported by the media, the role of which is crucial in public perception.

In the event, some 15 crop lines in active use in the U.S.A. are no longer regulated by the Animal and Plant Health Inspection Service (APHIS) of the United States Department of Agriculture (USDA). Areas under *Bt* crops amount to 7 million ha of maize, 850,000 of cotton, and 20,000 of potatoes, with an associated saving of one million litres of applied insecticide [S. Morris, University of Guelph, personal communication, Canada, 2001]; of course, herbicide usage has greatly increased on foot of herbicide-resistant GM crops. Overall, such data taken in conjunction with the potential of this technology to engineer crops resistant to fungi, bacteria, viruses including those transmitted by aphids, and nematodes, and capable of fixing nitrogen, and bioremediating contaminated soil, leave little doubt that this technology is here to stay and to expand in the industrialized world.

Third World Perspectives

Conflicting Third World forces are represented by a rapidly growing human population, 3 billion more people by 2050, with 80% of the world's population urban based, requiring an estimated additional 30 million tonnes of food p.a., and by declining availability of good quality land and water; these will increasingly necessitate larger crop yields on less land in regions where losses attributable to pests may amount to 40 % [10.5]. This is not in dispute, and it is increasingly put forward as a justification for producing GM foods. But the developed world cannot solve this problem by dumping its excess food in the Third World. Experience has shown that the primary effects of such dumping are to create a dependency culture, and to simultaneously under price the small proportion of indigenous growers who are efficient producers, and who could lead the way to enhanced production on a national scale.

Nor is there dispute over the concept that any enduring solution must constitute a sustainable system of food production. The enhanced yields that produced agricultural gluts in the West may be attributed, at least in part, to unsustainable inputs of water, fertilisers, pesticide and energy, with unacceptable side-effects of contaminated and nitrogen-rich water. Accordingly, the issue clarifies as to which approaches and methods are most appropriate to achieve the aim of sustainable development in the Third World. The two principal options are, in broad terms, biotechnology as expressed through GM foods, and sustainable ecological methods. As it is unlikely that one single path represents exclusively the best choice for all situations, the issue ultimately resolves as to the optimal mix, that will surely differ from region to region in response to various socio-economic and cultural considerations.

Biotechnology

The sophisticated potential of biotechnology may be gauged from its ability to affect both input and output traits of a crop [10.5]. Input traits comprise biotic and abiotic aspects: manipulation of the former has been achieved through use of herbicide- and insect-resistant crops; fungal- and nematode resistance should be achievable over time. Of course, it is unlikely that insect resistance, for example, will continue to rely on the efficacy of a single gene. More likely is gene pyramiding i.e. the introduction of two–three independent genes to provide a pesticidal cocktail. In terms of abiotic traits, there seems to be a real prospect of engineering resistance to difficult environmental conditions such as heat, cold, drought, water logging, and salt-laden soils [10.38].

Output traits appropriate for genetic engineering include the chemical composition of individual crops, especially in the broad terms of expression of proteins, starches and oils. Fine tuning will manipulate the expression of individual sugars (as for example in barley to facilitate brewing) and health-promoting constituents such as vitamins, enzymes to enhance digestibility, and proteins of pharmaceutical significance, especially vaccines in banana that is edible raw by infants [10.27].

Engineered GM crops were originally intended for the cash-rich, industrialized world. Its growers are willing to pay a significant cash premium for pest-resistant, pesticide-saving, labour-saving, cultivars and are willing to do so, year after year, which is enforceable by a terminator gene that precludes seeding. A significant cash premium is about the last thing that an impoverished peasant farmer of the Third World wishes to pay, or is capable of paying, and 80% of its crops are derived from seed of a previous harvest. The prospect of incorporating that terminator gene in seeds for such growers elicited protests from concerned scientists sufficient to prompt undertakings from major agrotech companies not to deploy it in the Third World. But how realistic is this?

Such companies require a profit on their investment or they cease to exist. So, profit on their investment foregone in the Third World must be recouped in the industrialized one, and by agreement between competing companies. This is an intrinsic improbability as it would require only one major company to withdraw from the agreement for the others to follow suit. Therefore, price-accessible GM crops for the Third World are more likely to be provided from public sources, or from public-private partnerships.

Fortunately, a paradigm is already in place by reference to those Centres that enabled the Green Revolution by promoting Host Plant Resistance (Chapter 8), that in a very broad sense subsumes the use of some GM crops. These Centres must be reinvigorated, redirected, and refinanced. Significant progress is claimed in public sector institutions advancing virus-resistant cassava and papaya, nutritionally-improved rice, and vaccine-bearing bananas. Input should also be harnessed from the Consultative Group on International Agricultural Research (CGIAR), the Institute for International Tropical Agriculture (IITA) based in Nigeria, and the Centre for Application of Molecular Biology to International Agriculture (CAMBIA) based in Canberra. In the latter, a particularly exciting project addresses apomixis, or reproduction of a plant by parthenogenesis from its seed. This would secure access by the small grower to elite seed, and could clear the burden of pathogens acquired during vegetative propagation. CAMBIA has also contributed to a 40 % increase in rice productivity by the incorporation of an antisenescence gene. Also significant is the Cassava Biotechnology Network (CBN) based in Colombia but extending to some 35 countries. Cassava, rich in edible starch and one of the world's most significant food crops for low-income people, of whom an estimated 500 million consume it daily, has proved amenable to genetic transformation. There is surely a need to provide an overarching structure with adequate funding to cohesively advance these activities.

The U.S.A. and EU, as the world's two richest trading entities may owe a duty of care to the people of Third World countries, and perhaps the most efficacious contribution would be a coalition of public-private partnerships designed to increase the yield of key crops.

Sustainable Ecology

Integrated Pest Management (IPM) comprises a varied toolbox ranging from the use of naturally occurring chemicals (Chapter 7) resistant cultivars (Chapter 8) and use of pheromones (Chapter 9), either individually or severally, and as such with or without the use of pesticides. The very significant gains achieved by pest-resistant cultivars was particularly emphasized.

That sustainable ecology can succeed in the Third World is revealed by the outstanding successes achieved by so-called primitive peoples as, for example, the Kayapó of Amazonia. They employ three major strategies: a sophisticated system of soil taxonomy (ethnopedology); complex planting patterns in time involving continuous planting, relay planting, and successional strategies; and complex patterns in space involving concentric ring planting and intercropping [10.15, and references therein].

Components and consequences of those three include: crop rotations; a rich spectrum of inputs including ashes, residues, mulches, dung; controlled, periodic in-field burning and use of fire-resistant crops; and sophisticated experimentation with a range of main crop cultivars. This system, on some of the world's poorest soil, facilitates active production for five yr, followed by decreasing levels for a further six yr.

Specifically, their ethnopedology enables prediction of which crops are best suited to particular soils. Concentric field planting takes account of small lateral differences in nutrient levels, and the broken pattern of crops in different circles impairs host location by pests. Manioc, yams and perennials are intercropped with maize, cucurbits, beans, rice, minor crops, and ritual plants. The advantages of crop rotation and use of mulches are obvious. Fire resistant crops include manioc, yams, sweet potatoes and marantaceas that also benefit from fertilizer supplied in ashes. The combination of short and long cycle cultigens and cultivars means that some food plants are available at all times.

Soil pH is improved by burning and although nitrogen levels remain low, this is less important for their root crops than for cereals. Specifically, rice requires an input of 23 kg nitrogen/ton while manioc and sweet potato remove only 3.7 and 0.3 kg/ton respectively. Root crops yields are adequate to meet carbohydrate and protein requirements if supplemented by meat from game. There is some evidence that a diet rich in sweet potatoes promotes nitrogen-fixing bacteria in the gut.

After five yr continuous production, Kayapó yields (kg/ha) of crops and of protein from plants are about three- and two-fold higher respectively than for colonist agriculture based on western methods; the values for protein from the former do not take account of game. After this interval the soil is exhausted by western methods but Kayapó production continues, albeit at a decreased rate, for another five-six yr. A comparison of Kayapó protein production with that for livestock indicates a superiority in favour of the former of 176-fold after five yr, and 20-fold after 10 yr. Given that the soils involved correspond to 80% of those represented in Amazonia, Kayapó practices may have widespread significance [10.15].

Against this background, it is really rather uncertain that a Third World agronomic society equipped with a knowledge of soil potential, relay planting, intercropping and indigenous resistant cultivars, truly requires transgenic technologies. It would seem more appropriate, in the first instance, to affirm and promote the former practices both as a way of improving the prevailing economic position, and consequentially of slowing migration to the cities. In the second instance, it would seem appropriate to affirm and promote the technologies of IPM, especially where these maximize indigenous knowledge of resistant cultivars and minimize pesticide usage. Accordingly, the reinvigoration, envisioned above of research centres associated with the Green Revolution, must also provide for the deployment of IPM, and indigenous knowledge and culture.

All this notwithstanding, it must be remembered that if the United States were to produce the yields in 1990 of its 17 most important crops by the methods of 1940, it would require 465 million additional acres (188 m ha) of similar quality. This would entail ploughing up 73% of that nations permanent pasture and ranges, or converting 61% of forests and woodlands to cropland. In fact, 465 m acres of such poor quality lands would be quite inadequate to bridge the gap [10.1]. So there can be no question of blocking scientific progress. The issue is to find the blend of traditional and new that is consistent with truly sustainable development. The foregoing text may provide some indicators.

And the words of Borlaug are surely apposite:

> The greatest evils which stalk our Earth are ignorance and oppression, and not science, technology and industry, whose instruments when adequately managed are indispensable tools of a future shaped by Humanity, by itself and for itself, in overcoming major problems like overpopulation, starvation, and world-wide diseases [10.1].

REFERENCES

10.1 Borlaug, N.E. (1997) Feeding a world of 10 billion people: The miracle ahead. *Plant Tis. Cult. Biotech.* **3**, 119-127.

10.2 Cho, H.-J., Choi, K.-P., Yamashita, M., Marikawa, H. and Murooka, Y. (1995) Introduction and expression of the *Streptomyces* cholesterol oxidase gene (*ChoA*), a potent insecticidal protein active against boll weevil larvae, into tobacco cells. *Appl. Microbiol. Biotech.* **44**, 133-138.

10.3 Chrispeels, M.J. and Raikhel, N.V (1991) Lectins, lectin genes, and their role in plant defense. *Plant Cell* **3**, 1-9.

10.4 Coates, C.J., Jasinskiene, N., Miyashiro, L. and James A.A. (1998) *Mariner* transposition and transformation of the yellow fever mosquito, *Aedes aegypti*. *Proc. Natl. Acad. Sci.* **95**, 3748-3751.

10.5 de Greef, W. (1999) A long term perspective on Ag-biotech. In Lutman, P.J.W. (chairman) *Gene Flow and Agriculture Relevance for Transgenic Crops*. British Crop Protection Council Symposium Proceedings No. 72, pp. 33-37.

10.6 Delannay, X., LaVallee, B.J., Proksch, R.K., Fuchs, R.L., Sims, S.R., Greenplate, J.T., Marrone, P.M., Dodson, R.B., Augustine, J.J., Layton, J.G. and Fischhoff, D.A. (1989) Field performance of transgenic tomato plants expressing the *Bacillus thuringiensis* var. *kurstaki* insect control protein. *Bio/Technology* **7**, 1265-1269.

10.7 Duan, X., Li, X., Qingzhong, X., Abo-El-Saad, M., Xu, D. and Wu, R. (1996) Transgenic rice plants harboring an introduced potato proteinase inhibitor II gene are insect resistant. *Nature Biotech.* **14**, 494-498.

10.8 Estruch, J.J., Carozzi, N.B., Desai, N., Duck, N.B., Warren, G.W. and Koziel, M.G. (1997) Transgenic plants: An emerging approach to pest control. *Nature Biotech.* **15**, 137-141.

10.9 Estruch, J.J., Warren, G.W., Mullins, M.A., Nye, G.J., Craig, J.A. and Koziel, M.G. (1996). Vip3A, a novel *Bacillus thuringiensis* vegetative insecticidal protein with a wide spectrum of activities against lepidopteran insects. *Proc. Natl. Acad. Sci.* **93**, 5389-5394.

10.10 Fischhoff, D.A., Bowdish, K.S., Perlak, F.J., Marrone, P.G., McCormick, S.M., Niedermeyer, J.G., Dean, D.A., Kusano-Kretzmer, K., Mayer, E.J., Rochester, D.E., Rogers, S.G. and Fraley, R.T. (1987) Insect tolerant transgenic tomato plants. *Bio/Technology* **5**, 807-813.

10.11 Frewer, C.I., Howard, C., Hedderlay, D. and Sheperd, R. (1996) What determines trust in information about food-related risks? Underlying psychological constructs. *Risk Anal.* **16**, 473-486.

10.12 Greene, A.E. and Allison, R.F. (1994) Recombination between viral RNA and transgenic plant transcripts. *Science* **263**, 1423-1425.

10.13 Greenplate, J.T., Duck, N.B., Pershing, J.C. and Purcell, J.P. (1995) Cholesterol oxidase: an oöstatic and larvicidal agent active against the cotton boll weevil *Anthonomus grandis*. *Entomol. Exp. Appl.* **74**, 253-258.

10.14 Handler, A.M. and James, A.A. (eds.) (2000) *Insect Trangenesis: Methods and Applications*. CRC Press, Boca Raton.

10.15 Hecht, S.A. and Posey, D.A. (1990) Indigenous soil management in the Latin American tropics: Some indicators for the Amazon basin. In Posey, D.A. and Overal, W.L. (organisers) *Ethnobiology: Implications and Applications*. Proc. First. Int. Cong. Ethnobiology (Belem, 1988), pp. 73-86,

10.16 Hilder, V.A., Gatehouse, A.M.R., Sheerman, S.E., Barker, R.F. and Boulter, D. (1987) A novel mechanism of insect resistance engineered into tobacco. *Nature* **330**, 160-163.

10.17 Horsch, R.B., Fry, J., Hinchee, M.A.W., Klee, H.J., Rogers, S.G. and Fraley, R.T. (1988) Strategies for practical gene transfer into agriculturally important crops. In Fraley, R.T., Frey, N.M. and Schell, J. (eds.) *Genetic Improvements of Asgriculturally Important Crops: Progress and Issues*. Cold Spring Harbor, New York, pp. 13-19.

10.18 Hoyle, R. (1995) EPA okays first pesticidal transgenic plants. *Bio/Technology* **13**, 434-435.

10.19 Johnson, R., Narvaez, J.,Gynheung, A. and Ryan, C. (1989) Expression of proteinase inhibitors I and II in transgenic tobacco plants: Effects on natural defense against *Manduca sexta* larvae. *Proc. Natl. Acad. Sci.* **86**, 9871-9875.

10.20 Kaiser, J. (1996) Pests overwhelm *Bt* cotton crop. *Science* **273**, 423.

10.21 Kirschbaum, J.B. (1985) Potential implication of genetic engineering and other biotechnologies to insect control. *Ann. Rev. Entomol.* **30**, 51-70.

10.22 Koziel, M.G., Carozzi, N.B., Currier, T.C., Warren, G.W. and Evola, S.V. (1993) The insecticidal crystal proteins of *Bacillus thuringiensis*: past, present and future uses. *Biotech. Gen. Eng. Rev.* **11**, 171-228.

10.23 Lecardonnel, A., Prévost, G., Beaujean, A., Sangwan, R.S. and Sangwan-Norreel, B.S. (1999) Genetic transformation of potato with nptII-gus marker genes enhances foliage consumption by Colorado beetle larvae. *Mol. Breed.***5**, 441-451.

10.24 Losey, J.E., Rayor, L.S. and Carter, M.E. (1999) Transgenic pollen harms monarch larvae. *Nature* **399**, 214.

10.25 Loukeris, T.G., Livadaras, I., Arcà, B. Zabalou, S. and Savakis, C. (1995). Gene transfer into the medfly, *Ceratitis capitata*, with a *Drosophila hydei* transposable element. *Science* **270**, 2002-2005.

10.26 Macilwain, C. (1996) Bollworms chew hole in gene-engineered cotton. *Nature* **382**, 289.

10.27 Moffat, A.S. (1995) Exploring transgenic plants as a new vaccine source. *Science* **268**, 658-660.

10.28 Perlak, F.J., Deaton, R.W., Armstrong, T.A., Fuchs, R.L., Sims, S.R., Greenplate, J.T. and Fischhoff, D.A. (1990) Insect resistant cotton plants. *Bio/Technology* **8**, 939-943.

10.29 Purcell, J.P., Greenplate, J.T., Jennings, M.G., Ryerse, J.S., Pershing, J.C., Sims, S.R., Prinsen, M.J., Corbin, D.R., Tran, M., Sammons, R.D. and Stonard, R.J. (1993) Cholesterol oxidase: a potent insecticidal protein active against boll weevil larvae. *Biochem. Biophys. Res. Comm.* **196**, 1406-14013.

10.30 Rissler, J. and Mallon, M. (1996) *The Ecological Risks of Engineered Crops*. MIT Press, London

10.31 Schuler, T.H., Poppy, G.M., Potting, R.P.J. Denholim, I. and Kerry, B.R. (1999). Interactions between insect tolerant genetically modified plants and natural enemies. In Lutman, P.J.W. (chairman) *Gene Flow and Agriculture Relevance for Transgenic Crops*. British Crop Protection Council Symposium Proceedings No. 72, pp. 197-202.

10.32 Sears, M.K., Hellmich, R.L., Stanley-Horn, D.E., Oberhauser, K.S., Pleasants, J.M., Mattila, H.R., Blair, D.S. and Dively, G.P. (2001) Impact of *Bt* corn pollen on monarch butterfly populations: A risk assessment. *Proc. Natl. Acad. Sci. USA.* **98**, 11937-11942.

10.33 Shelton, A. M., Robertson, J.L., Tang, J.D., Perez, C., Eigenbrode, S.D., Preisler, H.K., Wilsey, W.T. and Cooley, R.J. (1993) Resistance of diamond back moth (Lepidoptera: Plutellidae) to *Bacillus thuringiensis* subspecies in the field. *J. Econ. Ent.* **86**, 697-705.

10.34 Spradling, A.C. and Rubin, G.M. (1982) Transposition of cloned P elements into *Drosophila* germ line chromosomes. *Science* **218**, 341-347.
10.35 Stewart Jr., C.N. (1999) Insecticidal transgenes into nature: gene flow, ecological effects, relevancy, and monitoring. In Lutman, P.J.W. (chairman) *Gene Flow and Agriculture Relevance for Transgenic Crops.* British Crop Protection Council Symposium Proceedings No. 72, pp. 179-190.
10.36 Thompson, C.E., Squire, G., Mackay, G.R., Bradshaw, J.E., Crawford, J. and Ramsay, G. (1992) Regional patterns of gene flow and its consequence for GM oilseed rape. In Lutman, P.J.W. (chairman) *Gene Flow and Agriculture Relevance for Transgenic Crops.* British Crop Protection Council Symposium Proceedings No. 72, pp. 95-100.
10.37 Zangerl, A.R., McKenna, D., Wraight, C.L., Carroll, M., Ficarello, P., Warner, R. and Berenbaum, M.R. (2001) Effects of exposure to event 176 *Bacillus thuringiensis* corn pollen on monarch and black swallowtail caterpillars under field conditions. *Proc. Natl. Acad. Sci. USA.* **98**, 11908-11912.
10.38 Zhang, H.-X., Hodson, J.N., Williams, J.P. and Blumwald, E. (2001) Engineering salt-tolerant *Brassica* plants: Characterization of yield and seed oil quality in transgenic plants with increased vacuolar sodium accumulation. *Proc. Natl. Acad. Sci. USA.* **98**, 12832-12836.

SPECIES INDEX

SUBJECT INDEX

DATE DUE

AUG 2 3 2006	
JAN 2 3 2007	
OCT 1 8 2007	